To Dick:

With appreciation
for a very professional
job during the preparation
of this book.   Ralph Gonzalez

Knoxville,
December 5, 1977

Digital Image Processing

APPLIED MATHEMATICS AND COMPUTATION

A Series of Graduate Textbooks, Monographs, Reference Works

Series Editor: ROBERT KALABA, University of Southern California

Other Numbers in Preparation

Digital Image Processing

Rafael C. Gonzalez

Department of Electrical Engineering
University of Tennessee, Knoxville

Paul Wintz

School of Electrical Engineering
Purdue University, Lafayette, Indiana
and
Wintek Corporation, Lafayette, Indiana

 1977

Addison-Wesley Publishing Company
Advanced Book Program
Reading, Massachusetts

London · Amsterdam · Don Mills, Ontario · Sydney · Tokyo

Library of Congress Cataloging in Publication Data

Gonzalez, Rafael C
   Digital image processing.

   (Applied mathematics and computation ; no. 13)
   Includes bibliographies and index.
   1. Image processing. I. Wintz, Paul A., joint
author. II. Title.
TA1632.G66   621.38'0414   77-10317
ISBN: 0-201-02596-5
ISBN: 0-201-02597-3 pbk.

Reproduced by Addison-Wesley Publishing Company, Inc., Advanced Book Program, Reading, Massachusetts, from camera-ready copy prepared under the supervision of the authors.

American Mathematical Society (MOS) Subject Classification Scheme (1970): 68-00, 68A45, 62-04, 65C99, 42-00, 42A68, 42A76, 92-04, 93-00, 94-04, 94A05, 94A10.

Manufactured in the United States of America

ABCDEFGHIJ-HA-7987

To
Connie,
Ralph, and
Robert

CONTENTS

SERIES EDITOR'S FOREWORD

Execution times of modern digital computers are measured in nano-seconds. They can solve hundreds of simultaneous ordinary differential equations with speed and accuracy. But what does this immense capability imply with regard to solving the scientific, engineering, economic, and social problems confronting mankind? Clearly, much effort has to be expended in finding answers to that question.

In some fields, it is not yet possible to write mathematical equations which accurately describe processes of interest. Here, the computer may be used simply to simulate a process and, perhaps, to observe the efficacy of different control processes. In others, a mathematical description may be available, but the equations are frequently difficult to solve numerically. In such cases, the difficulties may be faced squarely and possibly overcome; alternatively, formulations may be sought which are more compatible with the inherent capabilities of computers. Mathematics itself nourishes and is nourished by such developments.

Each order of magnitude increase in speed and memory size of computers requires a reexamination of computational techniques and an assessment of the new problems which may be brought within the realm of solution. Volumes in this series will provide indications of current thinking regarding problem formulations, mathematical analysis, and computational treatment.

Digital image processing cuts across many areas — medical diagnosis, planetary physics, robotics, and industrial inspection, to name a few. But as useful as these applications are, the concepts that give rise to them are even more fascinating. In this upper-division and graduate-level volume, the authors demonstrate great skill in developing a thorough treatment of digital image processing concepts at an introductory level. The volume is particularly welcome in this series for the exemplary way in which it combines mathematical analysis and computation to obtain socially useful results.

Robert Kalaba

XI

PREFACE

Interest in digital image processing techniques dates back to the early 1920's when digitized pictures of world news events were first transmitted by submarine cable between New York and London. Application of digital image processing concepts, however, did not become widespread until the middle 1960's, when third-generation digital computers began to offer the speed and storage capabilities required for practical implementation of image processing algorithms. Since then, this area has experienced vigorous growth, having been a subject of interdisciplinary study and research in such fields as engineering, computer science, information science, statistics, physics, chemistry, biology, and medicine. The results of these efforts have established the value of image processing techniques in a variety of problems ranging from restoration and enhancement of space-probe pictures to processing of fingerprints for commercial transactions.

The principal objectives of this book are to provide an introduction to basic concepts and techniques for digital image processing and to lay a foundation which can be used as the basis for further study and research in this field. To achieve these objectives, we have focused attention on material which we feel is fundamental and where the scope of application is not limited to specialized problems. Most of the topics covered in the text have been taught by the authors in senior and first-year graduate courses at the University of Tennessee and at Purdue University. This book also contains revised material presented in short courses and seminars. The mathematical level is well within the grasp of seniors in a technical discipline such as engineering and computer science, which require introductory preparation in computer programming, matrix theory, probability, and mathematical analysis.

It has been our experience that one of the principal features which attracts students to a course in image processing is the opportunity to implement and to test with real data the concepts and algorithms developed in the classroom. The ideal environment for this is provided by an image processing system that includes an image digitizer, a general-purpose computer, and image display equipment. The appendices to this book provide an alternative route for instruction when such a system is not available. Appendix A contains a FORTRAN subroutine for displaying gray-tone images on an ordinary lineprinter, and Appendix B contains a set of coded images suitable for experimenting with the methods discussed in the text. This material can be utilized in conjunction with almost any general-purpose computer, thus allowing the reader to gain experience with image processing techniques through algorithm implementation and visual display of the results.

Digital Image Processing is one of three related books published by Addison-Wesley, Advanced Book Program. The first of these, *Pattern Recognition Principles* (Tou and Gonzalez, 1974) describes deterministic, statistical, and syntactic pattern recognition concepts. This latter topic is treated in greater depth in *Syntactic Pattern Recognition: An Introduction* (Gonzalez and Thomason, in preparation) which can be used by itself or as a supplement to *Pattern Recognition Principles*. The objective of these books is to provide a unified, introductory treatment of pattern recognition and image processing concepts with emphasis on fundamentals and consistency of notation.

Rafael C. Gonzalez
Paul Wintz

ACKNOWLEDGMENTS

We are indebted to a number of individuals who, either through discussions or by providing the facilities for our work, have contributed to the preparation of this book. In particular, we wish to extend our appreciation to E. L. Hall, J. M. Googe, F. N. Peebles, R. C. Kryter, M. T. Borelli, C. T. Huggins, M. G. Thomason, W. R. Wade, D. W. Bouldin, and W. Frei.

As is true with most projects carried out in a university environment, our students over the past few years have significantly influenced not only our thinking, but also the topics and material included in the text. The following individuals have worked with us in various aspects of digital image processing during the course of their graduate work at the University of Tennessee and at Purdue University: D. D. Thompson, J. D. Birdwell, B. A. Fittes, R. E. Woods, J. M. Harris, M. E. Casey, A. Miller, T. G. Saba, Po Chen, J. Duan, J. Essman, J. Gattis, J. Gupta, A. Habibi, C. Proctor, P. Ready, M. Tasto, W. Wilder, L. Wilkins, and T. Wallace.

We also wish to thank R. G. Gruber, G. S. Sodoski, and G. W. Roulette for their editorial assistance, G. C. Guerrant and R. E. Wright for the art work, and Vicki Bohanan, Diana Scott, Mary Bearden, and Margaret Barbour for their excellent typing of several drafts of the manuscript.

In addition, we express our appreciation to the National Aeronautics and Space Administration, the Office of Naval Research, the Oak Ridge National Laboratory, and the Advanced Research Projects Agency for their sponsorship of our research activities in image processing and pattern recognition.

Finally, we wish to acknowledge the following individuals and organizations for their permission to reproduce the material listed below.

Figures 1.1, 1.2, and 1.3 from M. D. McFarlane, "Digital Pictures Fifty Years Ago," *Proc. IEEE*, vol. 60, 1972, pp. 768–770. Figures 1.4(a) and (b) courtesy of the Jet Propulsion Laboratory. Figures 1.4(c) and (d) courtesy of E. L. Hall, University of Tennessee. Figures 1.4(e) and (f) courtesy of M. M. Sondhi, Bell Telephone Laboratories. Figure 2.6 from

T. N. Cornsweet, *Visual Perception*, Academic Press, 1970. Figures 2.9 and 2.10 from T. S. Huang, "PCM Picture Transmission," *IEEE Spectrum*, vol. 2, no. 12, 1965, pp. 57–63. Figure 3.25 from Cooley *et al*, "The Fast Fourier Transform and its Applications," *IEEE Trans. Educ.*, vol. E-12, no. 1, 1969, pp. 27–34. Figure 4.34 courtesy of E. L. Hall. Figure 4.37 from T. G. Stockham, Jr., "Image Processing in the Context of a Visual Model," *Proc. IEEE*, vol. 60, 1972, pp. 828–842. Plates I, II, III, and IV courtesy of General Electric, Lamp Business Division. Plate V courtesy of J. L. Blankenship, Oak Ridge National Laboratory. Plate VI from H. C. Andrews *et al*, "Image Processing by Digital Computer," *IEEE Spectrum*, vol. 9, 1972, pp. 20–32. Figure 5.3 from B. L. McGlamery, " Restoration of Turbulence-Degraded Images," *J. Opt. Soc. Amer.*, vol. 57, 1967, pp. 293–297. Figure 5.4 courtesy of M. M. Sondhi, Bell Telephone Laboratories. Figure 5.5 from J. L. Harris, "Potential and Limitations of Techniques for Processing Linear Motion-Degraded Images," NASA Publ. SP-193, 1968, pp. 131–138. Figure 5.6 from B. R. Hunt, "The Application of Constrained Least Squares Estimation to Image Restoration by Digital Computer," *IEEE Trans. Computers*, vol. C-22, 1973, pp. 805–812. Figures 5.8 through 5.11 courtesy of the Jet Propulsion Laboratory. Figures 7.5 through 7.7 from C. K. Chow and T. Kaneko, "Automatic Boundary Detection of the Left Ventricle from Cineangiograms," *Comp. and Biomed. Res.*, vol. 5, 1972, pp. 388–410. Figure 7.15 courtesy of W. Frei, University of Southern California. Figure 7.16 courtesy of E. L. Hall. Figure 7.22 courtesy of T. Wallace, Purdue University. Figure 7.23 courtesy of R. Y. Wong, University of Southern California. Figure 7.32 from R. S. Ledley, "High-Speed Automatic Analysis of Biomedical Pictures," *Science*, vol. 146, 1964, pp. 216–223. Figure 7.38 courtesy of E. L. Hall.

INTRODUCTION

One picture is worth more
than ten thousand words.
Anonymous

1.1 BACKGROUND

Interest in digital image processing methods stems from two principal application areas: improvement of pictorial information for human interpretation, and processing of scene data for autonomous machine perception. One of the first applications of image processing techniques in the first category was in improving digitized newspaper pictures sent by submarine cable between London and New York. Introduction of the Bartlane cable picture transmission system in the early 1920's reduced the time required to transport a picture across the Atlantic from more than a week to less than three hours. Pictures were coded for cable transmission and then reconstructed at the receiving end by specialized printing equipment. Figure 1.1 was transmitted in this way and reproduced on a telegraph printer fitted with type faces simulating a halftone pattern.

Some of the initial problems in improving the visual quality of these early digital pictures were related to the selection of printing procedures and the distribution of brightness levels. The printing method used to obtain Fig. 1.1 was abandoned toward the end of 1921 in favor of a technique based on photographic reproduction made from tapes perforated at the telegraph receiving terminal. Figure 1.2 shows an image obtained using this method. The improvements over Fig. 1.1 are evident, both in tonal quality and in resolution.

The early Bartlane systems were capable of coding images in five distinct brightness levels. This capability was increased to fifteen levels in 1929. Figure 1.3 is indicative of the type of image that could be obtained using the fifteen-tone equipment. During this period, the reproduction process was also improved considerably by the introduction of a system for developing a film plate via light beams which were modulated by the coded picture tape.

Figure 1.1. A digital picture produced in 1921 from a coded tape by a telegraph printer with special type faces. (From McFarlane [1972].)

Although improvements on processing methods for transmitted digital pictures continued to be made over the next thirty-five years, it took the combined advents of large-scale digital computers and the space program to bring into focus the potentials of image processing concepts. Work on using computer techniques for improving images from a space probe began at the Jet Propulsion Laboratory (Pasadena, California) in 1964, when pictures of the Moon transmitted by Ranger 7 were processed by a computer to correct various types of image distortion inherent in the on-board television camera. These techniques served as the basis for improved methods used in the enhancement and restoration of images from such familiar programs as the Surveyor missions to the Moon, the Mariner series of flyby missions to Mars, and the Apollo manned flights to the Moon.

From 1964 until this writing, the field of image processing has experienced vigorous growth. In addition to applications in the space program, digital image processing techniques are used today in a variety of

Figure 1.2. A digital picture made from a tape punched after the signals had crossed the Atlantic twice. Some errors are visible. (From McFarlane [1972].)

Figure 1.3. Unretouched cable picture of Generals Pershing and Foch, transmitted by 15-tone equipment from London to New York. (From McFarlane [1972].)

problems which, although often unrelated, share a common need for methods capable of enhancing pictorial information for human interpretation and analysis. In medicine, for instance, physicians are assisted by computer procedures that enhance the contrast or code the intensity levels into color for easier interpretation of x-rays and other biomedical images. The same or similar techniques are used by geographers in studying pollution patterns from aerial and satellite imagery. Image enhancement and restoration procedures have been used to process degraded images depicting unrecoverable objects or experimental results too expensive to duplicate. There have been instances in archeology, for example, where blurred pictures which were the only available records of rare artifacts lost or damaged after being photographed, have been successfully restored by image processing methods. In physics and related fields, images of experiments in such areas as high-energy plasmas and electron microscopy are routinely enhanced by computer techniques. Similar successful applications of image processing concepts can be found in astronomy, biology, nuclear medicine, law enforcement, defense, and industrial applications.

Some typical examples of the results obtainable with digital image processing techniques are shown in Fig. 1.4. The original images are shown on the left and the corresponding computer-processed images on the right. Figure 1.4(a) is a picture of the Martian surface which was corrupted by interference during transmission to Earth by a space probe. The interference, which in this case appears as a set of vertical, structured lines, can be almost completely removed by computer processing, as shown in Fig. 1.4(b). Figures 1.4(c) and (d) illustrate the considerable improvement that can be

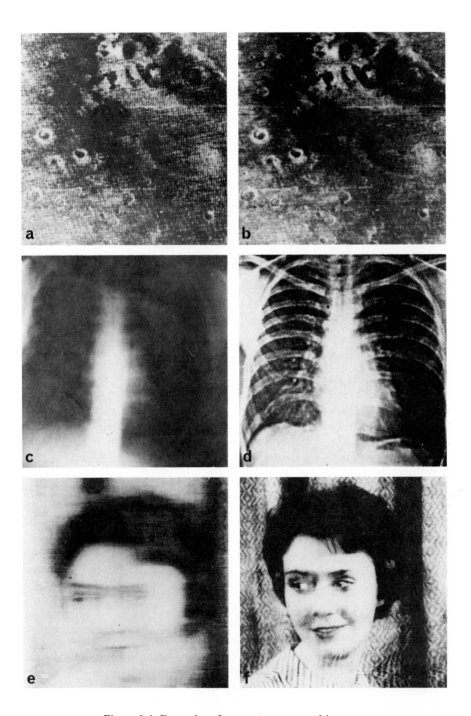

Figure 1.4. Examples of computer-processed images.

made on an x-ray image by contrast and edge enhancement. The image shown in Fig. 1.4(e) was blurred by uniform motion during exposure, and the image shown in Fig. 1.4(f) resulted after application of a deblurring algorithm. These illustrations are typical of those discussed in detail in Chapters 4 and 5.

The foregoing examples have in common the fact that processing results are intended for human interpretation. The second major application area of digital image processing techniques mentioned at the beginning of this section is in problems dealing with machine perception. In this case, interest is focused on procedures for extracting from an image information in a form suitable for computer processing. Often, this information bears little resemblance to visual features used by humans in interpreting the content of an image. Examples of the type of information used in machine perception are statistical moments, Fourier transform coefficients, and multidimensional distance measures.

Typical problems in machine perception which routinely employ image processing techniques are automatic character recognition, industrial robots for product assembly and inspection, military recognizance, automatic processing of fingerprints, screening of x-rays and blood samples, and machine processing of aerial and satellite imagery for weather prediction and crop assessment.

1.2 DIGITAL IMAGE REPRESENTATION

As used in this book, the term *monochrome image* or simply *image*, refers to a two-dimensional light intensity function $f(x, y)$, where x and y denote spatial coordinates and the value of f at any point (x, y) is proportional to the brightness (or *gray level*) of the image at that point. An example illustrating the axis convention used throughout the following chapters is shown in Fig. 1.5. It is sometimes useful to view an image function in perspective with the third axis being brightness. If Fig. 1.5 were viewed in this way it would appear as a series of active peaks in regions with numerous changes in brightness levels and smoother regions or plateaus where the brightness levels varied little or were constant. If we follow the convention of assigning proportionately higher values to brighter areas, the height of the components in the plot would be proportional to the corresponding brightness in the image.

A *digital image* is an image $f(x, y)$ which has been discretized both in spatial coordinates and in brightness. We may consider a digital image as a matrix whose row and column indices identify a point in the image and the corresponding matrix element value identifies the gray level at that point. The elements of such a digital array are called *image elements, picture elements, pixels,* or *pels,* with the last two names being commonly used abbreviations of "picture elements".

Although the size of a digital image varies with the application, it will become evident in the following chapters that there are numerous advantages in selecting square arrays with sizes and number of gray levels which are integer powers of 2. For example, a typical size comparable in quality to a monochrome TV image is a 512 × 512 array with 128 gray levels.

Figure 1.5. Axis convention used for digital image representation.

With the exception of a discussion in Chapter 4 of pseudo-color techniques for image enhancement, all the images considered in this book are digital monochrome images of the form described above. Thus, we will not be concerned with topics in three-dimensional scene analysis nor with optical techniques for image processing.

1.3 ELEMENTS OF A DIGITAL IMAGE PROCESSING SYSTEM

The components of a basic, general-purpose digital image processing system are shown in Fig. 1.6. The operation of such a system may be divided into three principal categories: digitization, processing, and display.

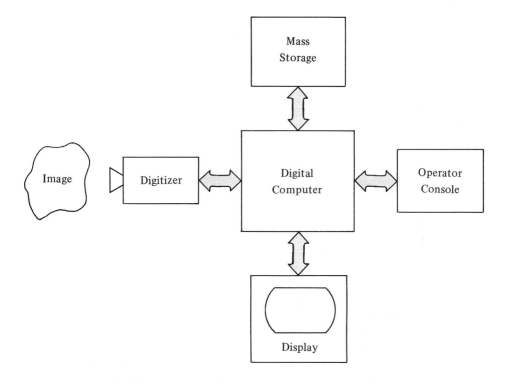

Figure 1.6. Elements of a digital image processing system.

1.3.1 Digitizers

A digitizer converts an image into a numerical representation suitable for input into a digital computer. Among the most commonly used input devices are microdensitometers, flying spot scanners, image dissectors, and TV camera digitizers. The first two devices require that the image to be digitized be in the form of a transparency (e.g., a film negative) or photograph. Image dissectors and TV cameras can accept images recorded in this manner, but they have the additional advantage of being able to digitize natural images that have sufficient light intensity to excite the detector.

In microdensitometers the transparency or photograph is mounted on a flat bed or wrapped around a drum. Scanning is accomplished by focusing a beam of light on the image and translating the bed or rotating the drum in relation to the beam. In the case of transparencies the beam passes through the film; in photographs it is reflected from the surface of the image. In both cases the beam is focused on a photodetector and the gray level at any point in the image is recorded by the detector based on the intensity of the beam. A digital image is obtained by allowing only discrete values of intensity and position in the output. Although microdensitometers are slow devices, they are capable of high degrees of position accuracy due to the essentially continuous nature of the mechanical translation used in the digitization process.

Flying spot scanners also operate on the principle of focusing a transmitted or reflected source beam on a photodetector. In this case, however, the image is stationary and the light source is a cathode-ray tube (CRT) in which a beam of electrons, deflected by electromagnets, impinges on a fluorescent phosphor surface. The beam thereby produces a spot of light that moves in a scanning pattern on the face of the tube. The fact that the beam is moved electronically allows high scanning speeds. Flying spot scanners are also ideally suited for applications in which it is desirable to control the beam scanning pattern externally (e.g., in tracing the boundaries of objects in an image). This flexibility is afforded by the fact that the position of the electron beam is quickly and easily established by external voltage signals applied to the electromagnets.

In image dissectors and TV cameras the image is focused directly on the surface of a photosensitive tube whose response is proportional to the incident light pattern. Dissector operation is based on the principle of electronic emission, where the image incident on the photosensitive surface produces an electron beam whose cross section is roughly the same as the geometry of the tube surface. Image pickup is accomplished by using electromagnets to deflect the entire beam past a pinhole located in the back of the dissector tube. The pinhole lets through only a small cross section of the beam and thus "looks" at one point in the image at a time. Since photoemissive materials are very inefficient, the time that the pinhole has to look at the point source in order to collect enough electrons tends to make image dissectors rather slow digitizers. Most devices integrate the emission of each input point over a specified time interval before yielding a signal which is proportional to the brightness of the point. This integration capability is beneficial in terms of noise reduction, thus making image dissectors attractive in applications where high signal-to-noise ratios are required. Like in flying spot scanners, control of the scanning pattern in image dissectors is easily varied by external voltage signals applied to the electromagnets.

Many general-purpose TV image digitizers employ a vidicon tube, whose operation is based on the principle of photoconductivity. An image focused on the tube surface produces a pattern of varying conductivity which matches the distribution of brightness in the optical image. An independent, finely focused electron beam scans the rear surface of the photoconductive target, and by charge neutralization, this beam creates a potential difference and produces on a collector a signal proportional to the input brightness pattern. A digital image is obtained by quantizing this signal, as well as the corresponding position of the scanning beam.

Although standard vidicon digitizers are in general less accurate than the systems discussed above, they have numerous advantages which in many applications outweigh their relative lack of precision. Vidicon systems, for example, are among the most inexpensive digitizers on the market. They also have the distinct advantage that the image being digitized can be viewed in its entirety on a TV monitor. This capability, not available in any of the systems discussed above, is ideal for general-purpose applications. Since vidicon systems employ electronic scanning, and photoconductive tubes are reasonably efficient, these digitizers are much faster than microdensitometers and image dissectors. They are not as fast or as flexible in terms of beam control as flying spot scanners because of the scanning constraints required to produce a video image which can be viewed on a standard TV monitor.

1.3.2 Image Processors

Systems used for image processing range from microprocessor devices for special-purpose applications to large computer systems capable of performing a variety of functions on high-resolution image arrays. The principal parameter influencing the structure of a computer system for image processing is the required data throughput. For general-purpose laboratory applications where fast data throughput is not essential, a moderately equipped minicomputer is often adequate. Since digitized image arrays are in most cases too large to fit completely in the memory of a small computer, the key in structuring such a system is to provide adequate and efficient bulk storage capabilities. The two most popular storage media are magnetic tapes and disk packs, both of which allow storage of numerous images per device, depending on the density of the medium and the size of the image arrays. Generally, magnetic tapes are used for archival storage, and disks are used during processing in order to improve data transfer speed between the computer memory and bulk storage media.

In terms of the above requirements a minimum, yet flexible computer system for image processing can be structured using a minicomputer with 32,000 to 64,000 words of core memory, two disk drives, a magnetic tape unit, and assorted peripherals such as scope terminals, cassette recorders, and a lineprinter or some other hard-copy output device. Although image processing programs are often coded in assembly language to gain speed, the flexibility of the system can be improved considerably by having at least one high-level language for use in program development.

A system with the components just described can be used without difficulty to implement most of the image processing methods discussed in the following chapters. The operating efficiency will, of course, depend on the size of the input images and the type of processing desired.

1.3.3 Display Devices

The function of the display unit in an image processing system is to convert the numerical arrays stored in the computer into a form suitable for human interpretation. The principal display media are CRTs, TV systems, and printing devices.

In CRT systems the horizontal and vertical positions of each element in the image array are converted into voltages which are used to deflect the CRT's electron beam, thus providing the two-dimensional drive necessary to produce an output image. At each deflection point, the intensity of the beam is modulated by using a voltage which is proportional to the value of the corresponding point in the numerical array, varying from zero intensity outputs for points whose numerical value corresponds to black, to maximum intensity for white points. The resulting variable-intensity light pattern is recorded by a photographic camera focused on the face of the cathode-ray tube. Some systems employ a long persistence phosphor tube which also allows viewing of the entire image after the scanning process is completed. Although images recorded by the photographic process can be of excellent quality, the same images generally appear of poor tonality when shown to an observer on a long persistence CRT because of limitations in the human visual system when responding to this type of display.

Television display systems convert an image stored in the computer into a video frame which can be displayed on a TV monitor. The advantage of these systems is that displays created on a video monitor have a tonality which closely resembles that of photographs, thus producing an output which is easily assimilated by the visual system. The disadvantage of TV displays is that they must refresh the monitor at a rate of about 30 frames (e.g., images) per second in order to avoid flicker. Since most general-purpose computers are not capable of transferring data at this rate, the principal problem in designing a TV display system is to provide some buffer storage medium for transferring data to a monitor at video rates. Most high-quality commercial systems accomplish this in one of two ways. One approach is to use a fast solid-state memory to store the entire image array; the screen is then refreshed at 30 frames per second by cycling through the memory and combining the stored binary information into an analog signal by means of conditioning circuits and fast digital-to-analog converters. The other method is to store the image array on a high-density disk. The storage arrangement is again in binary form, with each track in the disk containing a bit for all the pixels in the image. An $N \times N$ image with 2^n possible levels, for example, requires n storage tracks, each containing $N \times N$ bits. The necessary transfer speed is accomplished by rotating the disk past n sensing heads. For any disk position, the binary information in all n heads is combined to produce a voltage proportional to the gray level of a single pixel in the image. As the disk rotates, an analog signal is created by conditioning these voltage levels and inputing them into a fast digital-to-analog converter. In both of the approaches just discussed the analog signals out of the converters constitute the information carrying component of the video signal used to drive the TV display monitor.

Printing image display devices are useful primarily for low-resolution image processing work. One simple approach for generating gray-tone images directly on paper is to use the over-strike capability of a standard line-printer. The gray-level of any point in the printout can be controlled by the number and density of the characters overprinted at that point. By properly selecting the character set it is possible to achieve reasonable good gray-level distributions with a simple computer program and relatively few characters. An example of this approach is given in Appendix A. Other common means of recording an image directly on paper include heat sensitive paper devices and ink spray systems.

1.4 ORGANIZATION OF THE BOOK

Techniques for image processing may be divided into four principal categories: (1) image digitization; (2) image enhancement and restoration; (3) image encoding; and (4) image segmentation and representation. The material in the following chapters is organized in essentially the same order as these problem areas.

As discussed in Sections 1.2 and 1.3, the digitization problem is one of converting continuous brightness and spatial coordinates into discrete components. A preliminary discussion of digitization and its effect on image quality is given in Chapter 2, while a more theoretical treatment of the sampling process is developed in Chapter 3. Digitization considerations are a natural extension of the main theme of these two chapters, which is the introduction of concepts and mathematical tools used throughout the rest of the book.

Enhancement and restoration techniques deal with the improvement of a given image for human or machine perception. Image enhancement is the topic of Chapter 4, while image restoration methods are covered in Chapter 5. Image encoding procedures, discussed in Chapter 6, are used to reduce the number of bits in a digital image. The encoding process often plays a central role in image processing for the purpose of minimizing storage or transmission requirements. Segmentation and representation techniques are considered in Chapter 7. These procedures, which deal with the decomposition of an image into a set of simpler parts and the organization of these parts in a meaningful descriptive manner, are among the most important considerations in the development of autonomous image processing and recognition systems.

REFERENCES

The references cited below are of a general nature and cover the spectrum of available image processing techniques and their applications. References given at the end of later chapters are keyed to specific topics discussed in the text. All references are cited by author, book, or journal name followed by the year of publication. The bibliography at the end of the book is organized in the same way and contains all pertinent information for each reference.

Complementary reading for the material in this book may be found in the books by Rosenfeld [1969], Andrews [1970], Lipkin and Rosenfeld [1970], Duda and Hart [1973], Huang [1975], Andrews and Hunt [1976], and Rosenfeld and Kak [1976]. The following special issues of journals have been devoted to image processing: *Proc. IEEE* [1972], *IEEE Trans. Computers* [1972], *Computer* [1974], and *IEEE Trans. Circuits Syst.* [1975]. Other survey articles of interest are by Andrews, Tescher, and Kruger [1972], Rosenfeld [1972, 1973, 1974], Andrews [1974], and Fu and Rosenfeld [1976]. The pattern recognition literature often contains articles related to image processing. The books by Tou and Gonzalez [1974], and Gonzalez and Thomason (in preparation) contain a guide to the literature on pattern recognition and related topics.

2

DIGITAL IMAGE FUNDAMENTALS

Those who wish to succeed must ask the
right preliminary questions.
Aristotle

The purpose of this chapter is to introduce the reader to a number
of image concepts and to develop some of the notation that will be used
throughout the book. The first section is a brief summary of the mechanics
of the human visual system, including image formation in the eye and its
capabilities for brightness adaptation and discrimination. Section 2.2 presents an image model based on the illumination-reflection phenomenon
which gives rise to most images perceived in our normal visual activities. The
concepts of uniform image sampling and gray-level quantization are introduced in Section 2.3. The results of this section are then used as the basis for
Section 2.4, which deals with nonuniform sampling and quantization. Finally, Section 2.5 contains an introduction to photographic film and some of its
most important characteristics in terms of recording image processing results.

2.1 ELEMENTS OF VISUAL PERCEPTION

Since the ultimate goal of many of the techniques discussed in the
following chapters is to aid the observer in interpreting the content of an
image, it is important before proceeding to develop a basic understanding of
the visual-perception process. The following discussion is a brief account of
the human visual mechanism, with particular emphasis on concepts that will
serve as foundation to much of the material presented in later chapters.

2.1.1 Structure of the Human Eye

A horizontal cross section of the human eye is shown in Fig.
2.1. The eye is nearly spherical in form with an average diameter of

Rafael C. Gonzalez and Paul Wintz, Digital Image Processing   ISBN 0-201-02596-5; 0-201-02597-3(pbk.)

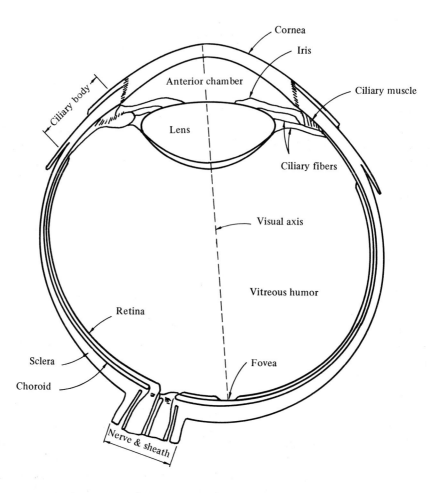

Figure 2.1. Simplified diagram of a cross section of the human eye.

approximately 20 mm. It is enclosed by three membranes: the *cornea* and *sclera* outer cover, the *choroid*, and the *retina*. The cornea is a tough, transparent tissue that covers the anterior surface of the eye. The sclera is continuous with the cornea; it is an opaque membrane that encloses the remainder of the optic globe.

The choroid lies directly below the sclera. This membrane contains a network of blood vessels which serve as the major source of nutrition to the eye. The choroid coat is heavily pigmented and hence helps to reduce the amount of extraneous light entering the eye and the backscatter within the optical globe. At its anterior extreme, the choroid is divided into the *ciliary body* and the *iris diaphragm*. The latter contracts or expands to control the amount of light that is permitted to enter the eye. The central opening of the iris (the *pupil*) is variable in diameter from approximately 2 mm up to 8 mm. The front of the iris contains the visible pigment of the eye, whereas the back contains a black pigment.

The innermost membrane of the eye is the retina, which lines the inside of the wall's entire posterior portion. When the eye is properly focused, light from an object outside the eye is imaged on the retina. Pattern vision is afforded by the distribution of discrete light receptors over the surface of the retina. There are two classes of receptors: cones and rods. The cones in each eye number between six and seven million. They are located primarily in the central portion of the retina, called the *fovea*, and are highly sensitive to color. Humans can resolve fine details with these cones largely because each one is connected to its own nerve end. Muscles controlling the eye rotate the eyeball until the image of an object of interest falls on the fovea. Cone vision is known as *photopic* or bright-light vision.

The number of rods is much larger, being in the order of 75 to 150 million distributed over the retinal surface. The larger area of distribution and the fact that several rods are connected to a single nerve end reduce the amount of detail discernible by these receptors. Rods serve to give a general, overall picture of the field of view. They are not involved in color vision and are sensitive to low levels of illumination. Objects which, for example, appear brightly colored in daylight, when seen by moonlight appear as colorless forms because only the rods are stimulated. This is known as *scotopic* or dim-light vision.

The *lens* is made up of concentric layers of fibrous cells and is suspended by fibers that attach to the ciliary body. It contains sixty to seventy percent water, about six percent fat, and more protein than any other tissue in the eye. The lens is colored by a slightly yellow pigmentation that increases with age. It absorbs approximately eight percent of the visible light spectrum, with relatively higher absorption at shorter wavelengths. Both infrared and ultraviolet light are absorbed appreciably by proteins within the lens structure and, in excessive amounts, can cause damage to the eye.

2.1.2 Image Formation in the Eye

The principal difference between the lens of the eye and an ordinary optical lens is that the former is flexible. As illustrated in Fig. 2.1, the radius of curvature of the anterior surface of the lens is greater than the radius of its posterior surface. The shape of the lens is controlled by the tension in the fibers of the ciliary body. To focus on distant objects, the controlling muscles cause the lens to be relatively flattened. Similarly, these muscles allow the lens to become thicker in order to focus on objects near the eye.

The distance between the focal center of the lens and the retina varies from approximately 17 mm down to about 14 mm, as the refractive power of the lens increases from its minimum to its maximum. When the eye is focused on an object farther than about 3 m away, the lens exhibits its lowest refractive power, and when focused on a very near object it is most strongly refractive. With this information, it is easy to calculate the size of the retinal image of any object. In Fig. 2.2, for example, the observer is looking at a tree 15 m high at a distance of 100 m. Letting x be the size of the retinal image in mm, we have from the geometry of Fig. 2.2 that $15/100 = x/17$ or $x = 2.55$ mm. As indicated in the previous section, the retinal image is reflected primarily in the area of the fovea. Perception then takes place by the relative excitation of light receptors which transform radiant energy into electrical impulses that are ultimately decoded by the brain.

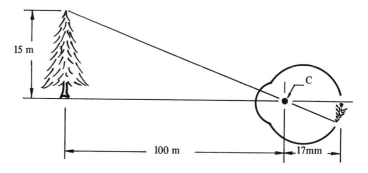

Figure 2.2. Optical representation of the eye looking at a tree. Point C is the optical center of the lens.

2.1.3 Brightness Adaptation and Discrimination

Since digital images are displayed as a discrete set of brightness points, the ability of the eye to discriminate between different brightness levels is an important consideration in presenting image processing results.

The range of light intensity levels to which the human visual system can adapt is enormous, being on the order of 10^{10} from the scotopic threshold to the glare limit. There is also considerable experimental evidence which indicates that subjective brightness (i.e., brightness as perceived by the human visual system) is a *logarithmic* function of the light intensity incident on the eye. This characteristic is illustrated in Fig. 2.3, which is a plot of light intensity versus subjective brightness. The long solid curve represents the range of intensities to which the visual system can adapt. In photopic vision alone, the range is about 10^6. The transition from scotopic to photopic vision is gradual over the approximate range from 0.001 to 0.1 milliambert (-3 to -1 mL in the log scale), as illustrated by the double branches of the adaptation curve in this range.

The key point in interpreting the impressive dynamic range depicted in Fig. 2.3 is that the visual system can by no means operate over such a range *simultaneously*. Rather, it accomplishes this large variation by changes in its overall sensitivity, a phenomenon known as *brightness adaptation*. The total range of intensity levels it can discriminate simultaneously is rather small when compared with the total adaptation range. For any given set of conditions, the current sensitivity level of the visual system is called the *brightness-adaptation level* which may correspond, for example, to brightness B_a in Fig. 2.3. The short intersecting curve represents the range of subjective brightness that the eye can perceive when adapted to this level. It is noted that this range is rather restricted, having a level B_b at and below which all stimuli are perceived as indistinguishable blacks. The upper (dashed) portion of the curve is not actually restricted but, if extended too far, loses its meaning because much higher intensities would simply raise the adaptation level to a higher value than B_a.

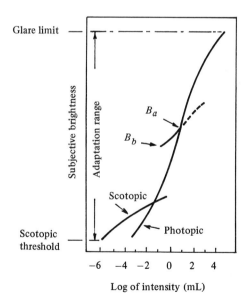

Figure 2.3. Range of subjective brightness sensations showing a particular adaptation level.

The contrast sensitivity of the eye can be measured by exposing an observer to a uniform field of light of brightness B, with a sharp-edged circular target in the center, of brightness $B + \Delta B$, as shown in Fig. 2.4(a). ΔB is increased from zero until it is just noticeable. The just-noticeable difference ΔB is measured as a function of B. The quantity $\Delta B/B$ is called the Weber ratio and is nearly constant at about 2 percent over a very wide range of brightness levels, as shown in Fig. 2.4(b). This phenomenon has given rise to the idea that the human eye has a much wider dynamic range than man-made imaging systems. However, this does not correspond to any ordinary seeing situation and more applicable results are obtained by using the pattern of Fig. 2.5(a). $\Delta B/B$ is again measured, but now B_o, the surrounding (adapting) brightness, is a parameter. The results are shown in Fig. 2.5(b). The dynamic range is about 2.2 log units centered about the adapting brightness which is comparable to what can be achieved with electronic imaging systems if they are correctly adjusted for the background brightness. The ease and rapidity with which the eye adapts itself—differently on different parts of the retina—is really the remarkable characteristic, rather than its overall dynamic range. What is meant by a dynamic range of 2.2 log units is that $\Delta B/B$ remains relatively constant in this range. As B becomes more and more different from the adapting brightness B_o, the appearance also changes. Thus, a brightness about 1.5 log units higher or lower than B_o appears white or black, respectively. If the central target is set at a constant level while B_o is varied over a wide range, the target appears to change from completely white to completely black.

In the case of a complex image, the visual system does not adapt to a single intensity level. Instead, it adapts to an average level which depends on the properties of the image. As the eye roams about the scene, the instanta-

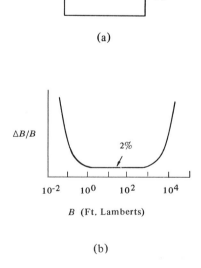

(a)

(b)

Figure 2.4. Contrast sensitivity with a constant background.

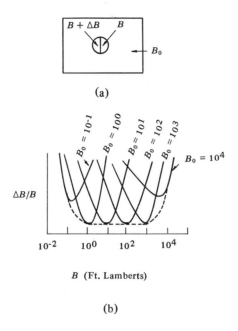

(a)

(b)

Figure 2.5. Contrast sensitivity with a varying background.

neous adaptation level fluctuates about this average. For any point or small area in the image, the Weber ratio is generally much larger than that obtained in an experimental environment because of the lack of sharply defined boundaries and intensity variations in the background. The result is that the eye can only detect in the neighborhood of one or two dozen intensity levels at any one point in a complex image. This does not mean, however, that an image need only be displayed in two dozen intensity levels to achieve satisfactory visual results. The above narrow discrimination range "tracks" the adaptation level as the latter changes in order to accommodate different intensity levels following eye movements about the scene. This allows a much larger range of *overall* intensity discrimination. To obtain displays that will appear reasonably smooth to the eye for a large class of image types, a range of over 100 intensity levels is generally required. This point will be considered in further detail in Section 2.3.

The brightness of a region, as perceived by the eye, depends upon factors other than simply the light radiating from that region. In terms of image processing applications, one of the most interesting phenomena related to brightness perception is that the response of the human visual system tends to "overshoot" around the boundary of regions of different intensity. The result of this overshoot is to make areas of constant intensity appear as if they had varying brightness. In Fig. 2.6(a), for example, the image shown was created by varying the intensity according to the intensity

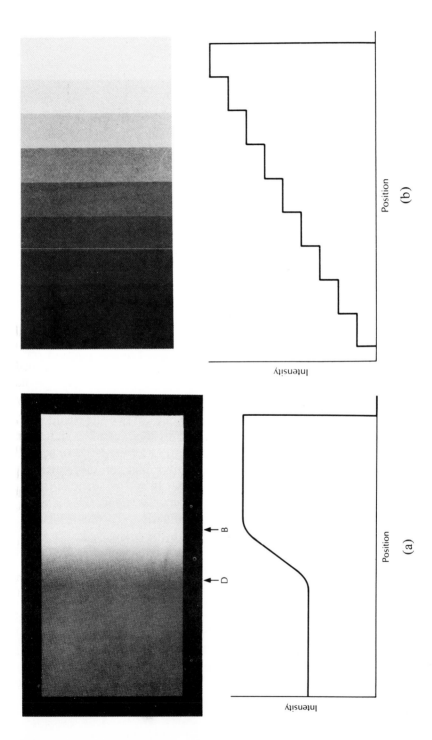

Figure 2.6. Examples of the Mach-band effect. (From Cornsweet [1970].)

profile shown below the photograph. Although the intensity variation is perfectly smooth, the eye perceives a brighter stripe in the region marked B and a darker stripe in the region marked D. These stripes are called *Mach bands*, after Ernst Mach who first described them in 1865. A more striking example of the Mach band effect is shown in Fig. 2.6(b). As indicated by the intensity profile, each band in the photograph was created by using a constant intensity. To the eye, however, the brightness pattern in the image appears strongly scalloped, particularly around the boundaries.

2.2 AN IMAGE MODEL

As used in this book, the term *image* refers to a two-dimensional light intensity function, denoted by $f(x, y)$, where the value or amplitude of f at spatial coordinates (x, y) gives the intensity (brightness) of the image at that point. Since light is a form of energy, $f(x, y)$ must be nonzero and finite, that is,

$$0 < f(x, y) < \infty \qquad (2.2\text{-}1)$$

The images we perceive in our everyday visual activities normally consist of light reflected from objects. The basic nature of $f(x, y)$ may be considered as being characterized by two components. One component is the amount of source light incident on the scene being viewed, while the other is the amount of light reflected by the objects in the scene. These components are appropriately called the *illumination* and *reflectance components*, and are denoted by $i(x, y)$ and $r(x, y)$, respectively. The functions $i(x, y)$ and $r(x, y)$ combine as a product to form $f(x, y)$:

$$f(x, y) = i(x, y)r(x, y) \qquad (2.2\text{-}2)$$

where

$$0 < i(x, y) < \infty \qquad (2.2\text{-}3)$$

and

$$0 < r(x, y) < 1 \qquad (2.2\text{-}4)$$

Equation (2.2-4) indicates the fact that reflectance is bounded by 0 (total absorption) and 1 (total reflectance). The nature of $i(x, y)$ is determined by the light source, while $r(x, y)$ is determined by the characteristics of the objects in a scene.

The values given in Eqs. (2.2-3) and (2.2-4) are theoretical bounds. The following *average* numerical figures illustrate some typical ranges of $i(x, y)$. On a clear day, the sun may produce in excess of 9000 foot-candles of illumination on the surface of the earth. This figure decreases to less than 1000 foot-candles on a cloudy day. On a clear evening, a full moon yields about 0.01 foot-candles of illumination. The typical illumination level in a commercial office is about 100 foot-candles. Similarly, the following are some typical values of $r(x, y)$: 0.01 for black velvet, 0.65 for stainless steel, 0.80 for flat-white wall paint, 0.90 for silver-plated metal, and 0.93 for snow.

Throughout this book, the intensity of a monochrome image f at coordinates (x, y) will be called the *gray level* (l) of the image at that point. From Eqs. (2.2-2) through (2.2-4), it is evident that l lies in the range

$$L_{min} \leqslant l \leqslant L_{max} \qquad\qquad (2.2\text{-}5)$$

In theory, the only requirement on L_{min} is that it be positive, and on L_{max} that it be finite. In practice, $L_{min} = i_{min}\ r_{min}$ and $L_{max} = i_{max}\ r_{max}$. Using the above values of illumination and reflectance as a guideline, one may expect the values $L_{min} \approx 0.005$ and $L_{max} \approx 100$ for indoors image processing applications.

The interval $[L_{min}, L_{max}]$ is called the *gray scale*. It is common practice to shift this interval numerically to the interval $[0, L]$, where $l = 0$ is considered black and $l = L$ is considered white in the scale. All intermediate values are shades of gray varying continuously from black to white.

2.3 UNIFORM SAMPLING AND QUANTIZATION

In order to be in a form suitable for computer processing, an image function $f(x, y)$ must be digitized both spatially and in amplitude. Digitization of the spatial coordinates (x, y) will be referred to as *image sampling*, while amplitude digitization will be called *gray-level quantization*.

Suppose that a continuous image $f(x, y)$ is approximated by equally-spaced samples arranged in the form of an $N \times N$ array† as shown in Eq. (2.3-1), where each element of the array is a discrete quantity:

---

†Digitization of an image need not be limited to square arrays. However, following discussions will often be simplified by the adoption of this convention.

$$f(x,y) \approx \begin{bmatrix} f(0,0) & f(0,1) & \cdots & f(0, N-1) \\ f(1,0) & f(1,1) & \cdots & f(1, N-1) \\ \vdots & & & \\ f(N-1,0) & f(N-1,1) & \cdots & f(N-1, N-1) \end{bmatrix} \qquad (2.3\text{-}1)$$

The right side of this equation represents what is commonly called a *digital image*, while each element of the array is referred to as an *image element*, *picture element*, *pixel*, or *pel*, as indicated in Section 1.2. The terms "image" and "pixels" will be used throughout the following discussions to denote a digital image and its elements.

The above digitization process requires that a decision be made on a value for N as well as on the number of discrete gray levels allowed for each pixel. It is common practice in digital image processing to let these quantities be integer powers of two; that is,

$$N = 2^n \qquad (2.3\text{-}2)$$

and

$$G = 2^m \qquad (2.3\text{-}3)$$

where G denotes the number of gray levels. It is assumed in this section that the discrete levels are equally spaced between 0 and L in the gray scale. Using Eqs. (2.3-2) and (2.3-3) the number, b, of bits required to store a digitized image is given by

$$b = N \times N \times m. \qquad (2.3\text{-}4)$$

For example, a 128 × 128 image with 64 gray levels requires 98,304 bits of storage. Table 2.1 summarizes values of b for some typical ranges of N and m. Table 2.2 gives the corresponding number of 8-bit bytes. Generally, it is not practical from a programming point of view to fill a byte completely if this implies a pixel overlap from one byte to the next. Thus, the figures in Table 2.2 represent the minimum number of bytes needed for each value of N and m when no overlap is allowed. For example, if $m = 5$, it is assumed

Table 2.1. Number of storage bits for various values of N and m.

m / N	1	2	3	4	5	6	7	8
32	1,024	2,048	3,072	4,096	5,120	6,144	7,168	8,192
64	4,096	8,192	12,288	16,384	20,480	24,576	28,672	32,768
128	16,384	32,768	49,152	65,536	81,920	98,304	114,688	131,072
256	65,536	131,072	196,608	262,144	327,680	393,216	458,752	524,288
512	262,144	524,288	786,432	1,048,576	1,310,720	1,572,864	1,835,008	2,097,152

that only one pixel is stored in a byte, even though this leaves three unused bits in the byte.

Since Eq. (2.3-1) is an approximation to a continuous image, a reasonable question to ask at this point is how many samples and gray levels are required for a good approximation. The *resolution* (i.e., the degree of discernable detail) of an image is strongly dependent on both N and m. The more these parameters are increased, the closer the digitized array will approximate the original image. However, Eq. (2.3-4) clearly points out the unfortunate fact that storage and, consequently, processing requirements increase rapidly as a function of N and m.

In view of the above comments, it is of interest to consider the effect that variations in N and m have on image quality. As might be suspected, a "good" image is difficult to define because quality requirements vary according to application. Figure 2.7 shows the effect of reducing the sampling-grid size on an image. Figure 2.7(a) is a 512 × 512, 256-level image showing Astronaut Aldrin during the first Moon landing (note the reflection of Neil Armstrong on the face plate). Figures 2.7(b) through (f) show the same image, but with $N = 256, 128, 64, 32,$ and 16. In all cases the maximum number of allowed gray levels was kept at 256. Since the display area used for each image was the same (i.e., 512 × 512 display points) pixels in the lower resolution images were duplicated in order to fill the entire display field. This produced a checker-board effect which is particularly noticeable

Table 2.2 Number of 8-bit bytes of storage for various values of N and m.

m / N	1	2	3	4	5	6	7	8
32	128	256	512	512	1,024	1,024	1,024	1,024
64	512	1,024	2,048	2,048	4,096	4,096	4,096	4,096
128	2,048	4,096	8,192	8,192	16,384	16,384	16,384	16,384
256	8,192	16,384	32,768	32,768	65,536	65,536	65,536	65,536
512	32,768	65,536	131,072	131,072	262,144	262,144	262,144	262,144

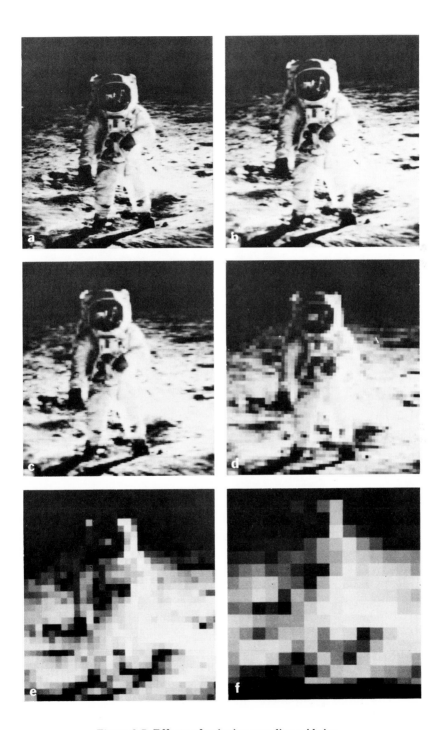

Figure 2.7. Effects of reducing sampling-grid size.

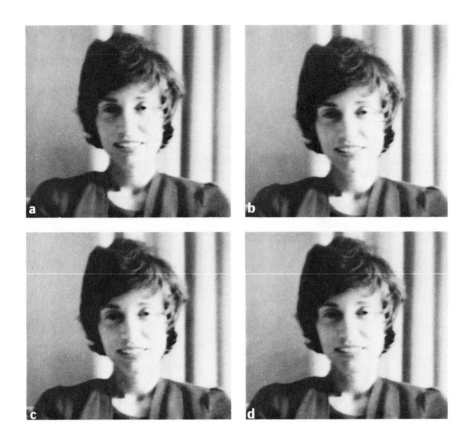

Figure 2.8. A 512 × 512 image displayed in 256, 128, 64, 32, 16, 8, 4, and 2 levels, respectively.

in the low-resolution images. It is noted that the 256 × 256 image is reasonably close to Fig. 2.7(a), but image quality deteriorated rapidly for the other values of N.

Figure 2.8 illustrates the effects produced by reducing the number of bits used to represent the gray levels in an image. Figure 2.8(a) is a picture of a subject digitized using a 512 × 512 array and 256 levels [$m = 8$ in Eq. (2.3-3)]. Figures 2.8(b) through (h) were obtained by reducing the number of bits from $m = 7$ to $m = 1$, respectively, while keeping the digitizing grid at 512 × 512. The 256-, 128-, and 64-level images are of acceptable quality. The 32-level image, however, has some mild "false contouring" in the smooth background area above the subject's right shoulder. This effect is considerably more pronounced in the image displayed in 16 levels, and increases sharply for the remaining images.

The number of samples and gray levels required to produce a faith-

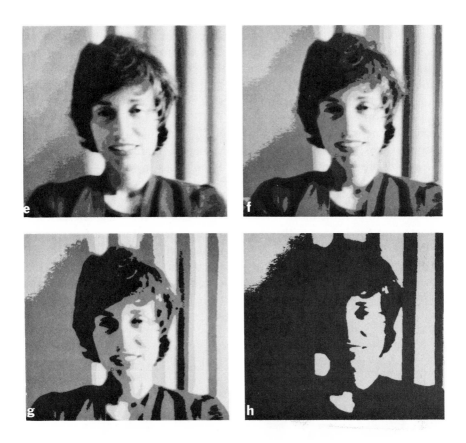

Figure 2.8. (Continued.)

ful reproduction of an original image depend on the image itself. As a basis for comparison, the requirements to obtain a quality comparable to that of monochrome TV pictures over a wide range of image types are on the order of 512 × 512 pixels with 128 gray levels. As a rule, a minimum system for general image processing work should be able to display 256 × 256 pixels with 64 gray levels.

The above results illustrate the effects produced on image quality by varying N and m independently. These results, however, only partially answer the question posed earlier since nothing has yet been said about the relation between these parameters. Huang [1965] considered this problem in an attempt to quantify experimentally the effects on image quality produced by varying N and m. The experiment consisted of a set of subjective tests. Three of the images used are shown in Fig. 2.9. The woman's face is representative of an image with relatively little detail; the cameraman picture contains an intermediate amount of detail; and the crowd picture contains, by comparison, a large amount of detail information.

Sets of these three images were generated by varying N and m and observers were then asked to rank them according to their subjective quality. The results are summarized in Fig. 2.10 in the form of *isopreference curves* in the N-m plane. Each point in this plane represents an image with values of N and m equal to the coordinates of that point. An isopreference curve is one in which the points represent images of equal subjective quality.

The isopreference curves of Fig. 2.10 are arranged, from left to right, in order of increasing subjective quality. These results suggest several empirical conclusions: (1) As expected, the quality of the images tends to increase as N and m are increased. There were a few cases in which, for fixed N, the quality improved by decreasing m. This is most likely due to the fact that a decrease in m generally increases the apparent contrast of an image. (2) The curves tend to become more vertical as the detail in the image increases. This suggests that for images with a large amount of detail only a few gray levels are needed. For example, it is noted in Fig. 2.10(c) that, for $N = 64$ or 128, image quality is not improved by an increase in m. This is not true for the curves in the other two figures. (3) The isopreference curves depart markedly from the curves of constant b, which are shown dotted in Fig. 2.10.

Figure 2.9. Test images used in evaluating subjective image quality. (From Huang [1965].)

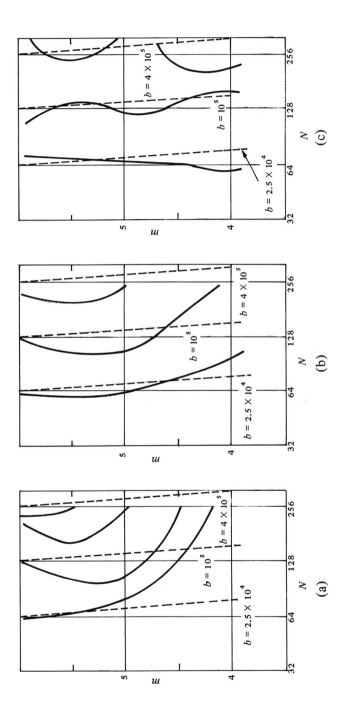

Figure 2.10. Isopreference curves for (a) face, (b) cameraman, and (c) crowd. (From Huang [1965].)

2.4 NONUNIFORM SAMPLING AND QUANTIZATION

2.4.1 Nonuniform Sampling

For a fixed value of N, it is possible in many cases to improve the appearance of an image by using an adaptive scheme where the sampling process depends on the characteristics of the image. In general, fine sampling is required in the neighborhood of sharp gray-level transitions, while coarse sampling may be employed in relatively smooth regions. Consider, for example, a simple image consisting of a face superimposed on a uniform background. Clearly, the background carries little detail information and can be quite adequately represented by coarse sampling. The face, on the other hand, contains considerably more detail. Using in this region of the image the additional samples not used in the background will tend to improve the overall result, particularly if N is small. In distributing the samples, greater sample concentration should be used in gray-level transition boundaries, such as the boundary between the face and the background in the preceding example.

It is important to note that the necessity of having to identify boundaries, even if only on a rough basis, is a definite drawback of the nonuniform sampling approach discussed above. Also, it should be kept in mind that this method is not practical for images containing relatively small uniform regions. For instance, nonuniform sampling would be difficult to justify for an image depicting a dense crowd.

2.4.2 Nonuniform Quantization

When the number of gray levels must be kept small, it is usually desirable to use unequally spaced levels in the quantization process. A method similar to the nonuniform sampling technique discussed in the previous section may be used for the distribution of gray levels in an image. However, since the eye is relatively poor at estimating shades of gray near abrupt level changes, the approach in this case is to use few gray levels in the neighborhood of boundaries. The remaining levels can then be used in those regions where gray-level variations are smooth, thus avoiding or reducing the false contours which often appear in these regions if they are too coarsely quantized.

The above method is subject to the same observations made in the previous section regarding boundary detection and detail content. An alternative technique which is particularly attractive for distributing gray levels consists of computing the frequency of occurrence of all allowed levels. If some gray levels in a certain range occur frequently, while the others occur rarely, the quantization levels are finely spaced in this range and coarsely spaced outside of it. This method is sometimes called *tapered quantization.*

2.5 PHOTOGRAPHIC FILM

Photographic film is an important element of image processing systems. It is often used as the medium where input images are recorded, and it is by far the most popular medium for recording output results. For these reasons, we consider in this section some basic properties of monochrome photographic film and their relation to image processing applications.

2.5.1 Film Structure and Exposure

A cross section of a typical photographic film as it would appear under magnification is shown in Fig. 2.11. It consists of the following layers and components: (1) a supercoat of gelatin used for protection against scratches and abrasion marks; (2) an emulsion layer consisting of minute silver halide crystals; (3) a substrate layer which promotes adhesion of the emulsion to the film base; (4) the film base or support, made of cellulose triacetate or a related polymer; and (5) a backing layer to prevent curling.

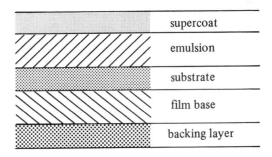

Figure 2.11. Structure of a modern black-and-white film.

When the film is exposed to light, the silver halide grains absorb optical energy and undergo a complex physical change. The grains that have absorbed a sufficient amount of energy contain tiny patches of metallic silver, called *development centers.* When the exposed film is developed, the existence of a single development center in a silver halide grain can precipitate the change of the entire grain to metallic silver. The grains that do not contain development centers do not undergo such a change. After development, the film is "fixed" by chemical removal of the remaining silver halide grains. The more light that reaches a given area of the film, the more silver halide is rendered developable and the denser the silver deposit that is formed there. Since the silver grains are largely opaque at optical frequencies, an image of gray tones is obtained where the brightness levels are reversed, thus producing the familiar film negative.

The process is repeated to obtain a positive picture. The negative is projected onto a sensitive paper carrying a silver halide emulsion similar to that used for the film. Exposure by a light source yields a latent image of the negative. After development, the paper bears a positive silver image. Enlargement of the negative is controlled by the choice of light source and size of positive paper used.

2.5.2 Film Characteristics

Of practical interest to the photographer are contrast, speed, graininess, and resolving power. An understanding of the effect of these parameters is particularly important in specialized applications such as photographing the results obtained in an image processing system.

2.5.2.1 Contrast

High-contrast films reproduce tone differences in the subject as large density differences in the photograph; low-contrast films translate tone differences as small density differences. The exposure, E, to which a film is subjected is defined as *energy per unit area* at each point on the photosensitive area. Exposure depends on the incident intensity, I, and the duration of the exposure, T. These quantities are related by the expression

$$E = IT \qquad\qquad (2.5\text{-}1)$$

The most widely used description of the photosensitive properties of photographic film is a plot of the density of the silver deposit on a film vs. the logarithm of E. These curves are called characteristic curves, D-log-E curves (density vs. log exposure), and H & D curves (after Hurter and Driffield who developed the method). Figure 2.12 shows a typical H & D curve for a photographic negative. When the exposure is below a certain level, the density is independent of exposure and equal to a minimum value called the *gross fog*. In the *toe* of the curve, density begins to increase with increasing exposure. There follows a region of the curve in which density is linearly proportional to logarithmic exposure. The slope of this linear region is referred to as the film *gamma* (γ). Finally, the curve saturates in a region called the *shoulder*, and again there is no change in density with increasing

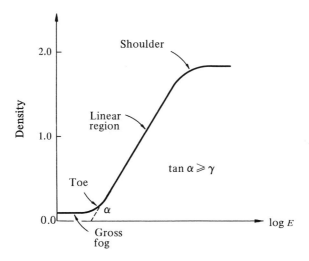

Figure 2.12. A typical H & D curve.

exposure. The value of γ is a measure of film contrast: the steeper the slope, the higher the contrast rendered. General-purpose films of medium contrast have gammas in the range 0.7 to 1.0. High-contrast films have gammas in the order of 1.5 to 10. As a rule, films with relatively low gammas are used for continuous-tone reproduction; high-contrast films are used for copying line originals and other specialized purposes.

2.5.2.2 Speed

The speed of a film determines how much light is needed to produce a given amount of silver on development. The lower the speed, the longer the film must be exposed to record a given image. The most widely used standard of speed is the ASA scale. This scale is arithmetic, with the speed number directly proportional to the sensitivity of the film. A film of ASA 200 is twice as fast (and for a given subject requires half as much exposure) as a film of ASA 100. Some speed scales, such as the DIN system used in Europe, are logarithmic. Every increase of three in the DIN speed number doubles the actual speed. An ASA 50 film is equivalent to a DIN 18, an ASA 100 to a DIN 21, and so on.

General purpose films for outdoor and some indoor photography have speeds between ASA 80 and ASA 160; fine-grain films for maximum image definition between ASA 20 and ASA 64; high-speed films for poor light and indoor photography between ASA 200 and ASA 500; and ultraspeed films for very low light levels from ASA 650 and up.

2.5.2.3 Graininess

The image derived from the silver halide crystals is discontinuous in structure. This gives an appearance of graininess in big enlargments. The effect is most prominent in fast films, which have comparatively large crystals; slower, fine-grain emulsions are therefore preferable in applications where fine detail is desired or where enlargement of the negatives is necessary.

2.5.2.4 Resolving Power

The fineness of detail that a film can resolve depends not only on its graininess, but also on the light scattering properties of the emulsion and on the contrast with which the film reproduces fine detail. Fine-grain films with thin emulsions yield the highest resolving power.

2.5.3 Diaphragm and Shutter Settings

Regardless of the type of film used, proper camera settings are essential in obtaining acceptable pictures. The principal settings are the lens diaphragm and shutter speed.

In the lens diaphragm, a series of leaves increase or decrease the size of the opening to control the amount of light passing through the lens to the film. The diaphragm control ring is calibrated with a scale of so-called f-numbers or stop numbers in a series such as: 1.4, 2, 2.8, 4, 5.6, 8, 11, 16, 22, and 32. The f-numbers are inversely proportional to the amount of light admitted. In the above series, each setting admits twice as much light as the next higher f-number (thus giving twice as much exposure), and half as much light as the next lower value. Shutter speed settings on present-day cameras also follow a standard double-or-half sequence. Typical speeds are 1, 1/2, 1/4, 1/8, 1/15, 1/30, 1/60, 1/125, 1/250, 1/500, and 1/1000 sec. The faster the shutter speed, the shorter the exposure time that is obtained.

The diaphragm and shutter control the amount of light reaching the film by adjusting the light intensity and the time during which it acts. Different aperture-shutter speed combinations can thus yield the same exposure. For example, diaphragm $f/2.8$ with 1/250 second, $f/4$ with 1/125 second, and $f/5.6$ with 1/60 second, all yield the same exposure. It should be noted, however, that the combination chosen for these two settings is not independent of the conditions under which a picture is taken. For example, when photographing a scene where depth of focus is of interest, the f-stop should be selected as high as possible to give the lens a "pin-hole" characteristic. For a given film, this requirement limits the range of shutter speeds which yield adequate exposures. In other applications, the shutter speed is the essential consideration. An example with image processing implications is the problem of photographing a television screen. In this case, the shutter speed must be set below the refreshing rate of the TV set (1/30 sec. per frame) or traceback information that is too fast for the human eye to see will be recorded on the film. Typically, 1/8 second is adequate, although slower speeds are often used. Many of the images in this book, for example, were photographed at 1/4 sec. with Kodak Panatomic-X fine-grain film (ASA 32). The diaphragm settings were determined by using an exposure meter to measure the light intensity of each image.

2.6 CONCLUDING REMARKS

The material in this chapter is primarily background information for subsequent discussions. Our treatment of the human visual system, although brief, should give the reader a basic idea of the capabilities of the eye in perceiving pictorial information. Similarly, the image model developed in Section 2.2 is used in Chapter 4 as the basis for an image enhancement technique called *homomorphic filtering*.

The sampling ideas introduced in Sections 2.3 and 2.4 are considered again in Section 3.3.9 after the necessary mathematical tools for a deeper analytical study of this problem are developed. Sampling and quantization considerations also play a central role in Chapter 6 in the context of image encoding applications, where the problem is one of compressing the large quantities of data that result from image digitization.

REFERENCES

The material presented in Sections 2.1.1 and 2.1.2 is based primarily on the books by Cornsweet [1970] and by Graham [1965]. Additional reading for Section 2.1.3 may be found in Sheppard [1968]; Sheppard, Stratton, and Gazley [1969]; and Stevens [1951]. The image model presented in Section 2.2 has been investigated by Oppenheim, Schafer, and Stockham [1968] in connection with image enhancement applications. References for the illumination and reflectance values used in that section are Moon [1961] and the *IES Lighting Handbook* [1972]. Some of the material presented in Section 2.3 is based on the work of Huang [1965]. The papers by Scoville [1965] and by Gaven, Tavitian, and Harabedian [1970] are also of interest. Additional reading on the material of Section 2.4 may be found in Rosenfeld and Kak [1976]. References for Section 2.5 are Mees [1966], Perrin [1960], Nelson [1971], and *Kodak Plates and Films for Scientific Photography* [1973].

3

IMAGE TRANSFORMS

And be not conformed to this world: but be
ye transformed by the renewing of your mind ...
Romans 12:2

The material in this chapter deals primarily with the development of two-dimensional transforms and their properties. Transform theory has played a key role in image processing for a number of years, and it continues to be a topic of interest in theoretical as well as applied work in this field. Two-dimensional transforms are used in the following chapters for image enhancement, restoration, encoding, and description.

Although other transforms are discussed in some detail in this chapter, emphasis is placed on the Fourier transform because of its wide range of applications in image processing problems. The Fourier transform of one and two continuous variables is introduced in Section 3.1. These concepts are then expressed in discrete form in Section 3.2. Several important properties of the two-dimensional Fourier transform are developed and illustrated in Section 3.3. This section is followed by the development of a fast Fourier transform algorithm which can be used to reduce the number of calculations to a fraction of that required to implement the discrete Fourier transform by direct methods. Finally, Section 3.5 deals with the development of the Walsh, Hadamard, discrete cosine, and Hotelling transforms.

3.1 INTRODUCTION TO THE FOURIER TRANSFORM

Let $f(x)$ be a continuous function of a real variable x. The *Fourier transform* of $f(x)$, denoted by $\mathfrak{F}\{f(x)\}$, is defined by the equation

$$\mathfrak{F}\{f(x)\} = F(u) = \int_{-\infty}^{\infty} f(x) \exp[-j2\pi ux]\, dx \qquad (3.1\text{-}1)$$

where $j = \sqrt{-1}$.

Given $F(u)$, $f(x)$ can be obtained by using the *inverse Fourier transform*

$$\mathcal{F}^{-1}\{F(u)\} = f(x)$$

$$= \int_{-\infty}^{\infty} F(u) \exp[j2\pi ux] \, du \qquad (3.1\text{-}2)$$

Equations (3.1-1) and (3.1-2), which are called the *Fourier transform pair*, can be shown to exist if $f(x)$ is continuous and integrable and $F(u)$ is integrable. These conditions are almost always satisfied in practice.

We will be concerned throughout this book with functions $f(x)$ which are real. The Fourier transform of a real function, however, is generally complex; that is,

$$F(u) = R(u) + jI(u) \qquad (3.1\text{-}3)$$

where $R(u)$ and $I(u)$ are, respectively, the real and imaginary components of $F(u)$. It is often convenient to express Eq. (3.1-3) in exponential form:

$$F(u) = |F(u)|e^{j\phi(u)} \qquad (3.1\text{-}4)$$

where

$$|F(u)| = \left[R^2(u) + I^2(u) \right]^{1/2} \qquad (3.1\text{-}5)$$

and

$$\phi(u) = \tan^{-1}\left[\frac{I(u)}{R(u)} \right] \qquad (3.1\text{-}6)$$

The magnitude function $|F(u)|$ is called the *Fourier spectrum* of $f(x)$, and $\phi(u)$ its *phase angle*. The square of the spectrum,

$$E(u) = |F(u)|^2$$

$$= R^2(u) + I^2(u) \qquad (3.1\text{-}7)$$

is commonly referred to as the *energy spectrum* of $f(x)$.

The variable u appearing in the Fourier transform is often called the *frequency variable*. This name arises from the fact that, using Euler's formula, the exponential term, $\exp[-j2\pi ux]$, may be expressed in the form:

$$\exp[-j2\pi ux] = \cos 2\pi ux - j \sin 2\pi ux \qquad (3.1\text{-}8)$$

If we interpret the integral in Eq. (3.1-1) as a limit-summation of discrete terms, it is evident that $F(u)$ is composed of an infinite sum of sine and cosine terms, and that each value of u determines the *frequency* of its corresponding sine-cosine pair.

Example: Consider the simple function shown in Fig. 3.1(a). Its Fourier transform is obtained from Eq. (3.1-1) as follows:

$$F(u) = \int_{-\infty}^{\infty} f(x) \exp[-j2\pi ux] \, dx$$

$$= \int_{0}^{X} A \exp[-j2\pi ux] \, dx$$

$$= \frac{-A}{j2\pi u} \left[e^{-j2\pi ux}\right]_{0}^{X} = \frac{-A}{j2\pi u} \left[e^{-j2\pi uX} - 1\right]$$

$$= \frac{A}{j2\pi u} \left[e^{j\pi uX} - e^{-j\pi uX}\right] e^{-j\pi uX}$$

$$= \frac{A}{\pi u} \sin(\pi uX) \, e^{-j\pi uX}$$

which is a complex function. The Fourier spectrum is given by

$$|F(u)| = \frac{A}{\pi u} |\sin(\pi uX)| \, |e^{-j\pi uX}|$$

$$= AX \left| \frac{\sin(\pi uX)}{(\pi uX)} \right|$$

A plot of $|F(u)|$ is shown in Fig. 3.1(b). ☐

The Fourier transform can be easily extended to a function $f(x, y)$ of two variables. If $f(x, y)$ is continuous and integrable, and $F(u, v)$ is integrable, we have that the following Fourier transform pair exists:

$$\mathfrak{F}\{f(x,y)\} = F(u, v) = \int\int_{-\infty}^{\infty} f(x, y) \exp[-j2\pi(ux + vy)] \, dx \, dy$$

$$(3.1\text{-}9)$$

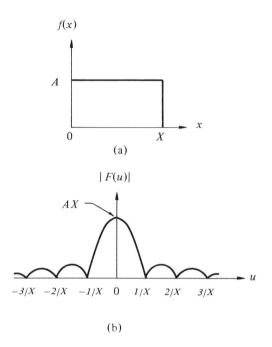

Figure 3.1. A simple function and its Fourier spectrum.

and

$$\mathcal{F}^{-1}\{F(u, v)\} = f(x, y) = \int\!\!\!\int_{-\infty}^{\infty} F(u, v)\, \exp[\,j2\pi(ux + vy)]\, du\, dv$$

(3.1-10)

where u and v are the frequency variables.

As in the one-dimensional case, the Fourier spectrum, phase, and energy spectrum are, respectively, given by the relations:

$$|F(u, v)| = [R^2(u, v) + I^2(u, v)]^{1/2}$$

(3.1-11)

$$\phi(u, v) = \tan^{-1}\left[\frac{I(u, v)}{R(u, v)}\right]$$

(3.1-12)

and

$$E(u, v) = R^2(u, v) + I^2(u, v)$$

(3.1-13)

Example: The Fourier transform of the function shown in Fig. 3.2(a) is given by

$$F(u, v) = \int\!\!\!\int_{-\infty}^{\infty} f(x, y) \exp\left[-j2\pi(ux + vy)\right] dx\, dy$$

$$= A \int_0^X \exp\left[-j2\pi ux\right] dx \int_0^Y \exp\left[-j2\pi vy\right] dy$$

$$= A \left[\frac{e^{-j2\pi ux}}{-j2\pi u}\right]_0^X \left[\frac{e^{-j2\pi vy}}{-j2\pi v}\right]_0^Y$$

$$= \frac{A}{-j2\pi u}\left[e^{-j2\pi uX} - 1\right]\frac{1}{-j2\pi v}\left[e^{-j2\pi vY} - 1\right]$$

$$= AXY\left[\frac{\sin(\pi uX)\, e^{-j\pi uX}}{(\pi uX)}\right]\left[\frac{\sin(\pi vY)\, e^{-j\pi vY}}{(\pi vY)}\right]$$

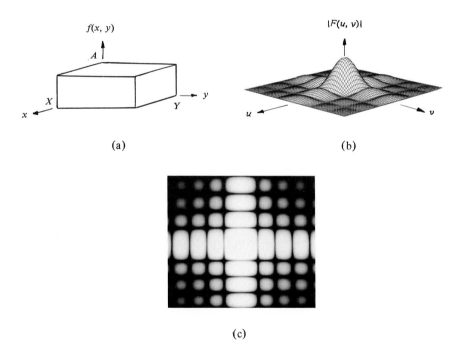

(a)

(b)

(c)

Figure 3.2. (a) A two-dimensional function, (b) its Fourier spectrum, and (c) the spectrum displayed as an intensity function.

The spectrum is given by

$$|F(u, v)| = AXY \left| \frac{\sin(\pi u X)}{(\pi u X)} \right| \left| \frac{\sin(\pi v Y)}{(\pi v Y)} \right|$$

A plot of this function is shown in Fig. 3.2(b) in three-dimensional perspective. Figure 3.2(c) shows the spectrum as an intensity function where brightness is proportional to the amplitude of $|F(u, v)|$. Other examples of two-dimensional functions and their spectra are shown in Fig. 3.3. In this case, both $f(x, y)$ and $|F(u, v)|$ are shown as images.   □

3.2 THE DISCRETE FOURIER TRANSFORM

Suppose that a continuous function $f(x)$ is discretized into a sequence $\{f(x_0), f(x_0 + \Delta x), f(x_0 + 2\Delta x), ..., f(x_0 + [N-1]\Delta x)\}$ by taking N samples Δx units apart, as shown in Fig. 3.4. It will be convenient in subsequent developments to use x as either a discrete or continuous variable, depending on the context of the discussion. We may do this by defining

$$f(x) = f(x_0 + x \, \Delta x) \tag{3.2-1}$$

where x now assumes the discrete values 0, 1, 2, ..., $N - 1$. In other words, the sequence $\{f(0), f(1), f(2), ..., f(N-1)\}$ will be used to denote *any* N uniformly-spaced samples from a corresponding continuous function.

With the above notation in mind, we have that the *discrete* Fourier transform pair that applies to sampled functions is given by[†]

$$F(u) = \frac{1}{N} \sum_{x=0}^{N-1} f(x) \exp[-j2\pi u x/N] \tag{3.2-2}$$

for $u = 0, 1, 2, ..., N - 1$, and

$$f(x) = \sum_{u=0}^{N-1} F(u) \exp[j2\pi u x/N] \tag{3.2-3}$$

for $x = 0, 1, 2, ..., N - 1$.

The values $u = 0, 1, 2, ..., N - 1$ in the discrete Fourier transform given in Eq. (3.2-2) correspond to samples of the continuous transform at values 0, Δu, $2\Delta u$, ..., $(N - 1)\Delta u$. In other words we are letting $F(u)$ represent $F(u\Delta u)$. This notation is similar to that used for the discrete $f(x)$, with the exception that the samples of $F(u)$ start at the origin of the frequency

---

† A proof of these results is outside the scope of this discussion. Proofs relating the continuous and discrete Fourier transforms can be found, for example, in Blackman and Tukey [1958]; Cooley, Lewis, and Welch [1967]; and Brigham [1974].

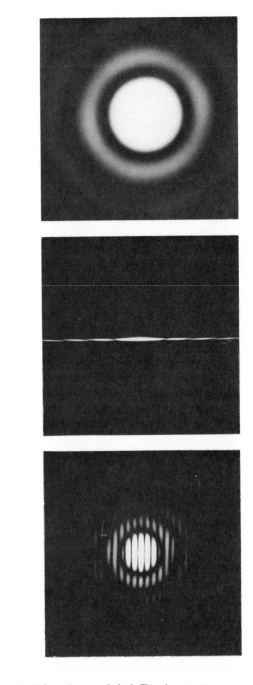

Figure 3.3. Some two-dimensional functions and their Fourier spectra.

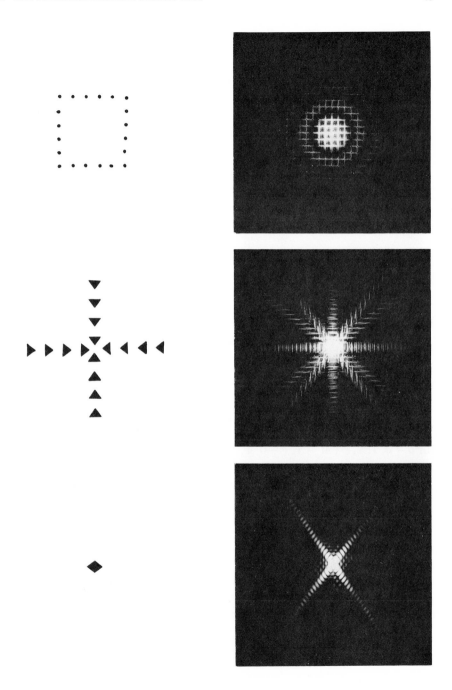

Figure 3.3. (Continued.)

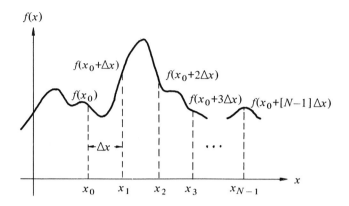

Figure 3.4 Sampling a continuous function.

axis. It can be shown that Δu and Δx are related by the expression

$$\Delta u = \frac{1}{N\,\Delta x} \tag{3.2-4}$$

In the two-variable case the discrete Fourier transform pair is given by the equations

$$F(u, v) = \frac{1}{MN} \sum_{x=0}^{M-1} \sum_{y=0}^{N-1} f(x, y)\, \exp\left[-j2\pi(ux/M + vy/N)\right] \tag{3.2-5}$$

for $u = 0, 1, 2, ..., M - 1, v = 0, 1, 2, ..., N - 1$, and

$$f(x, y) = \sum_{u=0}^{M-1} \sum_{v=0}^{N-1} F(u, v)\, \exp\left[j2\pi(ux/M + vy/N)\right] \tag{3.2-6}$$

for $x = 0, 1, 2, ..., M - 1$ and $y = 0, 1, 2, ..., N - 1$.

Sampling of a continuous function is now in a two-dimensional grid with divisions of width Δx and Δy in the x- and y-axis, respectively. As in the one-dimensional case, the discrete function $f(x, y)$ represents samples of the function $f(x_0 + x\Delta x, y_0 + y\Delta y)$ for $x = 0, 1, 2, ..., M - 1$ and $y = 0, 1, 2, ..., N - 1$. Similar comments hold for $F(u, v)$. The sampling increments in the spatial and frequency domains are related by

$$\Delta u = \frac{1}{M\Delta x} \tag{3.2-7}$$

and

$$\Delta v = \frac{1}{N\Delta y} \tag{3.2-8}$$

When images are sampled in a square array we have that $M = N$ and

$$F(u, v) = \frac{1}{N} \sum_{x=0}^{N-1} \sum_{y=0}^{N-1} f(x, y) \exp\left[-j2\pi(ux + vy)/N\right] \qquad (3.2\text{-}9)$$

for $u, v = 0, 1, 2, ..., N - 1$, and

$$f(x, y) = \frac{1}{N} \sum_{u=0}^{N-1} \sum_{v=0}^{N-1} F(u, v) \exp\left[j2\pi(ux + vy)/N\right] \qquad (3.2\text{-}10)$$

for $x, y = 0, 1, 2, ..., N-1$. It is noted that in this case we have included a $1/N$ term in both expressions. Since $F(u, v)$ and $f(x, y)$ are a Fourier transform pair, the grouping of these constant multiplicative terms is arbitrary. In practice, images are typically digitized in square arrays, so we will be mostly concerned with the Fourier transform pair given in Eqs. (3.2-9) and (3.2-10). The formulation given in Eqs. (3.2-5) and (3.2-6) will be used from time to time in situations where it is important to stress generality of the image size.

The Fourier spectrum, phase, and energy spectrum of one- and two-dimensional discrete functions are also given by Eqs. (3.1-4) through (3.1-6) and Eqs. (3.1-11) through (3.1-13), respectively. The only difference is that the independent variables are discrete.

Unlike the continuous case, we need not be concerned about the existence of the discrete Fourier transform since both $F(u)$ and $F(u, v)$ always exist in the discrete case. In the one-dimensional case, for example, this can be shown by direct substitution of Eq. (3.2-3) into Eq. (3.2-2):

$$F(u) = \frac{1}{N} \sum_{x=0}^{N-1} \left[\sum_{r=0}^{N-1} F(r) \exp\left[j2\pi rx/N\right] \exp\left[-j2\pi ux/N\right] \right]$$

$$= \frac{1}{N} \sum_{r=0}^{N-1} F(r) \left[\sum_{x=0}^{N-1} \exp\left[j2\pi rx/N\right] \exp\left[-j2\pi ux/N\right] \right]$$

$$= F(u) \qquad\qquad\qquad\qquad\qquad\qquad\qquad (3.2\text{-}11)$$

Identity (3.2-11) follows from the orthogonality condition

$$\sum_{x=0}^{N-1} \exp\left[j2\pi rx/N\right] \exp\left[-j2\pi ux/N\right] = \begin{cases} N & \text{if } r = u, \\ \\ 0 & \text{otherwise} \end{cases} \qquad (3.2\text{-}12)$$

It is noted that a change of variable from u to r was made in Eq. (3.2-2) to clarify the notation.

Substitution of Eq. (3.2-2) into Eq. (3.2-3) would also yield an identity on $f(x)$, thus indicating that the Fourier transform pair given by these equations always exists. A similar argument holds for the discrete, two-dimensional Fourier transform pair.

Example: As an illustration of Eqs. (3.2-2) and (3.2-3), consider the function shown in Fig. 3.5(a). If this function is sampled at the argument values $x_0 = 0.5$, $x_1 = 0.75$, $x_2 = 1.0$, $x_3 = 1.25$, and if the argument is redefined as discussed above, we obtain the discrete function shown in Fig. 3.5(b).

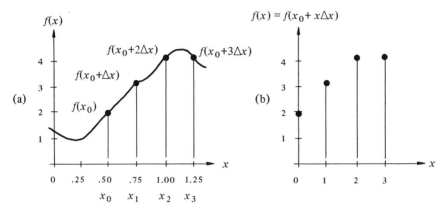

Figure 3.5 A simple function and samples in the x domain. In (a) x is a continuous variable; in (b) x is discrete.

Application of Eq. (3.2-2) to the resulting four samples yields the following sequence of steps:

$$F(0) = \frac{1}{4} \sum_{x=0}^{3} f(x) \exp[0]$$

$$= \frac{1}{4} \left[f(0) + f(1) + f(2) + f(3) \right]$$

$$= \frac{1}{4} \left[2 + 3 + 4 + 4 \right]$$

$$= 3.25$$

$$F(1) = \frac{1}{4} \sum_{x=0}^{3} f(x) \exp[-j2\pi x/4]$$

$$= \frac{1}{4} \left[2e^0 + 3e^{-j\pi/2} + 4e^{-j\pi} + 4e^{-j3\pi/2} \right]$$

$$= \frac{1}{4} \left[-2 + j \right]$$

where the last step follows from Euler's formula. Continuing with this procedure, we obtain:

$$F(2) = \frac{1}{4} \sum_{x=0}^{3} f(x) \exp[-j4\pi x/4]$$

$$= \frac{1}{4} \left[2e^0 + 3e^{-j\pi} + 4e^{-j2\pi} + 4e^{-j3\pi} \right]$$

$$= -\frac{1}{4} \left[1 + j0 \right]$$

and

$$F(3) = \frac{1}{4} \sum_{x=0}^{3} f(x) \exp[-j6\pi x/4]$$

$$= \frac{1}{4} \left[2e^0 + 3e^{-j3\pi/2} + 4e^{-j3\pi} + 4e^{-j9\pi/2} \right]$$

$$= -\frac{1}{4} \left[2 + j \right]$$

It is noted that all values of $f(x)$ contribute to each of the four terms of the discrete Fourier transform. Conversely, all terms of the transform contribute in forming the inverse transform via Eq. (3.2-3). The procedure for obtaining the inverse is analogous to the one described above for computing $F(u)$.

The Fourier spectrum is obtained from the magnitude of each of the transform terms; that is,

$$|F(0)| = 3.25$$

$$|F(1)| = \left[(2/4)^2 + (1/4)^2 \right]^{1/2} = \sqrt{5}/4$$

$$|F(2)| = \left[(1/4)^2 + (0/4)^2 \right]^{1/2} = 1/4$$

and

$$|F(3)| = \left[(2/4)^2 + (1/4)^2 \right]^{1/2} = \sqrt{5}/4 \qquad \square$$

3.3 SOME PROPERTIES OF THE TWO-DIMENSIONAL FOURIER TRANSFORM

Attention is focused in this section on properties of the Fourier transform which will be of value in subsequent discussions. Although our primary interest is on two-dimensional, discrete transforms, the underlying concepts of some of these properties are much easier to grasp if they are presented first in their one-dimensional, continuous form.

Since several of the topics considered in this section are illustrated by images and their Fourier spectra displayed as intensity functions, some comments concerning these displays are in order before beginning a discussion of Fourier transform properties. Many image spectra decrease rather rapidly as a function of increasing frequency and, therefore, their high-frequency terms have a tendency to become obscured when displayed in image form. A useful processing technique which compensates for this difficulty consists of displaying the function

$$D(u, v) = \log(1 + |F(u, v)|) \tag{3.3-1}$$

instead of $|F(u, v)|$. Use of this equation preserves the zero values in the frequency plane since $D(u, v) = 0$ when $|F(u, v)| = 0$. It is also noted that $D(u, v)$ is a nonnegative function.

As an illustration of the properties of the preceding logarithm transformation, consider the Fourier spectra shown in Fig. 3.6. Part (a) of this figure is the spectrum of a pulse of unit width and height (see Fig. 3.1). Figure 3.6(b) shows the effect of using Eq. (3.3-1) and rescaling the results to the amplitude interval [0, 1]. It is noted that, although the peak value is still one, the amplitude of the side lobes has increased slightly. The effect is more pronounced in Figs. 3.6(c) and (d), where the maximum amplitude of $|F(u)|$ is assumed to be 20 instead of 1. In Fig. 3.6(d), which was rescaled after taking the log, the ratio of the central peak to the peak of any of the side lobes is much smaller than the corresponding ratio in Fig. 3.6(b). If one considers the amplitude of these functions as being proportional to intensity, it is evident that the side lobes of Fig. 3.6(d) would be more visible than those in Fig. 3.6(b). The ideal case, of course, would be for all peaks to have the same amplitude since this would make them all equally bright in an intensity display.

The degree of amplitude equalization achieved by using Eq. (3.3-1) depends on the relative magnitude of the function $|F(u, v)|$, as illustrated in Fig. 3.6. In the one-dimensional case, the behavior of the ratio of the maximum to some minimum (nonzero) value of $F(u)$ as a function of amplitude can be examined by using the relation

$$R = \frac{\log(1 + KF_{max})}{\log(1 + KF_{min})} \tag{3.3-2}$$

where K is a scale factor.

Suppose, for example, that we consider the ratio of $F_{max} = 1.0$ to the peak of the first side lobe in Fig. 3.6(a). In this case, $F_{min} = 0.2$ and $F_{max}/F_{min} = 5.0$. In Fig. 3.6(b), on the other hand, use of the log operation yields the ratio $\log(1 + F_{max})/\log(1 + F_{min}) = 3.8$. Figure 3.6(c) is the same as Fig. 3.6(a), with the exception that all values have been multiplied by a scale factor of 20. Since the scale factor is the same for all values, the ratio F_{max}/F_{min} is again equal to 5.0. When the log is taken, however, we have the ratio $\log(1 + 20 F_{max})/\log(1 + 20 F_{min}) = 1.9$, which is substantially

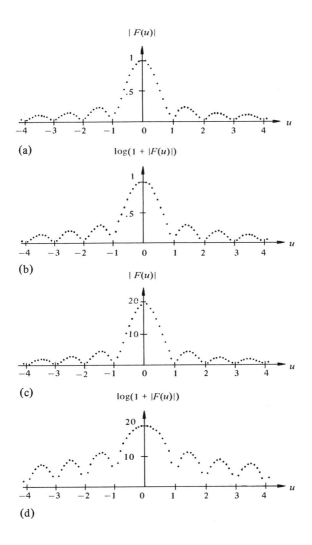

Figure 3.6. Effect of the log operation on a Fourier spectrum.

smaller. The general behavior of R as a function of K is illustrated in Fig. 3.7 for the above values of F_{max} and F_{min}. It is noted that the ratio decreases rather quickly at first, but it levels off for values of K greater than about 40. In the limit, R approaches 1 as K approaches infinity which, as mentioned above, is the ideal ratio for display purposes.

Use of Eq. (3.3-1) in a two-dimensional display of the spectrum greatly facilitates visual interpretation of the Fourier transform. An example of this is given in Fig. 3.8 which shows a picture of the planet Saturn, its

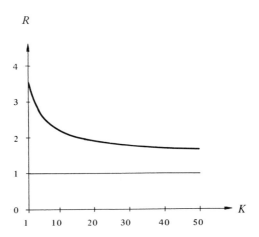

Figure 3.7. Behavior of Eq. (3.3-2) as a function of K for $F_{max} = 1.0$ and $F_{min} = 0.2$.

normal spectrum, and the spectrum processed by adding one to $|F(u, v)|$, taking the log, and rescaling the values to the same gray scale used in displaying $|F(u, v)|$. By equalizing the range of values in the spectrum, use of the log operation clearly brings out low-level information which, if shown simply as $|F(u, v)|$, is beyond the dynamic range of the display system. Most Fourier spectra shown in image form throughout this book were processed using Eq. (3.3-1).

3.3.1 Separability

The discrete Fourier transform pair given in Eqs. (3.2-9) and (3.2-10) can be expressed in the separable forms

$$F(u, v) = \frac{1}{N} \sum_{x=0}^{N-1} \exp[-j2\pi ux/N] \sum_{y=0}^{N-1} f(x, y) \exp[-j2\pi vy/N] \qquad (3.3\text{-}3)$$

for $u, v = 0, 1, ..., N - 1$, and

$$f(x, y) = \frac{1}{N} \sum_{u=0}^{N-1} \exp[j2\pi ux/N] \sum_{v=0}^{N-1} F(u, v) \exp[j2\pi vy/N] \qquad (3.3\text{-}4)$$

for $x, y = 0, 1, ..., N - 1$.

For our purposes, the principal advantage of the separability property is that $F(u, v)$ or $f(x, y)$ can be obtained in two steps by successive applications of the one-dimensional Fourier transform or its inverse. This becomes

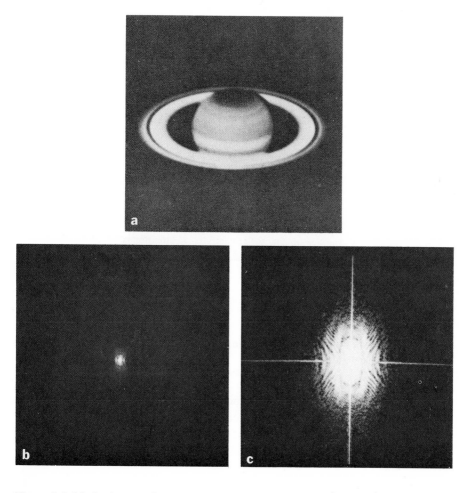

Figure 3.8 (a) A picture of the planet Saturn. (b) Display of $|F(u, v)|$. (c) Display of $\log(1 + |F(u, v)|)$.

evident if Eq. (3.3-3) is expressed in the form

$$F(u, v) = \frac{1}{N} \sum_{x=0}^{N-1} F(x, v) \exp[-j2\pi ux/N] \qquad (3.3\text{-}5)$$

where

$$F(x, v) = N\left[\frac{1}{N} \sum_{y=0}^{N-1} f(x, y) \exp[-j2\pi vy/N] \right] \qquad (3.3\text{-}6)$$

For *each* value of x, the expression inside the brackets is a one-dimensional transform with frequency values $v = 0, 1, ..., N - 1$. Therefore, the two-dimensional function $F(x, v)$ is obtained by taking a transform along *each* row of $f(x, y)$ and multiplying the result by N. The desired result, $F(u, v)$, is then obtained by taking a transform along each column of $F(x, v)$, as indicated by Eq. (3.3-5). The procedure is summarized in Fig. 3.9. It should be

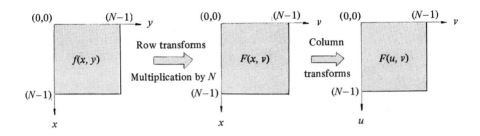

Figure 3.9. Computation of the two-dimensional Fourier transform as a series of one-dimensional transforms.

noted that the same results would be obtained by first taking transforms along the columns of $f(x, y)$ and then along the rows of the result. This is easily shown by reversing the order of the summations of Eq. (3.3-3). Identical comments hold for the implementation of Eq. (3.3-4).

3.3.2. Translation

The translation properties of the Fourier transform pair are given by

$$f(x, y) \exp[j2\pi(u_0 x + v_0 y)/N] \Leftrightarrow F(u - u_0, v - v_0) \qquad (3.3\text{-}7a)$$

and

$$f(x - x_0, y - y_0) \Leftrightarrow F(u, v) \exp[-j2\pi(ux_0 + vy_0)/N] \qquad (3.3\text{-}7b)$$

where the double arrow is used to indicate the correspondence between a function and its Fourier transform (and vice versa), as given in Eqs. (3.1-9) and (3.1-10) or Eqs. (3.2-9) and (3.2-10).

Equation (3.3-7a) shows that multiplying $f(x, y)$ by the indicated exponential term and taking the transform of the product results in a shift of the origin of the frequency plane to the point (u_0, v_0). Similarly, multiplying $F(u, v)$ by the exponential term shown and taking the inverse transform moves the origin of the spatial plane to (x_0, y_0).

Considerable use of Eq. (3.3-7a), with $u_0 = v_0 = N/2$, will be made in this and the next chapter. In this case it follows that

$$\exp\left[j2\pi(u_0 x + v_0 y)/N \right] = e^{j\pi(x+y)}$$

$$= (-1)^{x+y}$$

and

$$f(x, y)(-1)^{x+y} \Leftrightarrow F(u - N/2, v - N/2) \qquad (3.3\text{-}8)$$

Thus, the origin of the Fourier transform of $f(x, y)$ can be moved to the center of its corresponding $N \times N$ frequency square simply by multiplying $f(x, y)$ by $(-1)^{x+y}$. In the one-variable case this reduces to multiplication of $f(x)$ by the term $(-1)^x$.

It is interesting to note from Eq. (3.3-7b) that a shift in $f(x, y)$ does not affect the magnitude of its Fourier transform since

$$\left| F(u, v) \exp\left[-j2\pi(ux_0 + vy_0)/N \right] \right| = |F(u, v)| \qquad (3.3\text{-}9)$$

This should be kept in mind since visual examination of the transform is usually limited to a display of its magnitude.

3.3.3 Periodicity and Conjugate Symmetry

The discrete Fourier transform and its inverse are *periodic* with period N; that is,

$$F(u, v) = F(u + N, v) = F(u, v + N) = F(u + N, v + N) \qquad (3.3\text{-}10)$$

The validity of this property can be demonstrated by direct substitution of the variables of $(u + N)$ and $(v + N)$ in Eq. (3.2-9). Although Eq. (3.3-10) points out that $F(u, v)$ repeats itself for an infinite number of values of u and v, only the N values of each variable in any one period are required to obtain $f(x, y)$ from $F(u, v)$. In other words, only one period of the transform is necessary to completely specify $F(u, v)$ in the frequency domain. Similar comments hold for $f(x, y)$ in the spatial domain.

The Fourier transform also exhibits conjugate symmetry since

$$F(u, v) = F^*(-u, -v) \qquad (3.3\text{-}11)$$

or, more interestingly,

$$|F(u, v)| = |F(-u, -v)| \tag{3.3-12}$$

As mentioned earlier, it is often of interest to display the magnitude of the Fourier transform for interpretation purposes. In order to examine the implications of Eqs. (3.3-10) and (3.3-12) on a display of the transform magnitude, let us first consider the one-variable case where

$$F(u) = F(u + N)$$

and

$$|F(u)| = |F(-u)|.$$

The periodicity property indicates that $F(u)$ has a period of length N, and the symmetry property shows that the magnitude of the transform is centered about the origin, as shown in Fig. 3.10(a). It is evident from this figure and the above comments that the magnitudes of the transform values from $([N/2] + 1)$ to $(N - 1)$ are images of the values in the half period on the left side of the origin. Since the discrete Fourier transform has been formulated for values of u in the interval $[0, N - 1]$, we see that the result

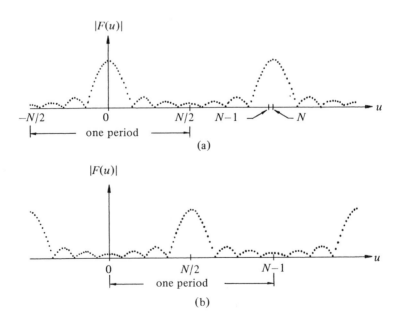

Figure 3.10 Illustration of the periodicity properties of the Fourier transform. (a) Fourier spectrum showing back-to-back half periods in the interval $[0, N - 1]$. (b) Shifted spectrum showing a full period in the same interval.

of this formulation yields two half-periods in this interval which are back to back. To display one full period, all that is necessary is to move the origin of the transform to the point $u = N/2$, as shown in Fig. 3.10(b). This is easily accomplished by multiplying $f(x)$ by $(-1)^x$ prior to taking the transform, as indicated earlier.

The same observations hold for the magnitude of the two-dimensional Fourier transform, with the exception that the results are considerably more difficult to interpret if the origin of the transform is not shifted to the frequency point $(N/2, N/2)$. This is shown in Figs. 3.11(a) and (b), where the latter figure was obtained by using the centering property of expression (3.3-8).

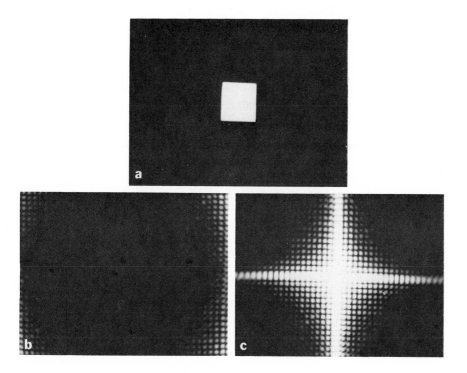

Figure 3.11. (a) A simple image. (b) Fourier spectrum without shifting. (c) Fourier spectrum shifted to the center of the frequency square.

3.3.4 Rotation

If we introduce the polar coordinates

$$x = r \cos \theta \qquad y = r \sin \theta \qquad u = \omega \cos \phi \qquad v = \omega \sin \phi$$

then $f(x, y)$ and $F(u, v)$ become $f(r, \theta)$ and $F(\omega, \phi)$, respectively. It can be shown by direct substitution in either the continuous or discrete Fourier transform pair that

$$f(r, \theta + \theta_0) \Leftrightarrow F(\omega, \phi + \theta_0) \qquad (3.3\text{-}13)$$

In other words, if $f(x, y)$ is rotated by an angle θ_0, then $F(u, v)$ is rotated by the same angle. Similarly, rotating $F(u, v)$ causes $f(x, y)$ to be rotated by the same angle. This property is illustrated in Fig. 3.12.

3.3.5 Distributivity and Scaling

It follows directly from the definition of the continuous or discrete transform pair that

$$\mathfrak{F}\{f_1(x, y) + f_2(x, y)\} = \mathfrak{F}\{f_1(x, y)\} + \mathfrak{F}\{f_2(x, y)\} \qquad (3.3\text{-}14)$$

and, in general, that

$$\mathfrak{F}\{f_1(x, y) \cdot f_2(x, y)\} \neq \mathfrak{F}\{f_1(x, y)\} \cdot \mathfrak{F}\{f_2(x, y)\} \qquad (3.3\text{-}15)$$

In other words, the Fourier transform and its inverse are distributive over addition, but not over multiplication.

It is also easy to show that for two scalars a and b,

$$af(x, y) \Leftrightarrow aF(u, v) \qquad (3.3\text{-}16)$$

and

$$f(ax, by) \Leftrightarrow \frac{1}{|ab|} F(u/a, v/b) \qquad (3.3\text{-}17)$$

3.3.6 Average Value

A widely-used definition of the average value of a two-dimensional discrete function is given by the expression

$$\bar{f}(x, y) = \frac{1}{N^2} \sum_{x=0}^{N-1} \sum_{y=0}^{N-1} f(x, y) \qquad (3.3\text{-}18)$$

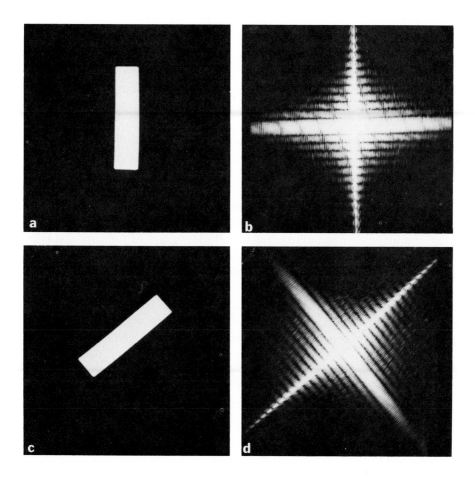

Figure 3.12. Rotational properties of the Fourier transform. (a) A simple image. (b) Spectrum. (c) Rotated image. (d) Resulting spectrum.

Substitution of $u = v = 0$ in Eq. (3.2-9) yields

$$F(0, 0) = \frac{1}{N} \sum_{x=0}^{N-1} \sum_{y=0}^{N-1} f(x, y) \tag{3.3-19}$$

We see, therefore, that $\bar{f}(x, y)$ is related to the Fourier transform of $f(x, y)$ by the equation

$$\bar{f}(x, y) = \frac{1}{N} F(0, 0) \qquad \text{(3.3-20)}$$

3.3.7 Laplacian

The Laplacian of a two-variable function $f(x, y)$ is defined as

$$\nabla^2 f(x, y) = \frac{\partial^2 f}{\partial x^2} + \frac{\partial^2 f}{\partial y^2} \qquad \text{(3.3-21)}$$

It follows from the definition of the two-dimensional Fourier transform that

$$\mathfrak{F}\{\nabla^2 f(x, y)\} \Leftrightarrow -(2\pi)^2(u^2 + v^2)F(u, v) \qquad \text{(3.3-22)}$$

The Laplacian operator is useful for outlining edges in an image.

3.3.8 Convolution and Correlation

In this section we consider two Fourier transform relationships which constitute a basic link between the spatial and frequency domains. These relationships, called convolution and correlation, are of fundamental importance in developing a firm understanding of image processing techniques based on the Fourier transform. In order to clarify the concepts involved, we begin the discussion by considering convolution in one dimension and with continuous arguments. The development is then extended to the discrete case and, finally, to the two-dimensional continuous and discrete cases. The same format is followed in developing the concept of correlation.

3.3.8.1 Convolution

The convolution of two functions $f(x)$ and $g(x)$, denoted by $f(x)*g(x)$, is defined by the integral

$$f(x)*g(x) = \int_{-\infty}^{\infty} f(\alpha) g(x - \alpha)\, d\alpha \qquad \text{(3.3-23)}$$

where α is a dummy variable of integration. Since the mechanics of the convolution integral are not particularly easy to visualize, let us illustrate graphically the use of Eq. (3.3-23) by two simple examples.

Example: The first example demonstrates convolution of the functions $f(x)$ and $g(x)$ shown in Figs. 3.13(a) and (b), respectively. Before carrying out the integration, it is necessary to form the function $g(x - \alpha)$. This is shown in two steps in Figs. 3.13(c) and (d). It is noted that this operation is simply one

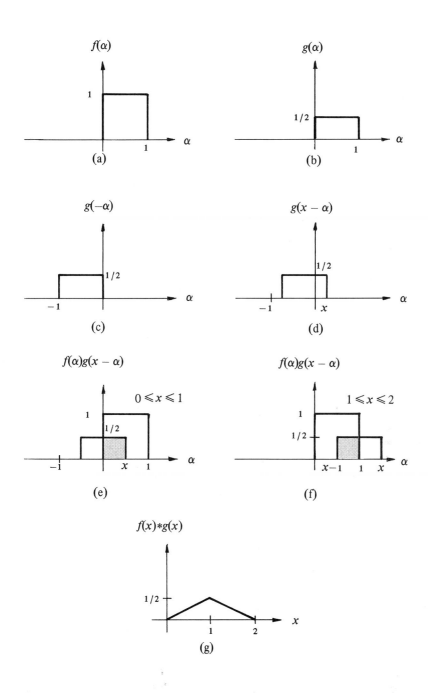

Figure 3.13. Graphical illustration of convolution. The shaded areas indicate regions where the product is not zero.

of folding $g(\alpha)$ about the origin to give $g(-\alpha)$ and then displacing this function by x. Then, for any given value of x, we multiply $f(\alpha)$ by the corresponding $g(x - \alpha)$ and integrate the product from $-\infty$ to ∞. The product of
$f(\alpha)$ and $g(x - \alpha)$ is the shaded portion of Fig. 3.13(e). This figure is valid
for $0 \leqslant x \leqslant 1$. Since the product is zero for values of α outside the interval
$[0, x]$, we have that $f(x)*g(x) = x/2$, which is simply the area of the shaded
region in Fig. 3.13(e). For x in the interval $[1, 2]$ we use Fig. 3.13(f). In this
case, $f(x)*g(x) = (1 - x/2)$. Thus, by noting that $f(\alpha)g(x - \alpha)$ is zero for
values of x outside the interval $[0, 2]$, we have

$$f(x)*g(x) = \begin{cases} x/2 & 0 \leqslant x \leqslant 1 \\ 1 - x/2 & 1 \leqslant x \leqslant 2 \\ 0 & \text{elsewhere} \end{cases}$$

The result is shown in Fig. 3.13(g).   □

One aspect of Eq. (3.3-23) which will be of use later in this section
involves the convolution of a   function $f(x)$ with the impulse function
$\delta(x - x_0)$, which is defined by the relation

$$\int_{-\infty}^{\infty} f(x)\delta(x - x_0)\, dx = f(x_0) \tag{3.3-24}$$

The function $\delta(x - x_0)$ may be viewed as having an area of unity in an
infinitesimal neighborhood about x_0 and being zero everywhere else; that is,

$$\int_{-\infty}^{\infty} \delta(x - x_0)\, dx = \int_{x_0^-}^{x_0^+} \delta(x - x_0)\, dx = 1 \tag{3.3-25}$$

For most purposes we can say that $\delta(x - x_0)$ is located at $x = x_0$, and that
the strength of the impulse is determined by the value of $f(x)$ at $x = x_0$.
For instance, if $f(x) = A$, we have that $A\delta(x - x_0)$ is an impulse of strength
A located at $x = x_0$. It is common practice to represent impulses graphically
by an arrow located at x_0 and with height equal to the impulse strength.
Figure 3.14 shows this representation for $A\delta(x - x_0)$.

Example: As a second illustration of the use of Eq. (3.3-23), suppose that
the function $f(x)$ shown in Fig. 3.15(a) is convolved with the function $g(x) =$
$\delta(x + T) + \delta(x) + \delta(x - T)$ shown in Fig. 3.15(b). By folding $g(x)$, sliding it
past $f(x)$, and making use of Eqs. (3.3-23) and (3.3-24), we obtain the result
shown in Fig. 3.15(c). It is noted that convolution in this case amounts

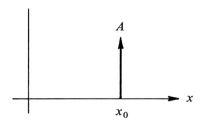

Figure 3.14. Graphical representation of $A\delta(x - x_0)$.

merely to "copying" $f(x)$ at the location of each impulse. $\qquad\qquad$ □

The importance of convolution in frequency-domain analysis lies in the fact that $f(x)*g(x)$ and $F(u)G(u)$ constitute a Fourier transform pair. In other words, if $f(x)$ has the Fourier transform $F(u)$ and $g(x)$ has the Fourier transform $G(u)$, then $f(x)*g(x)$ has the Fourier transform $F(u)G(u)$. This result, formally stated as

$$f(x)*g(x) \Leftrightarrow F(u)G(u) \qquad (3.3\text{-}26)$$

indicates that convolution in the x-domain can also be obtained by taking the inverse Fourier transform of the product $F(u)G(u)$. An analogous result is that convolution in the frequency domain reduces to multiplication in the x-domain; that is

$$f(x)g(x) \Leftrightarrow F(u)*G(u). \qquad (3.3\text{-}27)$$

These two results are commonly referred to as the *convolution theorem.*
Suppose that instead of being continuous, $f(x)$ and $g(x)$ are discretized into sampled arrays of size A and B, respectively: $\{f(0), f(1), f(2), ..., f(A - 1)\}$, and $\{g(0), g(1), g(2), ..., g(B - 1)\}$. As pointed out in Section 3.3.3, the discrete Fourier transform and its inverse are periodic functions. In order to formulate a discrete convolution theorem that is consistent with this periodicity property, we may *assume* that the discrete functions $f(x)$ and $g(x)$ are periodic with some period M. The resulting convolution will then be periodic with the same period. The problem is how to select a value for M. It can be shown (Brigham [1974]) that unless we choose

$$M \geqslant A + B - 1 \qquad (3.3\text{-}28)$$

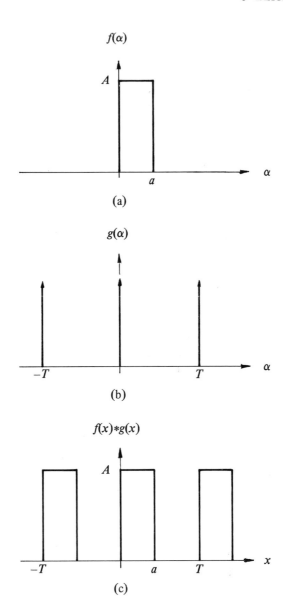

Figure 3.15. Convolution involving impulse functions.

the individual periods of the convolution will overlap; this overlap is commonly referred to as *wraparound error*. If $M = A + B - 1$, the periods will be adjacent; if $M > A + B - 1$, the periods will be spaced apart, with the degree of separation being equal to the difference between M and $A + B - 1$. Since the assumed period must be greater than either A or B, the length of the sampled sequences must be increased so that both are of length M. This can be done by appending zeros to the given samples to form the following *extended* sequences:

$$f_e(x) = \begin{cases} f(x) & 0 \leqslant x \leqslant A - 1 \\ \\ 0 & A \leqslant x \leqslant M - 1 \end{cases}$$

and

$$g_e(x) = \begin{cases} g(x) & 0 \leqslant x \leqslant B - 1 \\ \\ 0 & B \leqslant x \leqslant M - 1 \end{cases}$$

Based on this, we define the discrete convolution of $f_e(x)$ and $g_e(x)$ by the expression

$$f_e(x) * g_e(x) = \sum_{m=0}^{M-1} f_e(m) g_e(x - m) \tag{3.3-29}$$

for $x = 0, 1, 2, ..., M - 1$. The convolution function is a discrete, periodic array of length M, with the values $x = 0, 1, 2, ..., M - 1$ describing a full period of $f_e(x) * g_e(x)$.

The mechanics of discrete convolution are basically the same as for continuous convolution. The only differences are that displacements take place in discrete increments corresponding to the separation between samples, and that integration is replaced by a summation. Similarly, Eqs. (3.3-26) and (3.3-27) also hold in the discrete case where, to avoid wrap-around error, we use $f_e(x)$ and its transform. The discrete variables x and u assume values in the range $0, 1, 2, ..., M - 1$.

Example: The preceding considerations are illustrated graphically in Fig. 3.16 for continuous and discrete convolution. The diagrams for the discrete case show A samples for both $f(x)$ and $g(x)$ in the interval $[0, 1]$, as well as an assumed period of $M = A + B - 1 = 2A - 1$.

It is noted that the convolution function is periodic and that, since $M = 2A - 1$, the periods are adjacent. Choosing $M > 2A - 1$ would have

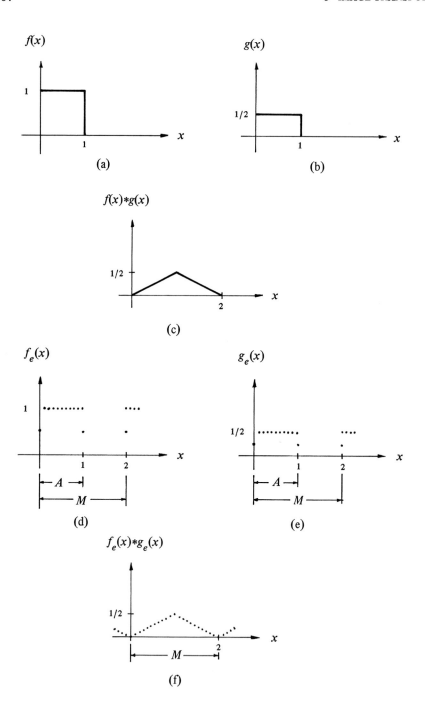

Figure 3.16. Comparison between continuous and discrete convolution.

produced a larger separation between these periods. It is also important to note that a period is completely described by M samples.   □

Two-dimensional convolution is analogous in form to Eq. (3.3-23). Thus, for two functions $f(x, y)$ and $g(x, y)$, we have

$$f(x, y) * g(x, y) = \int\limits_{-\infty}^{\infty} \int f(\alpha, \beta) g(x - \alpha, y - \beta) \, d\alpha \, d\beta \qquad (3.3\text{-}30)$$

The convolution theorem in two-dimensions then is given by the relations

$$f(x, y) * g(x, y) \Leftrightarrow F(u, v) G(u, v) \qquad (3.3\text{-}31)$$

and

$$f(x, y) g(x, y) \Leftrightarrow F(u, v) * G(u, v) \qquad (3.3\text{-}32)$$

Equation (3.3-30) is more difficult to illustrate graphically than Eq. (3.3-23). Figure 3.17 shows the basic folding, displacement, and multiplication operations required for two-dimensional convolution. The result of varying the displacement variables, x and y, would be a two-dimensional convolution surface with a shape dependent on the nature of the functions involved in the process.

The two-dimensional, discrete convolution is formulated by letting $f(x, y)$ and $g(x, y)$ be discrete arrays of size $A \times B$ and $C \times D$, respectively. As in the one-dimensional case, these arrays must be assumed periodic with some period M and N in the x and y directions, respectively. Wraparound error in the individual convolution periods is avoided by choosing

$$M \geqslant A + C - 1 \qquad (3.3\text{-}33)$$

and

$$N \geqslant B + D - 1 \qquad (3.3\text{-}34)$$

The periodic sequences are formed by extending $f(x, y)$ and $g(x, y)$ as follows:

$$f_e(x, y) = \begin{cases} f(x, y) & 0 \leqslant x \leqslant A - 1 \quad \text{and} \quad 0 \leqslant y \leqslant B - 1 \\ 0 & A \leqslant x \leqslant M - 1 \quad \text{or} \quad B \leqslant y \leqslant N - 1 \end{cases}$$

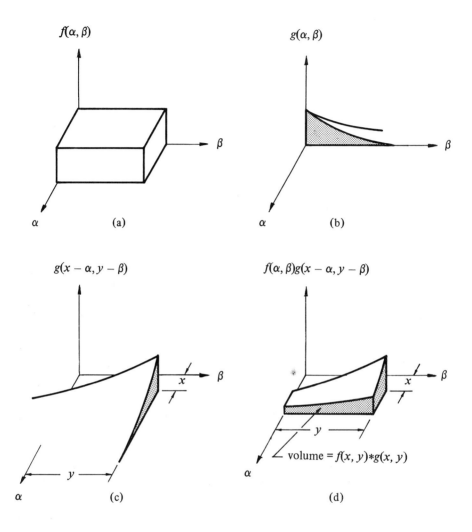

Figure 3.17. Illustration of the folding, displacement, and multiplication steps needed to perform two-dimensional convolution.

and

$$
g_e(x, y) = \begin{cases} g(x, y) & 0 \leqslant x \leqslant C - 1 \quad \text{and} \quad 0 \leqslant y \leqslant D - 1 \\ \\ 0 & C \leqslant x \leqslant M - 1 \quad \text{or} \quad D \leqslant y \leqslant N - 1 \end{cases}
$$

The two-dimensional convolution of $f_e(x, y)$ and $g_e(x, y)$ is given by the relation

$$f_e(x, y) * g_e(x, y) = \sum_{m=0}^{M-1} \sum_{n=0}^{N-1} f_e(m, n) g_e(x - m, y - n) \qquad (3.3\text{-}35)$$

for $x = 0, 1, 2, ..., M - 1$ and $y = 0, 1, 2, ..., N - 1$. The $M \times N$ array given by this equation is one period of the discrete, two-dimensional convolution. If M and N are chosen according to Eqs. (3.3-33) and (3.3-34), this array is guaranteed to be free of interference from other adjacent periods. As in the one-dimensional case, the continuous convolution theorem given in Eqs. (3.3-31) and (3.3-32) also applies to the discrete case with $u = 0, 1, 2, ..., M - 1$ and $v = 0, 1, 2, ..., N - 1$. All computations involve the extended functions $f_e(x, y)$ and $g_e(x, y)$.

The theoretical power of the convolution theorem will become evident in Section 3.3.9 when we discuss the sampling theorem. From a practical point of view, it is often more efficient to compute the discrete convolution in the frequency domain instead of using Eq. (3.3-35) directly. The procedure is to compute the Fourier transforms of $f_e(x, y)$ and $g_e(x, y)$ by using a fast Fourier transform (FFT) algorithm (see Section 3.4). The two transforms are then multiplied and the inverse Fourier transform of the product will yield the convolution function. A comparison by Brigham [1974] shows that, for one-dimensional arrays, the FFT approach is faster if the number of points is greater than 32. Although this figure is dependent on the particular machine and algorithms used, it is well below the number of points in a row or column of a typical image.

3.3.8.2 Correlation

The correlation[†] of two continuous functions $f(x)$ and $g(x)$, denoted by $f(x) \circ g(x)$, is defined by the relation

$$f(x) \circ g(x) = \int_{-\infty}^{\infty} f(\alpha) g(x + \alpha) \, d\alpha \qquad (3.3\text{-}36)$$

The forms of Eqs. (3.3-36) and (3.3-23) are similar, the only difference being that the function $g(x)$ is not folded about the origin. Thus, to perform correlation we simply slide $g(x)$ by $f(x)$ and integrate the product from $-\infty$ to ∞ for each value of displacement x. The procedure is illustrated in Fig. 3.18, which should be compared with Fig. 3.13.

[†]If $f(x)$ and $g(x)$ are the same function, Eq. (3.3-36) is usually called the *autocorrelation* function; if $f(x)$ and $g(x)$ are different, the term *crosscorrelation* is normally used.

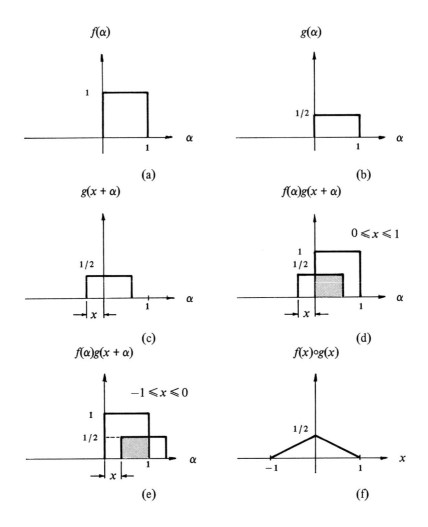

Figure 3.18. Graphical illustration of correlation. The shaded areas indicate regions where the product is not zero.

The discrete equivalent of Eq. (3.3-36) is defined as

$$f_e(x) \circ g_e(x) = \sum_{m=0}^{M-1} f_e(m) g_e(x+m) \tag{3.3-37}$$

for $x = 0, 1, 2, ..., M - 1$. The same comments made above with regard to $f_e(x)$ and $g_e(x)$, the assumed periodicity of these functions, and the choice of values for M, also apply to Eq. (3.3-37).

Similar expressions hold for two dimensions. Thus, if $f(x, y)$ and $g(x, y)$ are functions of continuous variables, their correlation is defined as

$$f(x, y) \circ g(x, y) = \int\int_{-\infty}^{\infty} f(\alpha, \beta) g(x + \alpha, y + \beta) \, d\alpha \, d\beta \qquad (3.3\text{-}38)$$

For the discrete case we have

$$f_e(x, y) \circ g_e(x, y) = \sum_{m=0}^{M-1} \sum_{n=0}^{N-1} f_e(m, n) g_e(x + m, y + n) \qquad (3.3\text{-}39)$$

for $x = 0, 1, 2, ..., M - 1$ and $y = 0, 1, 2, ..., N - 1$. As in the case of discrete convolution, $f_e(x, y)$ and $g_e(x, y)$ are extended functions, and M and N are chosen according to Eqs. (3.3-33) and (3.3-34) in order to avoid wraparound error in the periods of the correlation function.

It can be shown for both the continuous and discrete cases that the following *correlation theorem* holds:

$$f(x, y) \circ g(x, y) \Leftrightarrow F(u, v) G^*(u, v) \qquad (3.3\text{-}40)$$

and

$$f(x, y) g^*(x, y) \Leftrightarrow F(u, v) \circ G(u, v) \qquad (3.3\text{-}41)$$

where "$*$" represents the complex conjugate. It is understood that, when interpreted for discrete variables, all functions are assumed to be extended and periodic.

One of the principal applications of correlation in image processing is in the area of *template* or *prototype matching*, where the problem is to find the closest match between a given unknown image and a set of images of known origin. One approach to this problem is to compute the correlation between the unknown and each of the known images. The closest match can then be found by selecting the image that yields the correlation function with the largest value. Since the resultant correlations are two-dimensional functions, this involves searching for the largest amplitude of each function. As in the case of discrete convolution, the computation of $f_e(x, y) \circ g_e(x, y)$ is often more efficiently carried out in the frequency domain using an FFT algorithm to obtain the forward and inverse transforms.

When comparing the results of discrete *vs.* continuous convolution or correlation, it should be noted that the way we have defined the discrete cases amounts to evaluation of the continuous forms by rectangular integration. Thus, if one wishes to compare discrete and continuous results on an absolute basis, Eqs. (3.3-29) and (3.3-37) should be multiplied by Δx, and Eqs. (3.3-35) and (3.3-39) by $\Delta x \Delta y$, where Δx and Δy are the separations between samples, as defined in Section 3.2. In Fig. 3.16, for example, the continuous and discrete convolution functions have the same amplitude because the discrete result was multiplied by Δx. When one is only computing and evaluating the discrete forms, however, the inclusion of these scale factors is a matter of preference. It is also important to note that all convolution and correlation expressions hold if $f(x)$ and $g(x)$, along with their corresponding transforms, are interchanged. This is also true if the functions are two-dimensional.

3.3.9 Sampling

The basic idea of image sampling was introduced in Section 2.3 on an intuitive basis. The Fourier transform and the convolution theorem provide the tools for a deeper analytical study of this problem. In particular, we are interested in looking at the question of how many samples should be taken so that no information is lost in the sampling process. Expressed differently, the problem is one of establishing under what sampling conditions a continuous image can be fully recovered from a set of sampled values. We begin the analysis with the one-dimensional case.

3.3.9.1 One-Dimensional Functions

Consider the function shown in Fig. 3.19(a), which is assumed to extend from $-\infty$ to ∞, and suppose that the Fourier transform of $f(x)$ vanishes for values of u outside the interval $[-W, W]$. The transform might appear as shown in Fig. 3.19(b)[†]. A function whose transform has this property for any finite value of W is called a *band-limited* function.

To obtain a sampled version of $f(x)$, we simply multiply this function by a sampling function $s(x)$, which consists of a train of impulses Δx units apart, as shown in Fig. 3.19(c). Since, by the convolution theorem, multiplication in the x domain is equivalent to convolution in the frequency domain, we obtain the Fourier transform shown in Fig. 3.19(f) for the product $s(x)f(x)$. It is noted that the transform is periodic with period $1/\Delta x$, and that the individual repetitions of $F(u)$ can overlap. In the first period, for example, the center of the overlapped region will occur at $u = 1/2\Delta x$ if the quantity $1/2\Delta x$ is less than W. To avoid this problem, therefore, we select the sampling interval Δx so that $1/2\Delta x \geqslant W$, or

$$\Delta x \leqslant \frac{1}{2W} \tag{3.3-42}$$

[†]Recall that the Fourier transform is a complex function. In the following graphical illustrations we show only the magnitude of the transforms for simplicity. The ordinate axis, however, will be labeled with $F(u)$, $G(u)$, etc., to indicate that the concepts involved are valid for the complete transform and not just its magnitude.

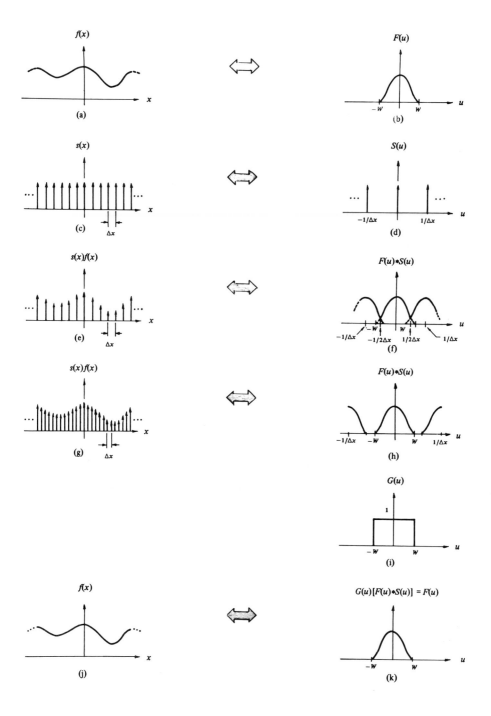

Figure 3.19 Graphical development of sampling concepts.

The result of decreasing Δx is shown in Figs. 3.19(g) and (h). The net effect is to separate the periods so that no overlap occurs. The importance of this operation lies in the fact that if one multiplies the transform of Fig. 3.19(h) by the function

$$G(u) = \begin{cases} 1 & -W \leqslant u \leqslant W \\ \\ 0 & \text{elsewhere} \end{cases} \tag{3.3-43}$$

it becomes possible to isolate $F(u)$ completely, as shown in Fig. 3.19(k). The inverse Fourier transform then yields the original *continuous* function $f(x)$. The result that a band-limited function can be recovered completely from samples whose spacing satisfies Eq. (3.3-42) is known as the *Whittaker-Shannon sampling theorem*.

It is important to keep in mind that all the frequency-domain information of a band-limited function is contained in the interval $[-W, W]$. If Eq. (3.3-42) is not satisfied, however, the transform in this interval is corrupted by contributions from adjacent periods. This phenomenon, frequently referred to as *aliasing*, precludes complete recovery of an under-sampled function.

The foregoing results apply to functions which are of unlimited duration in the x domain. Since this implies an infinite sampling interval, it is of interest to examine the practical case where a function is sampled only over a finite region. This situation is shown graphically in Fig. 3.20. Parts (a) through (f) are the same as in Fig. 3.19, with the exception that the separation between samples is assumed to satisfy the sampling theorem so that no aliasing is present. A finite sampling interval $[0, X]$ can be represented mathematically by multiplying the sampled result shown in Fig. 3.20(e) by the function

$$h(x) = \begin{cases} 1 & 0 \leqslant x \leqslant X \\ \\ 0 & \text{elsewhere} \end{cases} \tag{3.3-44}$$

This function, often called a *window*, and its Fourier transform are shown in Fig. 3.20(g) and (h), respectively. The results of multiplication are illustrated in Figs. 3.20(i) and (j). It is important to note that the final frequency domain result is obtained by convolving the function $S(u)*F(u)$ with $H(u)$, which is the Fourier transform of the window function $h(x)$. Since $H(u)$ has frequency components which extend to infinity, the convolution of these functions introduces a distortion in the frequency-domain representation of a function that has been sampled and limited to a finite region by $h(x)$, as shown in Fig. 3.20(j). Thus, even if the separation between samples satisfies

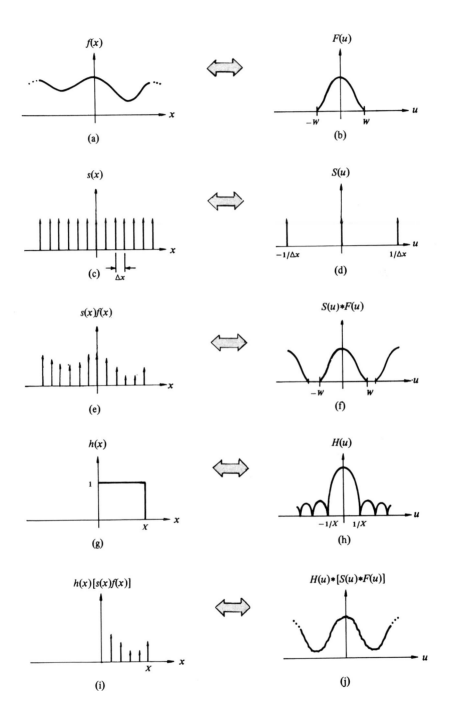

Figure 3.20. Graphical illustration of finite-sampling concepts.

the sampling theorem, it is generally impossible to recover completely a function that has been sampled only in a finite region of the x domain. This fact can be appreciated by noting that it becomes impossible under these conditions to isolate the original Fourier transform. The only exception to this is when $f(x)$ is band-limited and periodic with a period equal to X. In this case it can be shown that the corruptions due to $H(u)$ cancel out, thus allowing complete recovery of $f(x)$ if the sampling theorem is satisfied. It is important to note that the recovered function still extends from $-\infty$ to ∞ and is *not* zero outside the range in which $h(x)$ is zero. These considerations lead us to the important conclusion that no function $f(x)$ of finite duration can be band-limited. Conversely, a function that is band-limited must extend from $-\infty$ to ∞ in the x domain. This is an important practical result because it establishes a fundamental limitation in our treatment of digital functions.

Before leaving the discussion of one-dimensional functions, it will be of interest to use the above results to give an alternate reason for the periodicity of the discrete Fourier transform. We begin by noting that, thus far, all results in the frequency domain have been of a continuous nature. To obtain a discrete Fourier transform, we simply "sample" the continuous transform with a train of impulses which are Δu units apart. The situation is depicted in Fig. 3.21, where use has been made of Figs. 3.20(i) and (j). The notation $f(x)$ and $F(u)$ is used in Fig. 3.21 to facilitate comparison with the discussion in Section 3.2. It should be kept in mind, however, that Figs. 3.21(a) and (b) are assumed to be the result of the sequence of operations shown in Fig. 3.20.

As previously pointed out, sampling can be represented by multiplying the impulse train and the function of interest. In this case, we multiply $F(u)$ by $S(u)$ and obtain the result shown in Fig. 3.21(f). The equivalent operation in the x domain is convolution, which yields the function shown in Fig. 3.21 (e). This function is periodic with period $1/\Delta u$. If N samples of $f(x)$ and $F(u)$ are taken and the spacings between samples are selected so that a period in each domain is covered by N uniformly-spaced samples, we have that $N\Delta x = X$ in the x domain and $N\Delta u = 1/\Delta x$ in the frequency domain. The latter equation follows from the fact that the Fourier transform of a sampled function is periodic with period $1/\Delta x$, as shown earlier. It follows from this equation that

$$\Delta u = \frac{1}{N\Delta x} \qquad (3.3\text{-}45)$$

which agrees with Eq. (3.2-4). Choosing this spacing yields the function in Fig. 3.21(e), which is periodic with period $1/\Delta u$. From Eq. (3.3-45) we have that $1/\Delta u = N\Delta x = X$, which is the overall sampling interval in Fig. 3.21(a).

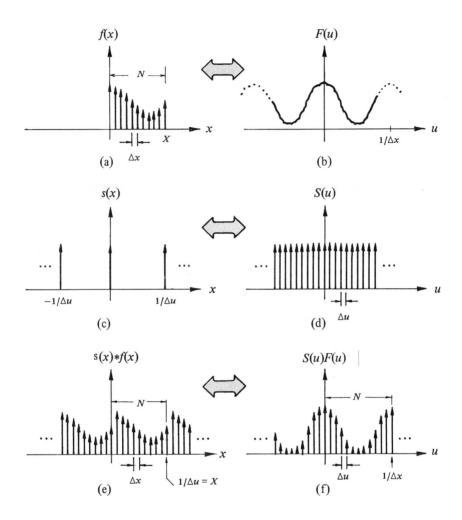

Figure 3.21. Graphical illustration of the discrete Fourier transform.

3.3.9.2 Two-Dimensional Functions

The sampling concepts developed above are, after some modifications in notation, directly applicable to two-dimensional functions. The sampling process for these functions can be formulated mathematically by making use of the two-dimensional impulse function $\delta(x, y)$ which we define as follows:

$$\int\int_{-\infty}^{\infty} f(x, y)\delta(x - x_0, y - y_0) \, dx \, dy = f(x_0, y_0) \qquad (3.3\text{-}46)$$

The interpretation of Eq. (3.3-46) is analogous to that given in connection with Eqs. (3.3-24) and (3.3-25). A two-dimensional sampling function consisting of a train of impulses separated Δx units in the x direction and Δy units in the y direction is shown in Fig. 3.22.

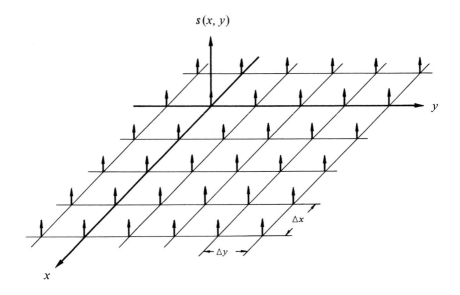

Figure 3.22. A two-dimensional sampling function.

Given a function $f(x, y)$, where x and y are continuous, a sampled function is obtained by forming the product $s(x, y)f(x, y)$. The equivalent operation in the frequency domain is convolution of $S(u, v)$ and $F(u, v)$, where $S(u, v)$ is a train of impulses with separation $1/\Delta x$ and $1/\Delta y$ in the u and v directions, respectively. If $f(x, y)$ is band-limited (i. e., its Fourier transform vanishes outside some finite region, R) the result of convolving $S(u, v)$ and $F(u, v)$ might look like the case shown in Fig. 3.23. It is noted that the function shown is periodic in two dimensions.

Let $2W_u$ and $2W_v$ represent the widths in the u and v directions, respectively, of the *smallest* rectangle that completely encloses the region R. Then, from Fig. 3.23, if $1/\Delta x > 2W_u$ and $1/\Delta y > 2W_v$ (i. e., there is no

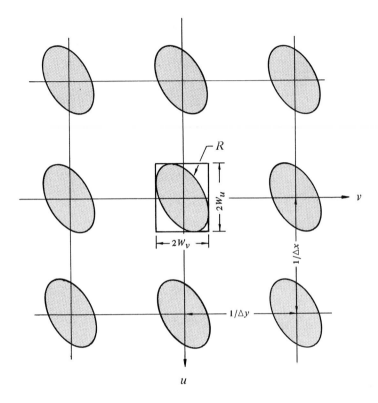

Figure 3.23. Frequency-domain representation of a sampled two-dimensional, band-limited function.

aliasing), one of the periods can be recovered completely if we multiply $S(u, v)*F(u, v)$ by the function

$$G(u, v) = \begin{cases} 1 & (u, v) \text{ inside one of the} \\ & \text{rectangles enclosing } R \\ \\ 0 & \text{elsewhere} \end{cases} \quad (3.3\text{-}47)$$

The inverse Fourier transform of $G(u, v)[S(u, v)*F(u, v)]$ yields $f(x, y)$.

The foregoing considerations lead us to a form of the two-dimensional sampling theorem, which states that a band-limited function $f(x, y)$ can be recovered completely from samples whose separation is given by

$$\Delta x \leqslant \frac{1}{2W_u} \tag{3.3-48a}$$

and

$$\Delta y \leqslant \frac{1}{2W_v} \tag{3.3-48b}$$

When $f(x, y)$ is spatially-limited by using a two-dimensional window $h(x, y)$ analogous to the function $h(x)$ used in Fig. 3.20, we again have the problem that the transform of the sampled function is distorted by the convolution of $H(u, v)$ and $S(u, v)*F(u, v)$. This distortion, which is due to the spatially-limited nature of digital images, precludes complete recovery of $f(x, y)$ from its samples. As in the one-dimensional case, periodic functions are an exception, but images satisfying this condition are rarely found in practice.

An argument similar to the one developed for the one-dimensional case can be carried out to show how periodicity arises in the two-dimensional, discrete Fourier transform. For an $N \times N$ image, this analysis would also yield the following results:

$$\Delta u = \frac{1}{N\Delta x} \tag{3.3-49a}$$

and

$$\Delta v = \frac{1}{N\Delta y} \tag{3.3-49b}$$

These relationships between the sample separations guarantee that a complete (two-dimensional) period will be covered by $N \times N$ uniformly-spaced values in both the spatial and frequency domains.

3.4 THE FAST FOURIER TRANSFORM

The number of complex multiplications and additions required to implement Eq. (3.2-2) is proportional to N^2. This can be seen easily by noting that, for *each* of the N values of u, expansion of the summation sign requires N complex multiplications of $f(x)$ by $\exp[-j2\pi ux/N]$ and $(N - 1)$ additions of the results. The terms of $\exp[-j2\pi ux/N]$ can be computed once and stored in a table for all subsequent applications. For this reason, the multiplication of u by x in these terms is usually not considered a direct part of the implementation.

In this section it is shown that, by properly decomposing Eq. (3.2-2), the number of multiply and add operations can be made proportional to $N \log_2 N$. The decomposition procedure is called the *fast Fourier transform* (FFT) *algorithm*. The reduction in proportionality from N^2 to $N \log_2 N$ multiply/add operations represents a significant saving in computation effort, as shown by the figures in Table 3.1. It is evident from this table that the FFT approach offers a considerable computational advantage over a direct implementation of the Fourier transform, particularly when N is relatively large. For example, a direct implementation of the transform for $N = 8192$ requires on the order of three-quarters of an hour in a machine such as an IBM 7094. By contrast, the same job can be done in this machine in about 5 sec. using an FFT algorithm.

Table 3.1. A comparison of N^2 vs. $N \log_2 N$ for various values of N.

N	N^2 (Conventional FT)	$N \log_2 N$ (FFT)	Computational Advantage ($N/\log_2 N$)
2	4	2	2.00
4	16	8	2.00
8	64	24	2.67
16	256	64	4.00
32	1,024	160	6.40
64	4,096	384	10.67
128	16,384	896	18.29
256	65,536	2,048	32.00
512	262,144	4,608	56.89
1024	1,048,576	10,240	102.40
2048	4,194,304	22,528	186.18
4096	16,777,216	49,152	341.33
8192	67,108,864	106,496	630.15

Attention is focused below on the development of an FFT algorithm of one variable. As pointed out in Section 3.3.1, a two-dimensional Fourier transform can be readily computed by a series of applications of the one-dimensional transform.

3.4.1 FFT Algorithm

The FFT algorithm developed in this section is based on the so-called "successive doubling" method. It will be convenient in the following discussion to express Eq. (3.2-2) in the form

$$F(u) = \frac{1}{N} \sum_{x=0}^{N-1} f(x) W_N^{ux} \tag{3.4-1}$$

where

$$W_N = \exp[-j2\pi/N] \tag{3.4-2}$$

and N is assumed to be of the form

$$N = 2^n \tag{3.4-3}$$

where n is a positive integer. Based on this, N can be expressed as

$$N = 2M \tag{3.4-4}$$

where M is also a positive integer. Substitution of Eq. (3.4-4) into Eq. (3.4-1) yields:

$$F(u) = \frac{1}{2M} \sum_{x=0}^{2M-1} f(x) W_{2M}^{ux}$$

$$= \frac{1}{2} \left\{ \frac{1}{M} \sum_{x=0}^{M-1} f(2x) W_{2M}^{u(2x)} + \frac{1}{M} \sum_{x=0}^{M-1} f(2x+1) W_{2M}^{u(2x+1)} \right\} \tag{3.4-5}$$

Since, from Eq. (3.4-2), $W_{2M}^{2ux} = W_M^{ux}$, Eq. (3.4-5) may be expressed in the form

$$F(u) = \frac{1}{2} \left\{ \frac{1}{M} \sum_{x=0}^{M-1} f(2x) W_M^{ux} + \frac{1}{M} \sum_{x=0}^{M-1} f(2x+1) W_M^{ux} W_{2M}^{u} \right\} \tag{3.4-6}$$

If we define

$$F_{\text{even}}(u) = \frac{1}{M} \sum_{x=0}^{M-1} f(2x) W_M^{ux} \tag{3.4-7}$$

for $u = 0, 1, 2, ..., M - 1$, and

$$F_{\text{odd}}(u) = \frac{1}{M} \sum_{x=0}^{M-1} f(2x+1) W_M^{ux} \tag{3.4-8}$$

for $u = 0, 1, 2, ..., M - 1$, Eq. (3.4-6) then becomes

$$F(u) = \frac{1}{2} \left\{ F_{\text{even}}(u) + F_{\text{odd}}(u) W_{2M}^{u} \right\} \tag{3.4-9}$$

Also, since $W_M^{u+M} = W_M^u$ and $W_{2M}^{u+M} = -W_{2M}^u$, it follows from Eqs. (3.4-7) through (3.4-9) that

$$F(u + M) = \frac{1}{2} \left\{ F_{\text{even}}(u) - F_{\text{odd}}(u) W_{2M}^u \right\} \qquad (3.4\text{-}10)$$

Careful analysis of Eqs. (3.4-7) through (3.4-10) reveals some interesting properties of these expressions. It is noted that an N-point transform can be computed by dividing the original expression into two parts, as indicated in Eqs. (3.4-9) and (3.4-10). Computation of the first half of $F(u)$ requires the evaluation of the two $(N/2)$-point transforms given in Eqs. (3.4-7) and (3.4-8). The resulting values of $F_{\text{even}}(u)$ and $F_{\text{odd}}(u)$ are then substituted into Eq. (3.4-9) to obtain $F(u)$ for $u = 0, 1, 2, ..., (N/2 - 1)$. The other half then follows directly from Eq. (3.4-10) without additional transform evaluations.

In order to examine the computational implications of the above procedure, let $m(n)$ and $a(n)$ represent the number of complex multiplications and additions, respectively, required to implement this method. As before, the number of samples is equal to 2^n where n is a positive integer. Suppose first that $n = 1$. A two-point transform requires the evaluation of $F(0)$; then $F(1)$ follows from Eq. (3.4-10). To obtain $F(0)$, we must first compute $F_{\text{even}}(0)$ and $F_{\text{odd}}(0)$. In this case $M = 1$ and Eqs. (3.4-7) and (3.4-8) are one-point transforms. Since the Fourier transform of a single point is the sample itself, however, no multiplications or additions are required to obtain $F_{\text{even}}(0)$ and $F_{\text{odd}}(0)$. One multiplication of $F_{\text{odd}}(0)$ by W_2^0 and one addition yield $F(0)$ from Eq. (3.4-9). $F(1)$ then follows from Eq. (3.4-10) with one more addition (we consider subtraction to be the same as addition). Since $F_{\text{odd}}(0)W_2^0$ was already computed, we have that the total number of operations required for a two-point transform consists of $m(1) = 1$ multiplication and $a(1) = 2$ additions.

The next allowed value for n is 2. According to the above development, a four-point transform can be divided into two parts. The first half of $F(u)$ requires evaluation of two, two-point transforms, as given in Eqs. (3.4-7) and (3.4-8) for $M = 2$. Since a two-point transform requires $m(1)$ and $a(1)$ multiplications and additions, respectively, it is evident that evaluation of these two equations requires a total of $2m(1)$ multiplications and $2a(1)$ additions. Two additional multiplications and two additions are necessary to obtain $F(0)$ and $F(1)$ from Eq. (3.4-9). Since $F_{\text{odd}}(u)W_{2M}^u$ was already computed for $u = \{0, 1\}$, we have that two more additions give $F(2)$ and $F(3)$. The total is then $m(2) = 2m(1) + 2$ and $a(2) = 2a(1) + 4$.

When n is equal to 3, we consider two, four-point transforms in the evaluation of $F_{even}(u)$ and $F_{odd}(u)$. These require $2m(2)$ multiplications and $2a(2)$ additions. Four more multiplications and eight additions yield the complete transform. The total is then $m(3) = 2m(2) + 4$ and $a(3) = 2a(2) + 8$.

By continuing this argument one would find that, for any positive integer value of n, the number of multiplications and additions required to implement the FFT is given by the recursive expressions:

$$m(n) = 2m(n-1) + 2^{n-1} \qquad n \geqslant 1 \qquad\qquad (3.4\text{-}11)$$

and

$$a(n) = 2a(n-1) + 2^n \qquad n \geqslant 1 \qquad\qquad (3.4\text{-}12)$$

where $m(0) = 0$ and $a(0) = 0$, since the transform of a single point does not require any additions or multiplications.

Implementation of Eqs. (3.4-7) through (3.4-10) constitutes the successive-doubling FFT algorithm. This name arises from the fact that a two-point transform is computed from two, one-point transforms, a four-point transform from two, two-point transforms, and so on for any N that is equal to an integer power of 2.

3.4.2 Number of Operations

In this section it is shown by induction that the number of complex multiplications and additions required to implement the above FFT algorithm is given by

$$m(n) = \frac{1}{2} 2^n \log_2 2^n$$

$$= \frac{1}{2} N \log_2 N$$

$$= \frac{1}{2} Nn \qquad n \geqslant 1 \qquad\qquad (3.4\text{-}13)$$

and

$$a(n) = 2^n \log_2 2^n$$

$$= N \log_2 N$$

$$= Nn \qquad n \geqslant 1 \qquad\qquad (3.4\text{-}14)$$

respectively.

First, it is necessary to prove that Eqs. (3.4-13) and (3.4-14) hold for $n = 1$. It was already shown that

$$m(1) = \tfrac{1}{2}(2)(1) = 1$$

and

$$a(1) = (2)(1) = 2$$

Next, it is assumed that the expressions hold for n. It is then required to prove that they are also true for $n + 1$.

From Eq. (3.4-11) it follows that

$$m(n + 1) = 2m(n) + 2^n$$

Substituting Eq. (3.4-13), which is assumed to be valid for n, yields

$$m(n + 1) = 2\left(\tfrac{1}{2}Nn\right) + 2^n$$

$$= 2\left(\tfrac{1}{2}2^n n\right) + 2^n$$

$$= 2^n(n + 1)$$

$$= \tfrac{1}{2}2^{n+1}(n + 1)$$

Equation (3.4-13) is, therefore, valid for all positive integer values of n.

From Eq. (3.4-12), we have

$$a(n + 1) = 2a(n) + 2^{n+1}$$

Substitution of Eq. (3.4-14) for $a(n)$ yields

$$a(n + 1) = 2Nn + 2^{n+1}$$

$$= 2(2^n n) + 2^{n+1}$$

$$= 2^{n+1}(n + 1)$$

This completes the proof.

3.4.3 The Inverse FFT

Thus far, little has been said concerning the inverse Fourier transform. It turns out that any algorithm for implementing the discrete forward transform can also be used (with minor modifications in the input) to compute the inverse. To see this, let us consider Eqs. (3.2-2) and (3.2-3), which are repeated below:

$$F(u) = \frac{1}{N} \sum_{x=0}^{N-1} f(x) \exp[-j2\pi ux/N] \qquad (3.4-15)$$

and

$$f(x) = \sum_{u=0}^{N-1} F(u) \exp[j2\pi ux/N] \qquad (3.4\text{-}16)$$

Taking the complex conjugate of Eq. (3.4-16) and dividing both sides by N yields

$$\frac{1}{N} f^*(x) = \frac{1}{N} \sum_{u=0}^{N-1} F^*(u) \exp[-j2\pi ux/N] \qquad (3.4\text{-}17)$$

By comparing this result with Eq. (3.4-15), we see that the right side of Eq. (3.4-17) is in the form of the forward Fourier transform. Thus, if we input $F^*(u)$ into an algorithm designed to compute the forward transform, the result will be the quantity $f^*(x)/N$. Taking the complex conjugate and multiplying by N yields the desired inverse $f(x)$.

For two-dimensional square arrays we take the complex conjugate of Eq. (3.2-10); that is

$$f^*(x, y) = \frac{1}{N} \sum_{u=0}^{N-1} \sum_{v=0}^{N-1} F^*(u, v) \exp[-j2\pi(ux + vy)/N] \qquad (3.4\text{-}18)$$

which we see is in the form of the two-dimensional forward transform given in Eq. (3.2-9). It follows, therefore, that if we input $F^*(u, v)$ into an algorithm designed to compute the forward transform, the result will be $f^*(x, y)$. By taking the complex conjugate of this result we obtain $f(x, y)$. In the case where $f(x)$ or $f(x, y)$ are real, the complex conjugate operation is unnecessary since $f(x) = f^*(x)$ and $f(x, y) = f^*(x, y)$ for real functions.

The fact that the two-dimensional transform is usually computed by successive passes of the one-dimensional transform is a frequent source of confusion when using the above technique to obtain the inverse. The reader should keep in mind the procedure outlined in Section 3.3.1 and avoid being misled by Eq. (3.4-17). In other words, when using a one-dimensional algorithm to compute the two-dimensional inverse, the method is *not* to compute the complex conjugate after each row or column is processed. Instead, the function $F^*(u, v)$ is treated as if it were $f(x, y)$ in the forward, two-dimensional transform procedure summarized in Fig. 3.9. The complex conjugate of the result (if necessary) will yield the proper inverse, $f(x, y)$.

3.4.4 Implementation

A computer implementation of the FFT algorithm developed in Section 3.4.1 is straightforward. The principal point to keep in mind is that the input data must be arranged in the order required for successive applications of Eqs. (3.4-7) and (3.4-8). The ordering procedure can be illustrated by a simple example. Suppose that we wish to use the successive doubling algorithm to compute the FFT of an eight-point function $\{f(0), f(1), ..., f(7)\}$. Equation (3.4-7) uses the samples with even arguments, $\{f(0), f(2), f(4), f(6)\}$, and Eq. (3.4-8) uses the samples with odd arguments, $\{f(1), f(3), f(5), f(7)\}$. However, each four-point transform is computed as two, two-point transforms. This also requires the use of Eqs. (3.4-7) and (3.4-8). Thus, to compute the FFT of the first set above, we must divide it into its even part $\{f(0), f(4)\}$ and odd part $\{f(2), f(6)\}$. Similarly, the second set is subdivided into $\{f(1), f(5)\}$ for Eq. (3.4-7) and $\{f(3), f(7)\}$ for Eq. (3.4-8). No further rearrangement is required since each two-element set is considered as having one even and one odd element. Combining these results, we have that the input array must be expressed in the form $\{f(0), f(4), f(2), f(6), f(1), f(5), f(3), f(7)\}$. The successive doubling algorithm operates on this array in the manner shown in Fig. 3.24. At the first level of computation

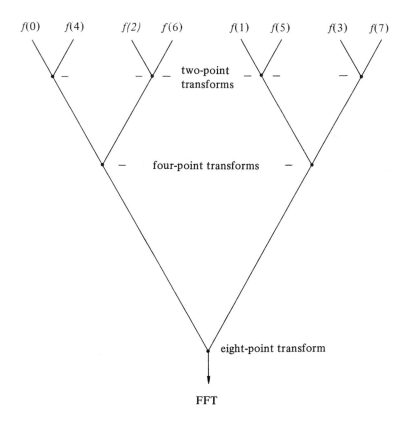

Figure 3.24. Ordered input array and its use in the successive-doubling method.

are four two-point transforms involving { $f(0)$, $f(4)$} , { $f(2)$, $f(6)$} ,
{$f(1)$, $f(5)$}, and {$f(3)$, $f(7)$}. The next level uses these results to form two,
four-point transforms, and the last level uses these two results to produce the
desired transform.

Fortunately, the general procedure for reordering an input array
follows a simple *bit-reversal* rule. If we let x represent any valid argument
value in $f(x)$, then the corresponding argument in the reordered array is given
by expressing x in binary and reversing the bits. For example, if $N = 2^3$, the
seventh element in the original array, $f(6)$, becomes the fourth element in
the reordered array since $6 = 110_2$ becomes $011_2 = 3$ when the bits are
reversed. It is noted that this is a left-to-right reversal of the binary number
and should not be confused with the binary complement. The procedure is
summarized in Table 3.2 for $N = 8$. If the reordered array is used in
computing the FFT, the answer will be the elements of the Fourier transform in the correct order. Conversely, it can be shown that if the array is
used in its natural order, the answer will be bit-reversed. Identical comments
hold for computing the inverse transform.

Table 3.2. Example of bit reversal and reordering of array for input into FFT algorithm.

Original Argument			Original Array	Bit-reversed Argument			Reordered Array
0	0	0	$f(0)$	0	0	0	$f(0)$
0	0	1	$f(1)$	1	0	0	$f(4)$
0	1	0	$f(2)$	0	1	0	$f(2)$
0	1	1	$f(3)$	1	1	0	$f(6)$
1	0	0	$f(4)$	0	0	1	$f(1)$
1	0	1	$f(5)$	1	0	1	$f(5)$
1	1	0	$f(6)$	0	1	1	$f(3)$
1	1	1	$f(7)$	1	1	1	$f(7)$

A FORTRAN subroutine for computing the FFT by the successive-
doubling method is shown in Fig. 3.25. The parameters in the subroutine
argument are as follows. On input, F is an array whose transform is desired

```
SUBROUTINE FFT(F,LN)
COMPLEX F(1024),U,W,T,CMPLX
PI=3.141593
N=2**LN
NV2=N/2
NM1=N-1
J=1
DO 3 I=1,NM1
IF(I.GE.J) GO TO 1
T=F(J)
F(J)=F(I)
F(I)=T
1 K=NV2
2 IF(K.GE.J) GO TO 3
J=J-K
K=K/2
GO TO 2
3 J=J+K
DO 5 L=1,LN
LE=2**L
LE1=LE/2
U=(1.0,0.0)
W=CMPLX(COS(PI/LE1),-SIN(PI/LE1))
DO 5 J=1,LE1
DO 4 I=J,N,LE
IP=I+LE1
T=F(IP)*U
F(IP)=F(I)-T
4 F(I)=F(I)+T
5 U=U*W
DO 6 I=1,N
6 F(I)=F(I)/FLOAT(N)
RETURN
END
```

Figure 3.25. A FORTRAN implementation of the successive-doubling FFT algorithm. (Adapted from Cooley *et al* [1969].)

and LN is equal to n. On output, the array F contains the Fourier transform. It is noted that F is a complex array so that if the input is a real function the imaginary part of F must be set to zero before calling the subroutine.

The first part of the program, including the "DO 3" loop, performs reordering of the input data. The second part, including the "DO 5" loop, does the successive-doubling calculations. The "DO 6" loop divides the results by N. It has been estimated that, for $N = 1024$, this simple program is only about 12% less efficient than a more optimally written FORTRAN program which uses a table for storing values of W_{2M}^{u}.

Equations (3.3-3) through (3.3-6), along with Figs. 3.9 and 3.25, provide the necessary information for implementing the two-dimensional, forward FFT. The same concepts are applicable to the inverse transform by using the complex conjugate of the Fourier transform as the input to the FFT subroutine, as indicated in Section 3.4.3.

It should be noted before leaving this section that it is possible to formulate the FFT using other integer bases greater than two. In fact, a base-three formulation requires slightly fewer operations than any other base (Cooley, Lewis, and Welch [1969]), but its disadvantage in terms of programming makes it an unattractive choice. A base-four is equal to a base-two implementation in terms of required operations, and all other bases are less efficient, requiring a progressively larger number of operations. Fast Fourier transform algorithms are usually formulated in a base-two format because it is easier to implement in assembly language.

3.5 OTHER IMAGE TRANSFORMS

Although the Fourier transform is the transform most often used in image processing applications, there are other transforms which are also of interest in this area. Four of these, the Walsh, Hadamard, discrete cosine, and Hotelling transforms are discussed in this section.

3.5.1 General Formulation

The one-dimensional, discrete Fourier transform is one of a class of important transforms which can be expressed in terms of the general relation

$$T(u) = \sum_{x=0}^{N-1} f(x) g(x, u) \tag{3.5-1}$$

where $T(u)$ is the transform of $f(x)$, $g(x, u)$ is the *forward transformation kernel*, and u assumes values in the range 0, 1, ..., $N - 1$. Similarly, the inverse transform is given by the relation

$$f(x) = \sum_{u=0}^{N-1} T(u) h(x, u) \tag{3.5-2}$$

where $h(x, u)$ is the *inverse transformation kernel* and x assumes values in the range 0, 1, ..., $N - 1$. The nature of a transform is determined by the properties of its transformation kernel.

For two-dimensional square arrays the forward and inverse transforms are given by the equations

$$T(u, v) = \sum_{x=0}^{N-1} \sum_{y=0}^{N-1} f(x, y) g(x, y, u, v) \qquad (3.5\text{-}3)$$

and

$$f(x, y) = \sum_{u=0}^{N-1} \sum_{v=0}^{N-1} T(u, v) h(x, y, u, v) \qquad (3.5\text{-}4)$$

where, as above, $g(x, y, u, v)$ and $h(x, y, u, v)$ are called the *forward* and *inverse transformation kernels*, respectively.

The forward kernel is said to be *separable* if

$$g(x, y, u, v) = g_1(x, u) g_2(y, v) \qquad (3.5\text{-}5)$$

The kernel is in addition *symmetric* if g_1 is functionally equal to g_2. In this case, Eq. (3.5-5) can be expressed in the form

$$g(x, y, u, v) = g_1(x, u) g_1(y, v) \qquad (3.5\text{-}6)$$

Identical comments hold for the inverse kernel if $g(x, y, u, v)$ is replaced by $h(x, y, u, v)$ in Eqs. (3.5-5) and (3.5-6).

The two-dimensional Fourier transform is a special case of Eq. (3.5-3). It has the kernel

$$g(x, y, u, v) = \frac{1}{N} \exp\left[-j2\pi(ux + vy)/N \right]$$

which is separable and symmetric since

$$g(x, y, u, v) = g_1(x, u) g_1(y, v)$$

$$= \frac{1}{\sqrt{N}} \exp\left[-j2\pi ux/N \right] \frac{1}{\sqrt{N}} \exp\left[-j2\pi vy/N \right] \qquad (3.5\text{-}7)$$

It is easily shown that the inverse Fourier kernel is also separable and symmetric.

A transform with a separable kernel can be computed in two steps, each requiring a one-dimensional transform. First, the one-dimensional transform is taken along each row of $f(x, y)$, yielding

$$T(x, v) = \sum_{y=0}^{N-1} f(x, y) g_2(y, v) \qquad (3.5\text{-}8)$$

for $x, v = 0, 1, 2, ..., N - 1$. Next, the one-dimensional transform is taken along each column of $T(x, v)$; this results in the expression

$$T(u, v) = \sum_{x=0}^{N-1} T(x, v) g_1(x, u) \qquad (3.5\text{-}9)$$

for u, $v = 0, 1, 2, ..., N - 1$. It is noted that this procedure agrees with the method given in Section 3.3.1 for the Fourier transform. The same final results are obtained if the transform is taken first along each column of $f(x, y)$ to obtain $T(y, u)$ and then along each row of the latter function to obtain $T(u, v)$. Similar comments hold for the inverse transform if $h(x, y, u, v)$ is separable.

If the kernel $g(x, y, u, v)$ is separable and symmetric, Eq. (3.5-3) can also be expressed in the following matrix form:

$$\mathbf{T} = \mathbf{AFA} \qquad (3.5\text{-}10)$$

where $\mathbf{F}$ is the $N \times N$ image matrix, $\mathbf{A}$ is an $N \times N$ symmetric transformation matrix with elements $a_{ij} = g_1(i, j)$, and $\mathbf{T}$ is the resulting $N \times N$ transform for values of u and v in the range $0, 1, 2, ..., N - 1$.

To obtain the inverse transform we pre- and post-multiply Eq. (3.5-10) by an inverse transformation matrix $\mathbf{B}$. This operation yields the expression

$$\mathbf{BTB} = \mathbf{BAFAB} \qquad (3.5\text{-}11)$$

If $\mathbf{B} = \mathbf{A}^{-1}$, it then follows that

$$\mathbf{F} = \mathbf{BTB} \qquad (3.5\text{-}12)$$

which indicates that the digital image $\mathbf{F}$ can be recovered completely from its transform. If $\mathbf{B}$ is not equal to $\mathbf{A}^{-1}$, then use of Eq. (3.5-11) yields an approximation to $\mathbf{F}$, given by the relation

$$\hat{\mathbf{F}} = \mathbf{BAFAB} \qquad (3.5\text{-}13)$$

A number of transforms, including the Fourier, Walsh, Hadamard, and discrete cosine transforms, can be expressed in the forms of Eqs. (3.5-10) and (3.5-12). An important property of the resulting transformation matrices is that they can be decomposed into products of matrices with fewer nonzero entries than the original matrix. This result, first formulated by Good [1958] for the Fourier transform, reduces redundancy and, consequently, the number of operations required to implement a two-dimensional transform. The degree of reduction is equivalent to that achieved by an FFT algorithm, being on the order of $N \log_2 N$ multiply/add operations for each row or column of an $N \times N$ image. Although attention is focused in this book on computational procedures which are based on successive applications of one-dimensional algorithms to obtain the forward and inverse transforms of an image, it should be kept in mind that equivalent computational results can be obtained via a matrix formulation of the problem. The interested reader should consult the book by Andrews [1970] for additional details on this topic.

3.5.2 Walsh Transform

When $N = 2^n$, the discrete Walsh transform of a function $f(x)$, denoted by $W(u)$, is obtained by substituting the kernel

$$g(x, u) = \frac{1}{N} \prod_{i=0}^{n-1} (-1)^{b_i(x)b_{n-1-i}(u)} \qquad (3.5\text{-}14)$$

into Eq. (3.5-1). In other words,

$$W(u) = \frac{1}{N} \sum_{x=0}^{N-1} f(x) \prod_{i=0}^{N-1} (-1)^{b_i(x)b_{n-1-i}(u)} \qquad (3.5\text{-}15)$$

where $b_k(z)$ is the kth bit in the binary representation of z. For example, if $n = 3$ and $z = 6$ (110 in binary), we have that $b_0(z) = 0$, $b_1(z) = 1$, and $b_2(z) = 1$.

The values of $g(x, u)$, excluding the $1/N$ constant term, are listed in Table 3.3 for $N = 8$. The array formed by the Walsh transformation kernel is

Table 3.3. Values of the Walsh transformation kernel for $N = 8$.

u \ x	0	1	2	3	4	5	6	7
0	+	+	+	+	+	+	+	+
1	+	+	+	+	−	−	−	−
2	+	+	−	−	+	+	−	−
3	+	+	−	−	−	−	+	+
4	+	−	+	−	+	−	+	−
5	+	−	+	−	−	+	−	+
6	+	−	−	+	+	−	−	+
7	+	−	−	+	−	+	+	−

a symmetric matrix whose rows and columns are orthogonal. These properties, which hold in general, lead to an inverse kernel which is identical to the forward kernel, except for a constant multiplicative factor of $1/N$; that is,

$$h(x, u) = \prod_{i=0}^{n-1} (-1)^{b_i(x)b_{n-1-i}(u)} \qquad (3.5\text{-}16)$$

Thus, the inverse Walsh transform is given by

$$f(x) = \sum_{u=0}^{N-1} W(u) \prod_{i=0}^{n-1} (-1)^{b_i(x)b_{n-1-i}(u)} \tag{3.5-17}$$

It is noted that, unlike the Fourier transform which is based on trigonometric terms, the Walsh transform consists of a series expansion of basis functions whose values are either plus or minus one.

The validity of Eq. (3.5-17) is easily established by substituting Eq. (3.5-15) for $W(u)$ and making use of the orthogonality condition mentioned above. It is also of interest to note from Eqs. (3.5-15) and (3.5-17) that the forward and inverse Walsh transforms differ only by the $1/N$ term. Thus, any algorithm for computing the forward transform can be used directly to obtain the inverse transform simply by multiplying the result of the algorithm by N.

The two-dimensional forward and inverse Walsh kernels are given by the relations

$$g(x, y, u, v) = \frac{1}{N} \prod_{i=0}^{n-1} (-1)^{[b_i(x)b_{n-1-i}(u) + b_i(y)b_{n-1-i}(v)]} \tag{3.5-18}$$

and

$$h(x, y, u, v) = \frac{1}{N} \prod_{i=0}^{n-1} (-1)^{[b_i(x)b_{n-1-i}(u) + b_i(y)b_{n-1-i}(v)]} \tag{3.5-19}$$

Although it is just as valid to group both $1/N$ terms in front of $g(x, y, u, v)$ or $h(x, y, u, v)$, the forms given in Eqs. (3.5-18) and (3.5-19) are preferable in image processing applications where there is equal interest in taking the forward and inverse transforms. Since the formulation given in these equations yields identical kernels, it follows from Eqs. (3.5-3) and (3.5-4) that the forward and inverse Walsh transforms are also equal in form; that is

$$W(u, v) = \frac{1}{N} \sum_{x=0}^{N-1} \sum_{y=0}^{N-1} f(x, y) \prod_{i=0}^{n-1} (-1)^{[b_i(x)b_{n-1-i}(u) + b_i(y)b_{n-1-i}(v)]} \tag{3.5-20}$$

and

$$f(x, y) = \frac{1}{N} \sum_{u=0}^{N-1} \sum_{v=0}^{N-1} W(u, v) \prod_{i=0}^{n-1} (-1)^{[b_i(x)b_{n-1-i}(u) + b_i(y)b_{n-1-i}(v)]} \tag{3.5-21}$$

Thus, any algorithm which is used to compute the two-dimensional forward Walsh transform can also be used without modification to compute the inverse transform.

The Walsh transform kernels are separable and symmetric since

$$g(x,y,u,v) = g_1(x,u) g_1(y,v)$$

$$= h_1(x,u) h_1(y,v)$$

$$g(x,y,u,v) = \left[\frac{1}{\sqrt{N}} \prod_{i=0}^{n-1} (-1)^{b_i(x)b_{n-1-i}(u)} \right] \left[\frac{1}{\sqrt{N}} \prod_{i=0}^{n-1} (-1)^{b_i(y)b_{n-1-i}(v)} \right] \qquad (3.5\text{-}22)$$

It follows, therefore, that $W(u, v)$ and its inverse can be computed by successive applications of the one-dimensional Walsh transform given in Eq. (3.5-15). The procedure followed in the computation is the same as the one given in Section 3.3.1 and Fig. 3.9 for the Fourier transform.

The Walsh transform can be computed by a fast algorithm identical in form to the successive-doubling method given in Section 3.4.1 for the FFT. The only difference is that all exponential terms W_N are set equal to one in the case of the fast Walsh transform (FWT)[†]. Equations (3.4-9) and (3.4-10), which are the basic relations leading to the FFT, then become

$$W(u) = \tfrac{1}{2} \{ W_{even}(u) + W_{odd}(u) \} \qquad (3.5\text{-}23)$$

and

$$W(u + M) = \tfrac{1}{2} \{ W_{even}(u) - W_{odd}(u) \} \qquad (3.5\text{-}24)$$

where $M = N/2$, $u = 0, 1, ..., M - 1$, and $W(u)$ denotes the one-dimensional Walsh transform. Rather than giving a general proof of this result, let us illustrate the use of Eq. (3.5-15) and the validity of Eqs. (3.5-23) and (3.5-24) by means of a simple example. Further details on this subject may be found in Shanks [1969].

Example: If $N = 4$, use of Eq. (3.5-15) results in the following sequence of steps:

$$W(0) = \frac{1}{4} \sum_{x=0}^{3} \left[f(x) \prod_{i=0}^{1} (-1)^{b_i(x)b_{1-i}(0)} \right]$$

$$= \frac{1}{4} \left[f(0) + f(1) + f(2) + f(3) \right]$$

$$W(1) = \frac{1}{4} \sum_{x=0}^{3} \left[f(x) \prod_{i=0}^{1} (-1)^{b_i(x)b_{1-i}(1)} \right]$$

$$= \frac{1}{4} \left[f(0) + f(1) - f(2) - f(3) \right]$$

[†]The use of W in this section to denote the Walsh transform should not be confused with our use of the same symbol in Section 3.4.1 to denote exponential terms.

$$W(2) = \frac{1}{4} \sum_{x=0}^{3} \left[f(x) \prod_{j=0}^{1} (-1)^{b_i(x)b_{1-i}(2)} \right]$$

$$= \frac{1}{4} \left[f(0) - f(1) + f(2) - f(3) \right]$$

$$W(3) = \frac{1}{4} \sum_{x=0}^{3} \left[f(x) \prod_{i=0}^{1} (-1)^{b_i(x)b_{1-i}(3)} \right]$$

$$= \frac{1}{4} \left[f(0) - f(1) - f(2) + f(3) \right]$$

In order to show the validity of Eqs. (3.5-23) and (3.5-24) we subdivide these results into two groups; that is,

$$W_{even}(0) = \frac{1}{2} \left[f(0) + f(2) \right] \qquad W_{odd}(0) = \frac{1}{2} \left[f(1) + f(3) \right]$$

$$W_{even}(1) = \frac{1}{2} \left[f(0) - f(2) \right] \qquad W_{odd}(1) = \frac{1}{2} \left[f(1) - f(3) \right]$$

From Eq. (3.5-23) we have

$$W(0) = \frac{1}{2} \left[W_{even}(0) + W_{odd}(0) \right]$$

$$= \frac{1}{4} \left[f(0) + f(1) + f(2) + f(3) \right]$$

and

$$W(1) = \frac{1}{2} \left[W_{even}(1) + W_{odd}(1) \right]$$

$$= \frac{1}{4} \left[f(0) + f(1) - f(2) - f(3) \right]$$

The next two terms are computed from these results using Eq. (3.5-24):

$$W(2) = \frac{1}{2} \left[W_{even}(0) - W_{odd}(0) \right]$$

$$= \frac{1}{4} \left[f(0) - f(1) + f(2) - f(3) \right]$$

and

$$W(3) = \tfrac{1}{2} \left[W_{even}(1) - W_{odd}(1) \right]$$

$$= \tfrac{1}{4} \left[f(0) - f(1) - f(2) + f(3) \right]$$

Thus, computation of $W(u)$ by Eq. (3.5-15) or by Eqs. (3.5-23) and (3.5-24) yields identical results. □

As indicated above, an algorithm used to compute the FFT by the successive-doubling method can be easily modified for computing a fast Walsh transform simply by setting all trigonometric terms equal to one. Figure 3.26 illustrates the required modifications for the FFT program given

```
        SUBROUTINE FWT(F,LN)
        REAL*4 F(1024),T
        N=2**LN
        NV2=N/2
        NM1=N-1
        J=1
        DO 3 I=1,NM1
        IF (I.GE.J) GO TO 1
        T=F(J)
        F(J)=F(I)
        F(I)=T
    1   K=NV2
    2   IF (K.GE.J) GO TO 3
        J=J-K
        K=K/2
        GO TO 2
    3   J=J+K
        DO 5 L=1,LN
        LE=2**L
        LE1=LE/2
        DO 5 J=1,LE1
        DO 4 I=J,N,LE
        IP=I+LE1
        T=F(IP)
        F(IP)=F(I)-T
    4   F(I)=F(I)+T
    5   CONTINUE
        DO 6 I=1,N
    6   F(I)=F(I)/FLOAT(N)
        RETURN
        END
```

Figure 3.26. Modification of the successive-doubling FFT algorithm for computing the fast Walsh transform.

in Fig. 3.25. It is noted that the Walsh transform is real, thus requiring less computer storage for a given problem than the Fourier transform, which is generally complex.

3.5.3 Hadamard Transform

One of several known formulations for the one-dimensional, forward Hadamard kernel is given by the relation

$$g(x, u) = \frac{1}{N} (-1)^{\sum_{i=0}^{n-1} b_i(x) b_i(u)} \tag{3.5-25}$$

where the summation in the exponent is performed in modulo 2 arithmetic and, as in Eq. (3.5-14), $b_k(z)$ is the kth bit in the binary representation of z. Substitution of Eq. (3.5-25) into Eq. (3.5-1) yields the following expression for the one-dimensional Hadamard transform:

$$H(u) = \frac{1}{N} \sum_{x=0}^{N-1} f(x)(-1)^{\sum_{i=0}^{n-1} b_i(x) b_i(u)} \tag{3.5-26}$$

where $N = 2^n$, and u assumes values in the range 0, 1, 2, ..., $N - 1$.

As in the case of the Walsh transform, the Hadamard kernel forms a matrix whose rows (and columns) are orthogonal. This condition again leads to an inverse kernel which, except for the $1/N$ term, is equal to the forward Hadamard kernel; that is,

$$h(x, u) = (-1)^{\sum_{i=0}^{n-1} b_i(x) b_i(u)} \tag{3.5-27}$$

Substitution of this kernel into Eq. (3.5-2) yields the following expression for the inverse Hadamard transform:

$$f(x) = \sum_{u=0}^{N-1} H(u)(-1)^{\sum_{i=0}^{n-1} b_i(x) b_i(u)} \tag{3.5-28}$$

for $x = 0, 1, 2, ..., N - 1$.

The two-dimensional kernels are similarly given by the relations

$$g(x, y, u, v) = \frac{1}{N} (-1)^{\sum_{i=0}^{n-1} [b_i(x) b_i(u) + b_i(y) b_i(v)]} \tag{3.5-29}$$

and

$$h(x, y, u, v) = \frac{1}{N} (-1)^{\sum_{i=0}^{n-1} [b_i(x) b_i(u) + b_i(y) b_i(v)]} \tag{3.5-30}$$

where the summation in the exponent is again carried out in modulo 2 arithmetic. It is noted that, as in the case of the Walsh transform, the two-dimensional Hadamard kernels are identical.

Substitution of Eqs. (3.5-29) and (3.5-30) into Eqs. (3.5-3) and (3.5-4) yields the following two-dimensional Hadamard transform pair:

$$H(u, v) = \frac{1}{N} \sum_{x=0}^{N-1} \sum_{y=0}^{N-1} f(x, y)(-1)^{\sum_{i=0}^{n-1}[b_i(x)b_i(u) + b_i(y)b_i(v)]} \qquad (3.5\text{-}31)$$

and

$$f(x, y) = \frac{1}{N} \sum_{u=0}^{N-1} \sum_{v=0}^{N-1} F(u, v)(-1)^{\sum_{i=0}^{n-1}[b_i(x)b_i(u) + b_i(y)b_i(v)]} \qquad (3.5\text{-}32)$$

Since the forward and inverse transforms are identical, an algorithm used for computing $H(u, v)$ can be used without modification to obtain $f(x, y)$, and vice versa. It can also be shown by keeping in mind the modulo 2 summation that the Hadamard kernels are separable and symmetric. It follows, therefore, that

$$g(x,y,u,v) = g_1(x,u)\, g_1(y,v)$$

$$= h_1(x,u)\, h_1(y,v)$$

$$= \left[\frac{1}{\sqrt{N}} (-1)^{\sum_{i=0}^{n-1} b_i(x)b_i(u)} \right]\left[\frac{1}{\sqrt{N}} (-1)^{\sum_{i=0}^{n-1} b_i(y)b_i(v)} \right] \qquad (3.5\text{-}33)$$

With the exception of the $1/\sqrt{N}$ term, g_1 and h_1 are identical to Eq. (3.5-25). It is also noted that, since the two-dimensional Hadamard kernels are separable, the two-dimensional transform pair can be obtained by successive applications of any one-dimensional Hadamard transform algorithm.

The matrix of values produced by the one-dimensional Hadamard kernel given in Eq. (3.5-25) is shown in Table 3.4 for $N = 8$, where the constant $1/N$ term has been omitted for simplicity. It is noted that, although the entries in this table are the same as for the Walsh transform, the order of the rows and columns is different. In fact, when $N = 2^n$, this is the only difference between these two transforms. When N is not equal to an integer power of 2, the difference is more important. While the Walsh transform can be formulated for any positive integer value of N, existence of the Hadamard transform for values of N other than integer powers of 2 has been shown only up to $N = 200$.

Since most of the applications of transforms in image processing are based on $N = 2^n$ samples per row or column of an image, the use (and terminology) of the Walsh and Hadamard transforms is intermixed in the image processing literature, where the term Walsh-Hadamard transform is often used to denote either transform.

Two important features which might influence the choice of one of these transforms over the other are worth noting. As indicated in Section 3.5.2, the formulation given in Eq. (3.5-15) has the advantage that it can be

Table 3.4. Values of the Hadamard transformation kernel for $N = 8$.

u \ x	0	1	2	3	4	5	6	7
0	+	+	+	+	+	+	+	+
1	+	−	+	−	+	−	+	−
2	+	+	−	−	+	+	−	−
3	+	−	−	+	+	−	−	+
4	+	+	+	+	−	−	−	−
5	+	−	+	−	−	+	−	+
6	+	+	−	−	−	−	+	+
7	+	−	−	+	−	+	+	−

expressed directly in a successive-doubling format. This allows computation of the FWT by a straightforward modification of the FFT algorithm developed in Section 3.4.1. Further modifications of this algorithm would be required to compute the fast Hadamard transform (FHT) to take into account the difference in ordering. An alternative approach is to use the FWT algorithm of Fig. 3.24 and then reorder the results to obtain the Hadamard transform.

Although the Hadamard ordering has disadvantages in terms of a successive-doubling implementation, it leads to a simple recursive relationship for the generation of the transformation matrices needed to implement Eqs. (3.5-10) and (3.5-12). The Hadamard matrix of lowest order (i.e., $N = 2$) is given by

$$\mathbf{H}_2 = \begin{bmatrix} 1 & 1 \\ 1 & -1 \end{bmatrix} \tag{3.5-34}$$

Then, letting $\mathbf{H}_N$ represent the matrix of order N, the recursive relationship is given by the expression

$$\mathbf{H}_{2N} = \begin{bmatrix} \mathbf{H}_N & \mathbf{H}_N \\ \mathbf{H}_N & -\mathbf{H}_N \end{bmatrix} \tag{3.5-35}$$

where H_{2N} is the Hadamard matrix of order $2N$, and it is assumed that $N = 2^n$.

The transformation matrix for use in Eq. (3.5-10) is obtained by normalizing the corresponding Hadamard matrix by the square root of the matrix order. Thus, in the $N \times N$ case, these two matrices are related by the equation

$$A = \frac{1}{\sqrt{N}} H_N \qquad (3.5-36)$$

The expressions for the inverse Hadamard matrix are identical to Eqs. (3.5-34) through (3.5-36).

Example: Use of Eq. (3.5-34) and (3.5-35) leads to the following Hadamard matrices of order four and eight:

$$H_4 = \begin{bmatrix} H_2 & H_2 \\ H_2 & -H_2 \end{bmatrix}$$

$$= \begin{bmatrix} + & + & + & + \\ + & - & + & - \\ + & + & - & - \\ + & - & - & + \end{bmatrix}$$

and

$$H_8 = \begin{bmatrix} H_4 & H_4 \\ H_4 & -H_4 \end{bmatrix}$$

$$= \begin{bmatrix} + & + & + & + & + & + & + & + \\ + & - & + & - & + & - & + & - \\ + & + & - & - & + & + & - & - \\ + & - & - & + & + & - & - & + \\ + & + & + & + & - & - & - & - \\ + & - & + & - & - & + & - & + \\ + & + & - & - & - & - & + & + \\ + & - & - & + & - & + & + & - \end{bmatrix}$$

where $+$ and $-$ indicate $+1$ and -1, respectively.

As shown in Eqs. (3.5-25) and (3.5-33), $g(x, u)$ and $g_1(x, u)$ differ only by a constant multiplier term. Since $a_{ij} = g_1(i, j)$, it follows that the entries in the $\mathbf{A}$ matrix have the same form as the expansion of $g(x, u)$. This is easily seen for example by comparing Table 3.4 and the expression for $\mathbf{A} = \dfrac{1}{\sqrt{8}}\mathbf{H}_8$.   □

The number of sign changes along a column of the Hadamard matrix is often called the *sequency* of that column[†]. Since the elements of this matrix are derived from the kernel values, the sequency concept applies to the expansion of $g_1(x, u)$ for $x, u = 0, 1, ..., N - 1$. For instance, the sequencies of the eight columns of $\mathbf{H}_8$ and Table 3.4 are 0, 7, 3, 4, 1, 6, 2, and 5.

It is often of interest to express the Hadamard kernels so that the sequency increases as a function of increasing u. This formulation is analogous to the Fourier transform where frequency also increases as a function of increasing u. The one-dimensional Hadamard kernel which achieves this particular ordering is given by the relation

$$g(x, u) = \frac{1}{N}(-1)^{\sum\limits_{i=0}^{n-1} b_i(x)p_i(u)} \tag{3.5-37}$$

where

$$p_0(u) = b_{n-1}(u)$$

$$p_1(u) = b_{n-1}(u) + b_{n-2}(u)$$

$$p_2(u) = b_{n-2}(u) + b_{n-3}(u) \tag{3.5-38}$$

$$\vdots$$

$$p_{n-1}(u) = b_1(u) + b_0(u)$$

As before, the summations in Eqs. (3.5-37) and (3.5-38) are performed in modulo 2 arithmetic. The expansion of Eq. (3.5-37) is shown in Table 3.5 for $N = 8$, where the constant multiplier term has been omitted for simplicity and the $+$ and $-$ entries indicate $+1$ and -1, respectively. It is noted that the columns and, by symmetry, the rows of this table are in order of increasing sequency.

The inverse, ordered Hadamard kernel is given by the expression

$$h(x, u) = (-1)^{\sum\limits_{i=0}^{n-1} b_i(x)p_i(u)} \tag{3.5-39}$$

[†]As in the case of the Fourier transform where u is a frequency variable, the concept of sequency is normally restricted to this variable. Thus, the association of sequency with the columns of the Hadamard matrix is based on the assumption that the columns vary as a function of u and the rows as a function of x. This convention is used in Table 3.4.

Table 3.5. Values of the ordered Hadamard kernel for $N = 8$.

u \ x	0	1	2	3	4	5	6	7
0	+	+	+	+	+	+	+	+
1	+	+	+	+	−	−	−	−
2	+	+	−	−	−	−	+	+
3	+	+	−	−	+	+	−	−
4	+	−	−	+	+	−	−	+
5	+	−	−	+	−	+	+	−
6	+	−	+	−	−	+	−	+
7	+	−	+	−	+	−	+	−

where $p_i(u)$ is computed using Eq. (3.5-38). Substitution of the forward and inverse kernels into Eqs. (3.5-1) and (3.5-2) yields the following ordered Hadamard transform pair:

$$H(u) = \frac{1}{N} \sum_{x=0}^{N-1} f(x)(-1)^{\sum_{i=0}^{n-1} b_i(x)p_i(u)} \qquad (3.5\text{-}40)$$

and

$$f(x) = \sum_{u=0}^{N-1} H(u)(-1)^{\sum_{i=0}^{n-1} b_i(x)p_i(u)} \qquad (3.5\text{-}41)$$

The two-dimensional kernels are, as in the unordered case, separable and identical; they are given by the relations

$$g(x, y, u, v) = h(x, y, u, v)$$

$$= \frac{1}{N} (-1)^{\sum_{i=0}^{n-1} [b_i(x)p_i(u) + b_i(y)p_i(v)]} \qquad (3.5\text{-}42)$$

Substitution of these kernels into Eqs. (3.5-3) and (3.5-4) yields the following two-dimensional, ordered Hadamard transform pair:

$$H(u, v) = \frac{1}{N} \sum_{x=0}^{N-1} \sum_{y=0}^{N-1} f(x, y)(-1)^{\sum_{i=0}^{n-1} [b_i(x)p_i(u) + b_i(y)p_i(v)]} \qquad (3.5\text{-}43)$$

and

$$f(x, y) = \frac{1}{N} \sum_{u=0}^{N-1} \sum_{v=0}^{N-1} H(u, v)(-1)^{\sum_{i=0}^{n-1} [b_i(x)p_i(u) + b_i(y)p_i(v)]} \qquad (3.5\text{-}44)$$

3.5.4 Discrete Cosine Transform

The forward kernel of the one-dimensional, discrete cosine transform (DCT) is given by the relations

$$g(x, 0) = \frac{1}{\sqrt{N}} \qquad (3.5\text{-}45a)$$

$$g(x, u) = \sqrt{\frac{2}{N}} \cos \frac{(2x + 1)u\pi}{2N} \qquad (3.5\text{-}45b)$$

for $x = 0, 1, ..., N - 1$ and $u = 1, 2, ..., N - 1$. Substitution of these expressions into Eq. (3.5-1) yields

$$C(0) = \frac{1}{\sqrt{N}} \sum_{x=0}^{N-1} f(x) \qquad (3.5\text{-}46a)$$

$$C(u) = \sqrt{\frac{2}{N}} \sum_{x=0}^{N-1} f(x) \cos \frac{(2x + 1)u\pi}{2N} \qquad (3.5\text{-}46b)$$

where $C(u)$, $u = 0, 1, 2, ..., N - 1$, is the DCT of $f(x)$.

The inverse kernel is of the same form as in Eq. (3.5-45), and the inverse DCT is defined by the equation

$$f(x) = \frac{1}{\sqrt{N}} C(0) + \sqrt{\frac{2}{N}} \sum_{u=1}^{N-1} C(u) \cos \frac{(2x + 1)u\pi}{2N} \qquad (3.5\text{-}47)$$

for $x = 0, 1, 2, ..., N - 1$.

The two-dimensional forward DCT kernel is defined as

$$g(x, y, 0, 0) = \frac{1}{N} \qquad (3.5\text{-}48a)$$

$$g(x, y, u, v) = \frac{1}{2N^3} \left[\cos(2x + 1)u\pi \right]\left[\cos(2y + 1)v\pi \right] \qquad (3.5\text{-}48b)$$

for $x, y = 0, 1, ..., N - 1$, and $u, v = 1, 2, ..., N - 1$. The inverse kernel is also of this form. Thus, it follows from Eqs. (3.5-3) and (3.5-4) that the two-dimensional DCT pair is given by the expressions

$$C(0, 0) = \frac{1}{N} \sum_{x=0}^{N-1} \sum_{y=0}^{N-1} f(x, y) \qquad (3.5\text{-}49a)$$

$$C(u, v) = \frac{1}{2N^3} \sum_{x=0}^{N-1} \sum_{y=0}^{N-1} f(x, y) \left[\cos(2x + 1)u\pi \right] \left[\cos(2y + 1)v\pi \right]$$

$$(3.5\text{-}49\text{b})$$

for $u, v = 1, 2, ..., N - 1$, and

$$f(x, y) = \frac{1}{N} C(0, 0)$$

$$+ \frac{1}{2N^3} \sum_{u=1}^{N-1} \sum_{v=1}^{N-1} C(u, v) \left[\cos(2x + 1)u\pi \right] \left[\cos(2y + 1)v\pi \right]$$

$$(3.5\text{-}50)$$

for $x, y = 0, 1, ..., N - 1$.

By comparing Eqs. (3.5-45) and (3.5-48) we note that the DCT transformation kernels are separable, so that the two-dimensional forward or inverse transform can be computed by successive applications of a one-dimensional DCT algorithm. In fact, an interesting property of the DCT is that it can be obtained directly from an FFT algorithm. This can be seen by expressing Eq. (3.5-46) in the equivalent form

$$C(0) = \frac{1}{\sqrt{N}} \sum_{x=0}^{N-1} f(x) \qquad (3.5\text{-}51\text{a})$$

and

$$C(u) = \sqrt{\frac{2}{N}} \; \text{Re}\left\{ \left[\exp\left(\frac{-j2\pi}{2N} \right) \right] \sum_{x=0}^{2N-1} f(x) \exp\left(\frac{-j2\pi ux}{N} \right) \right\} \qquad (3.5\text{-}51\text{b})$$

$$u = 1, 2, \ldots, N - 1$$

where $f(x) = 0$ for $x = N, N + 1, ..., 2N - 1$, and $\text{Re}\{\cdot\}$ denotes the real part of the term enclosed. The summation term is recognized as a $2N$-point discrete Fourier transform. Similarly, a $2N$-point inverse FFT can be used to obtain $f(x)$ from $C(u)$.

3.5.5 Hotelling Transform

Unlike the transforms discussed previously, the *Hotelling transform*[†] developed in this section is based on statistical properties of an image. The principal uses of this transform are in data compression and rotation applications.

As an introduction to the problem, suppose that an $N \times N$ image $f(x, y)$ is transmitted M times over some communication channel. Since any

---

[†]This transform is also commonly referred to as the *eigenvector, principal component,* or *discrete Karhunen-Loève transform.*

physical channel is subject to random disturbances, the set of received images, $\{f_1(x, y), f_2(x, y), ..., f_M(x, y)\}$, will in general represent a statistical ensemble whose properties are determined by the channel characteristics and the nature of the disturbance. An example of such an ensemble is the set of images of the same scene transmitted by a space probe. In this case, the images are corrupted by atmospheric disturbances and electrical noise in the transmitter and receiver.

3.5.5.1 Formulation

Each sample image $f_i(x, y)$ can be expressed in the form of an N^2-dimensional vector x_i as follows:

$$
\mathbf{x}_i =
\begin{bmatrix}
x_{i1} \\
x_{i2} \\
\vdots \\
x_{ij} \\
\vdots \\
x_{iN^2}
\end{bmatrix}
\qquad (3.5\text{-}52)
$$

where x_{ij} denotes the jth component of vector $\mathbf{x}_i$. One way to construct such a vector is to let the first N components of $\mathbf{x}_i$ be formed from the first row of $f_i(x, y)$ [i.e., $x_{i1} = f(0, 0)$, $x_{i2} = f(0, 1)$, ..., $x_{iN} = f(0, N - 1)$]; the second set of N components from the second row; and so on. Another way is to use the columns of $f_i(x, y)$ instead of the rows. Of course, other representations are possible, but these two are the ones used most often.

The covariance matrix of the x vectors is defined as

$$
\mathbf{C}_x = E\{(\mathbf{x} - \mathbf{m})(\mathbf{x} - \mathbf{m})'\} \qquad (3.5\text{-}53)
$$

where

$$
\mathbf{m}_x = E\{\mathbf{x}\} \qquad (3.5\text{-}54)
$$

is the mean vector, E is the expected value, and the prime ($'$) indicates transpositon. Equations (3.5-53) and (3.5-54) can be approximated from the samples by using the relations

$$
\mathbf{m}_x \cong \frac{1}{M} \sum_{i=1}^{M} \mathbf{x}_i \qquad (3.5\text{-}55)
$$

and

$$
\mathbf{C}_x \cong \frac{1}{M} \sum_{i=1}^{M} (\mathbf{x}_i - \mathbf{m}_x)(\mathbf{x}_i - \mathbf{m}_x)'
$$

$$C_x \cong \frac{1}{M} \left[\sum_{i=1}^{M} x_i x_i' \right] - m_x m_x' \tag{3.5-56}$$

The mean vector is of dimensionality N^2 and C_x is an $N^2 \times N^2$ matrix.

Let e_i and λ_i, $i = 1, 2, ..., N^2$, be the eigenvectors and corresponding eigenvalues of C_x. It is assumed for convenience in notation that the eigenvalues have been arranged in decreasing order so that $\lambda_1 \geqslant \lambda_2 \geqslant \cdots \geqslant \lambda_{N^2}$. A transformation matrix whose rows are the eigenvalues of C_x is given by

$$A = \begin{bmatrix} e_{11} & e_{12} \cdots e_{1N^2} \\ e_{21} & e_{22} \cdots e_{2N^2} \\ \cdot \\ \cdot \\ \cdot \\ e_{N^21} & e_{N^22} \cdots e_{N^2N^2} \end{bmatrix} \tag{3.5-57}$$

where e_{ij} is the jth component of the ith eigenvector. The Hotelling transform then consists simply of multiplying a centralized image vector, $(x - m_x)$, by A to obtain a new image vector y, that is,

$$y = A(x - m_x) \tag{3.5-58}$$

Equation (3.5-58) has several important properties. Let us first examine the covariance matrix of the y vectors. This matrix is given by

$$C_y = E\{(y - m_y)(y - m_y)'\} \tag{3.5-59}$$

where m_y is equal to the zero vector, 0, as can be shown directly from Eqs. (3.5-54) and (3.5-58):

$$m_y = E\{y\}$$

$$= E\{A(x - m_x)\}$$

$$= AE\{x\} - Am_x$$

$$= 0 \tag{3.5-60}$$

Substitution of Eqs. (3.5-58) and (3.5-60) into Eq. (3.5-59) yields the following expressions for C_y in terms of C_x:

$$C_y = E\{(Ax - Am_x)(Ax - Am_x)'\}$$

$$= E\{A(x - m_x)(x - m_x)'A'\}$$

$$= AE\{(x - m_x)(x - m_x)'\}A'$$

$$\mathbf{C_y} = \mathbf{AC_xA'} \tag{3.5-61}$$

where the last step follows from the definition of $\mathbf{C_x}$ given in Eq. (3.5-53).

It can be shown (Lawley and Maxwell [1963]) that $\mathbf{C_y}$ is a diagonal matrix with elements equal to the eigenvalues of $\mathbf{C_x}$, that is,

$$\mathbf{C_y} = \begin{bmatrix} \lambda_1 & & & \\ & \lambda_2 & & 0 \\ & & \ddots & \\ 0 & & & \\ & & & \lambda_{N^2} \end{bmatrix} \tag{3.5-62}$$

The importance of this property is that, since the terms of the main diagonal are zero, the elements of $\mathbf{y}$ are *uncorrelated*. In addition, each eigenvalue λ_i is equal to the variance of the ith element of $\mathbf{y}$ along eigenvector $\mathbf{e}_i$.

Another property, which will be shown in Chapter 6 to be useful for image coding, deals with the reconstruction of $\mathbf{x}$ and $\mathbf{y}$. Since $\mathbf{C_x}$ is a real, symmetric matrix it is always possible to find a set of orthonormal eigenvectors (Noble [1969]). It follows, therefore, that $\mathbf{A}^{-1} = \mathbf{A'}$ and $\mathbf{x}$ can be reconstructed from $\mathbf{y}$ by using the relation

$$\mathbf{x} = \mathbf{A'y} + \mathbf{m_x} \tag{3.5-63}$$

Suppose, however, that instead of using all the eigenvectors of $\mathbf{C_x}$, we form $\mathbf{A}$ from the K eigenvectors corresponding to the largest eigenvalues. The $\mathbf{y}$ vectors will then be K-dimensional and the reconstruction given by Eq. (3.5-63) will no longer be exact. Let

$$\hat{\mathbf{x}} = \mathbf{A'_K y} + \mathbf{m_x} \tag{3.5-64}$$

represent the approximation to $\mathbf{x}$ obtained with a transformation matrix $\mathbf{A}_K$ composed of the first K eigenvectors of $\mathbf{C_x}$. It can be shown that the mean square error, R, between $\mathbf{x}$ and $\hat{\mathbf{x}}$ is given by the expression

$$R = \sum_{j=1}^{N^2} \lambda_j - \sum_{j=1}^{K} \lambda_j \tag{3.5-65a}$$

$$= \sum_{j=K+1}^{N^2} \lambda_j \tag{3.5-65b}$$

From Eq. (3.5-65a) we see that the error is zero if $K = N^2$ (i. e., if all the eigenvectors are used in the transformation). Since the λ_j decrease monotonically, either form of Eq. (3.5-65) shows that the error is minimized by selecting the eigenvectors associated with the largest eigenvalues. Thus, the Hotelling transform is optimum in the least-square-error sense.

The development of the Hotelling transform given above is quite different from the general approach presented in Section 3.5.1. The reason for this is that the Hotelling transformation matrix is not separable, thus precluding the use of Eq. (3.5-10). Another important difference between the Hotelling transform and the separable transforms studied thus far is that no fast algorithm exists for computing the former transform. This is a serious drawback because of the amount of computation involved in obtaining the eigenvalues and eigenvectors of the $N^2 \times N^2$ covariance matrix of any set of images of practical size. This problem is addressed again in Chapter 6.

3.5.5.2 Application to Image Rotation

One of the basic concepts underlying the derivation of the Hotelling transform is that the choice of base vectors (i.e., rows of A) is made such that these vectors point in the direction of maximum variance of the data, subject to the constraint that all vectors be mutually orthogonal and the transformed components be uncorrelated. These requirements lead to the solution stated above that the bases are the eigenvectors of the covariance matrix. The required orientation of the eigenvectors leads to the optimal properties of the Hotelling transform since, when fewer than N^2 vectors are used in A, the basis vectors selected correspond to the direction of largest variance in the data.

These properties can be used to advantage in image rotation. Let us consider the problem where an object has been extracted from an image and it is desired to rotate the object so that it is aligned in some standard or invariant direction. Many objects of interest in image processing applications (particularly man-made objects) can be easily standardized by performing a rotation which aligns the coordinate axes with the axes of maximum variance of the pixels in the object.

Consider the coordinate system with x_1- and x_2-axis shown in Fig. 3.27. If new coordinates are chosen having a different orientation, but the same origin as the original system, then we shall say that there has been a rotation of axes in the plane. Let θ be the angle of rotation from the positive half of the x_1-axis to the positive half of the y_1-axis. It then follows from elementary trigonometry that the new and old axes are related by the equations

$$y_1 = x_1 \cos \theta + x_2 \sin \theta$$

$$y_2 = -x_1 \sin \theta + x_2 \cos \theta$$

(3.5-66)

or

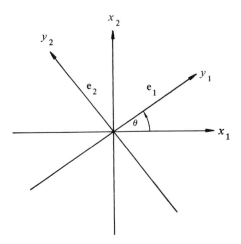

Figure 3.27. Rotation of coordinate system.

$$\begin{bmatrix} y_1 \\ y_2 \end{bmatrix} = \begin{bmatrix} \cos\theta & \sin\theta \\ -\sin\theta & \cos\theta \end{bmatrix} \begin{bmatrix} x_1 \\ x_2 \end{bmatrix} \tag{3.5-67}$$

The original coordinates of each pixel in the object can be interpreted as two-dimensional random variables with mean

$$\mathbf{m_x} \cong \frac{1}{P} \sum_{i=1}^{P} \mathbf{x}_i \tag{3.5-68}$$

and covariance matrix

$$\mathbf{C_x} \cong \frac{1}{P} \left[\sum_{i=1}^{P} \mathbf{x}_i \mathbf{x}_i' \right] - \mathbf{m_x} \mathbf{m_x'} \tag{3.5-69}$$

where P is the number of pixels in the object to be rotated and $\mathbf{x}_i$ is the vector composed of the coordinates of the ith pixel. It is noted that $\mathbf{m_x}$ is a two-dimensional vector and $\mathbf{C_x}$ a 2×2 matrix.

Since the eigenvectors of $\mathbf{C_x}$ point in the directions of maximum variance (subject to the constraint that they be orthogonal), a logical choice is to select the new coordinate system so that it will be aligned with these eigenvectors.

If we let the y_1- and y_2-axis be aligned with the *normalized* eigenvectors $\mathbf{e}_1$ and $\mathbf{e}_2$, it then follows from Fig. 3.27 that $\cos\theta = e_{11}$, $\sin\theta = e_{12}$, $-\sin\theta = e_{21}$, and $\cos\theta = e_{22}$, where e_{11} and e_{21} are the projections of $\mathbf{e}_1$ and $\mathbf{e}_2$ along the x_1-axis, and e_{12} and e_{22} are the projections of these two vectors along the x_2-axis. The rotation from the original coordinate system to this new system is then given by the relation.

$$
\begin{bmatrix} y_1 \\ \\ y_2 \end{bmatrix} = \begin{bmatrix} e_{11} & e_{12} \\ \\ e_{21} & e_{22} \end{bmatrix} \begin{bmatrix} x_1 \\ \\ x_2 \end{bmatrix}
\qquad (3.5\text{-}70)
$$

which is seen to be in the familiar form $\mathbf{y} = \mathbf{Ax}$, where

$$
\mathbf{A} = \begin{bmatrix} e_{11} & e_{12} \\ \\ e_{21} & e_{22} \end{bmatrix}
\qquad (3.5\text{-}71\text{a})
$$

$$
= \begin{bmatrix} \cos\theta & \sin\theta \\ \\ -\sin\theta & \cos\theta \end{bmatrix}
\qquad (3.5\text{-}71\text{b})
$$

It is noted that Eq. (3.5-70) is analogous to a two-dimensional Hotelling transform where the mean has not been subtracted from the original vectors. Subtraction of the mean vector simply centralizes the object so that its center of gravity is at the origin of the new coordinate system. In this case, the transformation assumes the form $\mathbf{y} = \mathbf{A}(\mathbf{x} - \mathbf{m_x})$.

Interpretation of Eq. (3.5-70) is different in another respect from our earlier discussion of the Hotelling transform where each vector was formed from pixels in an entire image. In the present application, we are concerned with the *coordinates* of pixels in a *single* image, and the ensemble of vectors used to arrive at the transformation matrix is derived from an object in the image.

The rotation-translation procedure is summarized in Fig. 3.28. Part (a) of this figure shows a simple object and the y_1- and y_2-axis chosen in the direction of its normalized eigenvectors. Figure 3.28(b) shows the object in the new coordinate system. Finally, Fig. 3.28(c) shows the result of rotation after the mean was subtracted from the data.

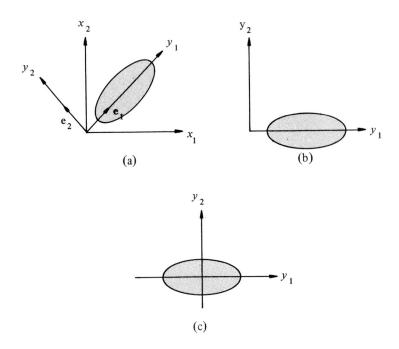

Figure 3.28. Rotation of a two-dimensional object. (a) Original data scatter showing direction of unit eigenvectors. (b) Data rotated by using the transformation $y = Ax$. (c) Data rotated and centralized by using the transformation $y = A(x - m_x)$.

It is important to note that, if e_1 and e_2 are valid eigenvectors, the pairs $(e_1, -e_2)$, $(-e_1, e_2)$, and $(-e_1, -e_2)$ are also valid. Since the foregoing rotation procedure is based on a right-handed coordinate system, the choice of direction of the eigenvectors must be taken into account. Failure to do so can produce rotated results which are reflected about the origin in one or both of the new axes. Figure 3.29 illustrates the results of rotation for all possible directions of the eigenvectors[†];   the rotation in each case was obtained by aligning the y_1-axis with e_1, which is the eigenvector associated with the largest variance. It is noted that Fig. 3.29(a) shows the expected results.   Figure 3.29(b) is based on a left-handed coordinate system and yields data which are reflected about the origin in the y_1-axis.

[†]Since the eigenvectors of the covariance matrix point in the directions of maximum variance, the only possible deviation of each vector is a change in sign.

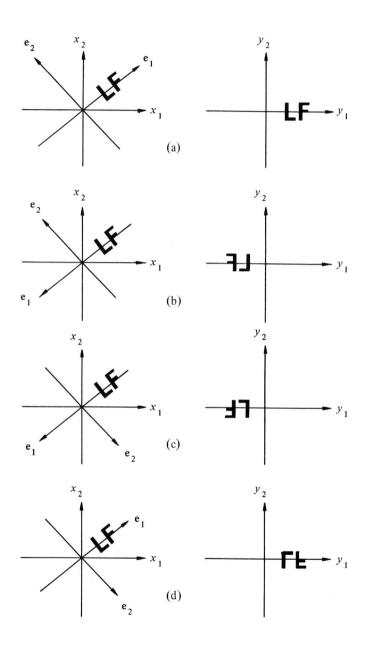

Figure 3.29. Example of possible directions of eigenvectors and corresponding rotation results.

Figure 3.29(c) is based on a right-handed system, but it also shows reflection about both axes. Actually this result is not incorrect; it merely points out the fact that it is not uncommon in practice to obtain the negative of the expected eigenvectors from a computational algorithm. Reversing the direction of both vectors would yield the same result as in Fig. 3.29(a). Finally, Fig. 3.29(d) again shows the reflection due to an improper coordinate system. It is noted that a right-handed system may or may not yield reflected results, but that a left-handed system will always be accompanied by this problem. In practice, the most common approach is to use a right-handed system and make provisions to handle both possible orientations of e_1 and e_2 in this system.

3.6 CONCLUDING REMARKS

The principal purpose of this chapter has been to present a theoretical foundation of image transforms and their properties. Within this framework, the essential points necessary for a basic understanding of these concepts have been developed and illustrated.

The emphasis placed on the Fourier transform reflects its wide scope of application in image processing problems. The material on the fast Fourier transform is of particular importance because of its computational implications. The separability, centralization, and convolution properties of the Fourier transform will also be used extensively in the following chapters.

Transform theory has played a central role in the development of image processing as a formal discipline, as will be evident in subsequent discussions. In the next chapters, we consider some uses of the Fourier transform for image enhancement and restoration. Further discussions of the Walsh-Hadamard, discrete cosine, and Hotelling transforms is deferred until Chapter 6, where use is made of these transforms for image encoding.

REFERENCES

Our treatment of the Fourier transform is of an introductory nature. The classic texts by Titchmarsh [1948] and Papoulis [1962] offer a comprehensive theoretical treatment of the continuous Fourier transform and its properties. The reader will have little difficulty finding additional references in this area. Most engineering circuits and communications texts offer a variety of developments and explanations of the Fourier transform. The books by Van Valkenburg [1955], Carlson [1968], and Thomas [1969] are representative.

The derivation of the discrete Fourier transform from its continuous form is also covered extensively in the literature. Three good references on this topic are Blackman and Tukey [1958], Cooley, Lewis, and Welch [1967], and Brigham [1974]. The first and last references are particularly suited for introductory reading.

Formulation of the fast Fourier transform is often credited to Cooley and Tukey [1965]. The FFT, however, has an interesting history worth sketching here. In response to the Cooley-Tukey paper, Rudnick [1966] reported that he was using a similar technique, whose number of operations was also proportional to $N \log_2 N$, and which was based on a method published by Danielson and Lanczos [1942]. These authors, in turn, referenced Runge [1903, 1905] as the source of their technique. The latter two papers together with the lecture notes of Runge and König [1939] contain the essential computational advantages of present FFT algorithms. Similar techniques were also published by Yates [1937], Stumpff [1939], Good [1958], and Thomas [1963]. A paper by Cooley, Lewis, and Welch [1967a] presents a historical summary and an interesting comparison of results prior to the 1965 Cooley-Tukey paper.

The FFT algorithm presented in this chapter is by no means a unique formulation. For example, the so-called Sande-Tukey algorithm (Gentleman and Sande [1966]) is based on an alternative decomposition of the input data. The book by Brigham [1974] contains a comprehensive discussion of this algorithm as well as numerous other formulations of the FFT including procedures for bases other than 2.

Although our attention has been focused exclusively on digital techniques, the reader should be aware that two-dimensional Fourier transforms can also be obtained by optical methods. The books by Papoulis [1968], Goodman [1968], and Hech and Zajac [1975] span the theoretical and applied aspects of optics and optical transforms at an introductory level.

Further reading in the matrix formulation of image transforms can be found in the book by Andrews [1970], which also develops the concept of matrix decomposition and discusses other image transforms in addition to the ones covered in this chapter. The papers by Good [1958], Gentleman [1968], and Kahaner [1970] are also of interest.

The original paper on the Walsh transform (Walsh [1923]) is worth reading from a historical point of view. Additional references on this transform are Fine [1949, 1950], Hammond and Johnson [1962], Henderson [1964], Shanks [1969], and Andrews [1970].

Further reading on the Hadamard transform can be found in the original paper by Hadamard [1893], and in Williamson [1944], Whelchel [1968], and Andrews [1970]. Two interesting notes dealing with the search for Hadamard matrices based on other than integer powers of 2 are Baumert [1962], and Golomb [1963]. The concept of sequency appears to have been introduced by Harmuth [1968]. References for the discrete cosine transform are Ahmed [1974], and Ahmed and Rao [1975]. The latter reference also contains a comprehensive discussion of other orthogonal transforms.

Hotelling [1933] was the first to derive and publish the transformation that transforms discrete variables into uncorrelated coefficients. He referred to this technique as *the method of principal components*. His paper gives considerable insight into the method and is worth reading. Hotelling's transformation was rediscovered by Kramer and Mathews [1956] and Huang and Schultheiss [1963]. See Lawley and Maxwell [1963] for a general discussion of this topic.

The analogous transformation for transforming continuous data into a set of uncorrelated coefficients was discovered by Karhunen [1947] and Loève [1948] and is called the *Karhunen-Loève expansion*. See Selin [1965] for an excellent discussion. The result that the Karhunen-Loève expansion minimizes the mean-square truncation error was first published by Koschman [1954] and rediscovered by Brown [1960].

IMAGE ENHANCEMENT

It makes all the difference whether one
sees darkness though the light or brightness
through the shadows.
David Lindsay

The principal objective of enhancement techniques is to process a given image so that the result is more suitable than the original image for a specific application. The word "specific" is important because it establishes at the onset that the techniques discussed in this chapter are very much problem-oriented. Thus, for example, a method that is quite useful for enhancing x-ray images may not necessarily be the best approach for enhancing pictures of Mars transmitted by a space probe.

The approaches discussed in this chapter may be divided into two broad categories: frequency-domain methods and spatial-domain methods. Processing techniques in the first category are based on modifying the Fourier transform of an image. The spatial-domain, on the other hand, refers to the image plane itself, and approaches in this category are based on direct manipulation of the pixels in an image.

The basic methodology underlying the material developed in this chapter is presented in Section 4.1. Section 4.2 deals with image enhancement by histogram modification techniques. Sections 4.3 and 4.4 contain a number of approaches for image smoothing and sharpening, respectively. This discussion is followed in Section 4.5 by an enhancement technique based on the illumination-reflectance model introduced in Section 2.2. Finally, Section 4.6 contains an introduction to color fundamentals, as well as several applications of pseudo-color concepts to image enhancement.

4.1 BACKGROUND

The foundation of frequency-domain techniques is the convolution theorem. Let $g(x, y)$ be an image formed by the convolution of an image

$f(x, y)$ and a *position-invariant operator* $h(x, y)^{\dagger}$, that is,

$$g(x, y) = h(x, y)*f(x, y) \tag{4.1-1}$$

Then, from the convolution theorem (Section 3.3.8), we have that the following frequency-domain relation holds:

$$G(u, v) = H(u, v)F(u, v) \tag{4.1-2}$$

where G, H, and F are the Fourier transforms of g, h, and f, respectively. The transform $H(u, v)$ is sometimes called the *transfer function* of the process.

It will be shown in this chapter that numerous enhancement and restoration problems can be expressed in the forms of Eqs. (4.1-1) and (4.1-2). Attention will be focused on the latter equation because, as pointed out in Section 3.3.8, discrete convolution is often more efficiently carried out in the frequency domain via an FFT algorithm.

As indicated in Section 3.3.8, discrete convolution is characterized by wraparound error unless the functions are assumed to be periodic with periods chosen according to Eqs. (3.3-33) and (3.3-34). In an image, wraparound error manifests itself as a distortion around the edges. In practice, however, this error is often not objectionable, even when the images are not extended using the procedure given in Section 3.3.8. The results given later in this chapter, for example, were obtained by direct FFT computations on the given images without extension.

In a typical image enhancement problem $f(x, y)$ is given and the goal, after computation of $F(u, v)$, is to select $H(u, v)$ so that the desired image, given by,

$$g(x, y) = \mathcal{F}^{-1}\{H(u, v)F(u, v)\} \tag{4.1-3}$$

exhibits some highlighted feature of $f(x, y)$. For instance, edges in $f(x, y)$ can be accentuated by using a function $H(u, v)$ which emphasizes the high-frequency components of $F(u, v)$.

Spatial domain techniques are based on gray-level mappings, where the type of mapping used depends on the criterion chosen for enhancement. As an illustration of this approach, consider the problem of enhancing the contrast of an image. Let r and s denote any gray level in the original and enhanced image, respectively, and suppose that for every pixel with level r in the original image, we create a pixel in the enhanced image with level $s = T(r)$. If $T(r)$ has the form shown in Fig. 4.1(a), the effect of this transformation will be to produce an image of higher contrast than the original by darkening the levels below a value m and brightening the levels above m in the original pixel spectrum. In this technique, known as *contrast*

---

†A position-invariant operator is one whose result depends only on the *value* of $f(x, y)$ at a given point in the image and not on the *position* of the point. Position invariance is an implicit requirement in the definition of the convolution integrals given in Eqs. (3.3-23) and (3.3-30).

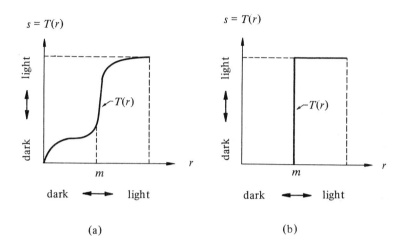

Figure 4.1. Gray-level transformation functions for contrast enhancement.

stretching, the levels of *r* below *m* are compressed by the transformation function into a narrow range of *s* toward the dark end of the spectrum; the opposite effect takes place for values of *r* above *m*. In the limiting case shown in Fig. 4.41(b), $T(r)$ produces a two-level (i.e., binary) image. As shown in the following sections, some fairly simple, yet powerful processing approaches can be formulated in the spatial domain of an image.

Although it may already be obvious, let us emphasize before leaving this section that there is no general theory of image enhancement. When an image is processed for visual interpretation, the viewer is the ultimate judge of how well a particular method works. Visual evaluation of image quality is a highly subjective process, thus making the definition of a "good image" an elusive standard by which to compare algorithm performance. When the problem is one of processing images for machine perception, the evaluation task is somewhat easier. If, for example, one were dealing with a character recognition application, the best image processing method would be the one yielding the best machine recognition results. It is noteworthy, however, that even in situations where a clear-cut criterion of performance can be imposed on the problem, one usually is faced with a certain amount of trial-and-error before being able to settle on a particular image processing approach.

4.2 IMAGE ENHANCEMENT BY HISTOGRAM MODIFICATION TECHNIQUES

A histogram of gray level content provides a global description of the appearance of an image. The methods discussed in this section achieve enhancement by modifying the histogram of a given image in a specified manner. The type and degree of enhancement obtained depends on the nature of the specified histogram.

4.2.1 Foundation

Let the variable r represent the gray level of the pixels in the image to be enhanced. For simplicity, it will be assumed in the following discussion that the pixel values have been normalized so that they lie in the range

$$0 \leqslant r \leqslant 1 \qquad\qquad (4.2\text{-}1)$$

with $r = 0$ representing black and $r = 1$ representing white in the gray scale.

For any r in the interval [0, 1], attention will be focused on transformations of the form

$$s = T(r) \qquad\qquad (4.2\text{-}2)$$

which produce a level s for every pixel value r in the original image. It is assumed that the transformation function given in Eq. (4.2-2) satisfies the conditions:

(a) $T(r)$ is a single-valued and monotonically increasing in the interval $0 \leqslant r \leqslant 1$, and

(b) $0 \leqslant T(r) \leqslant 1$ for $0 \leqslant r \leqslant 1$.

Condition (a) preserves the order from black to white in the gray scale, while condition (b) guarantees a mapping that is consistent with the allowed range of pixel values. A transformation function satisfying these conditions is illustrated in Fig. 4.2.

The inverse transformation from s back to r will be denoted by

$$r = T^{-1}(s) \qquad 0 \leqslant s \leqslant 1 \qquad\qquad (4.2\text{-}3)$$

where it is assumed that $T^{-1}(s)$ also satisfies conditions (a) and (b) with respect to the variable s.

The gray levels in an image are random quantities in the interval [0, 1]. Assuming for a moment that they are continuous variables, the original and transformed gray levels can be characterized by their probability density functions $p_r(r)$ and $p_s(s)$, respectively. A great deal can be said about the general characteristics of an image from the density function of its gray levels. For example, an image whose gray levels have a density

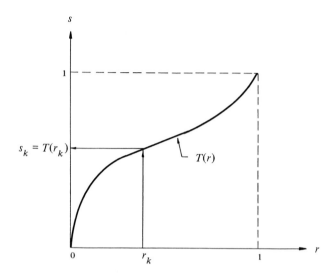

Figure 4.2. A gray-level transformation function.

function like the one shown in Fig. 4.3(a) would have fairly dark characteristics since the majority of its levels are concentrated in the dark region of the gray scale. An image whose gray levels have a density function like the one shown in Fig. 4.3(b), on the other hand, would have predominant light tones since the majority of its pixels are light gray.

It follows from elementary probability theory that if $p_r(r)$ and $T(r)$ are known, and $T^{-1}(s)$ satisfies condition (a), then the probability density function of the transformed gray levels is given by the relation

$$p_s(s) = \left[p_r(r) \frac{dr}{ds} \right]_{r = T^{-1}(s)} \tag{4.2-4}$$

The following enhancement techniques are based on modifying the appearance of an image by controlling the probability density function of its gray levels via the transformation function $T(r)$.

4.2.2 Histogram Equalization

Consider the transformation function

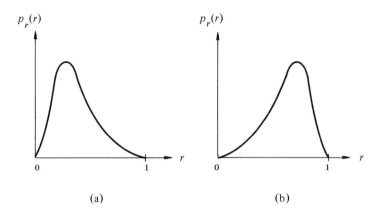

Figure 4.3. Gray-level probability density functions of (a) a "dark" image, and (b) a "light" image.

$$s = T(r) = \int_0^r p_r(w) \, dw \qquad 0 \leqslant r \leqslant 1 \tag{4.2-5}$$

where w is a dummy variable of integration. The rightmost side of Eq. (4.2-5) is recognized as the cumulative distribution function (CDF) of r. The two conditions set forth in the previous section are satisfied by this transformation function since the CDF increases monotonically from 0 to 1 as a function of r.

From Eq. (4.2-5) the derivative of s with respect to r is given by

$$\frac{ds}{dr} = p_r(r) \tag{4.2-6}$$

Substituting dr/ds into Eq. (4.2-4) yields

$$p_s(s) = \left[p_r(r) \frac{1}{p_r(r)} \right]_{r=T^{-1}(s)}$$

$$= [1]_{r=T^{-1}(s)}$$

$$= 1 \qquad 0 \leqslant s \leqslant 1 \tag{4.2-7}$$

which is a uniform density in the interval of definition of the transformed variable s. It is noted that this result is independent of the inverse transformation function. This is important because it is not always easy to obtain $T^{-1}(s)$ analytically.

The foregoing development indicates that using a transformation function equal to the cumulative distribution of r produces an image whose gray levels have a uniform density. In terms of enhancement, this implies an increase in the dynamic range of the pixels which, as will be seen below, can have a considerable effect in the appearance of an image.

Example: Before proceeding with a discussion of discrete variables, let us consider a simple illustration of the use of Eqs. (4.2-4) and (4.2-5). Assume that the levels r have the probability density function shown in Fig. 4.4(a).

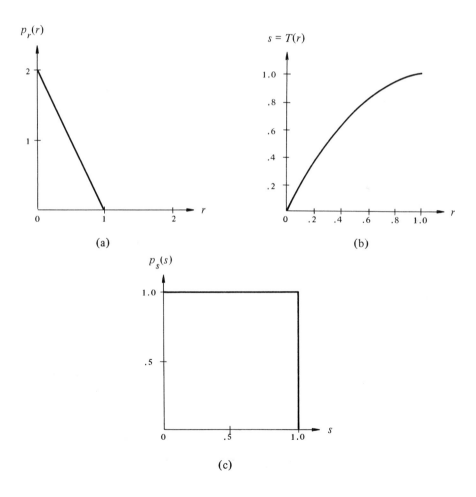

Figure 4.4. Illustration of the uniform-density transformation method. (a) Original probability density function. (b) Transformation function. (c) Resulting uniform density.

In this case $p_r(r)$ is given by

$$p_r(r) = \begin{cases} -2r + 2 & 0 \leqslant r \leqslant 1 \\ \\ 0 & \text{elsewhere} \end{cases}$$

Substitution of this expression in Eq. (4.2-5) yields the transformation function

$$s = T(r) = \int_0^r (-2w + 2)\, dw$$

$$= -r^2 + 2r$$

Although we only need $T(r)$ for histogram equalization, it will be instructive to show that the resulting density $p_s(s)$ is in fact uniform since this requires that $T^{-1}(s)$ be obtained. In practice this step is not required because Eq. (4.2-7) is independent of the inverse transformation function. Solving for r in terms of s yields

$$r = T^{-1}(s) = 1 \pm \sqrt{1 - s}$$

Since r lies in the interval $[0, 1]$, only the solution

$$r = T^{-1}(s) = 1 - \sqrt{1 - s}$$

is valid.

The probability density function of s is obtained by substituting the above results into Eq. (4.2-4):

$$p_s(s) = \left[p_r(r)\, \frac{dr}{ds} \right]_{r = T^{-1}(s)}$$

$$= \left[(-2r + 2)\, \frac{dr}{ds} \right]_{r = 1 - \sqrt{1 - s}}$$

$$= \left[(-2\sqrt{1 - s}\,)\, \frac{d}{ds} \left(\sqrt{1 - s}\, \right) \right]$$

$$= 1 \qquad 0 \leqslant s \leqslant 1$$

which is a uniform density in the desired level. The transformation function $T(r)$ is shown in Fig. 4.4(b), and $p_s(s)$ is shown in Fig. 4.4(c).   □

In order to be useful for digital image processing, the concepts developed above must be formulated in discrete form. For gray levels which assume discrete values, we deal with probabilities given by the relation

$$p_r(r_k) = \frac{n_k}{n} \qquad \begin{array}{l} 0 \leqslant r_k \leqslant 1 \\ k = 0, 1, \ldots, L-1 \end{array} \qquad (4.2\text{-}8)$$

where L is the number of levels, $p_r(r_k)$ is the probability of the kth gray level, n_k is the number of times this level appears in the image, and n is the total number of pixels in the image. A plot of $p_r(r_k)$ vs. r_k is usually called a *histogram*, and the technique used for obtaining a uniform histogram is known as *histogram equalization* or *histogram linearization*.

The discrete form of Eq. (4.2-5) is given by the relation

$$s_k = T(r_k) = \sum_{j=0}^{k} \frac{n_j}{n}$$

$$= \sum_{j=0}^{k} p_r(r_j) \qquad \begin{array}{l} 0 \leqslant r_k \leqslant 1 \\ k = 0, 1, \ldots, L-1 \end{array} \qquad (4.2\text{-}9)$$

The inverse transformation is denoted by

$$r_k = T^{-1}(s_k) \qquad 0 \leqslant s_k \leqslant 1$$

where both $T(r_k)$ and $T^{-1}(s_k)$ are assumed to satisfy conditions (a) and (b) stated in the previous section. It is noted that the transformation function $T(r_k)$ can be computed directly from the image in question by using Eq. (4.2-9). Although the inverse function $T^{-1}(s_k)$ is not used in histogram equalization, it plays a central role in the method discussed in the next section.

Example: Suppose that a 64 × 64, 8-level image has the gray-level distribution shown in Table. 4.1.

Table 4.1

r_k	n_k	$p_r(r_k) = n_k/n$
$r_0 = 0$	790	0.19
$r_1 = 1/7$	1023	0.25
$r_2 = 2/7$	850	0.21
$r_3 = 3/7$	656	0.16
$r_4 = 4/7$	329	0.08
$r_5 = 5/7$	245	0.06
$r_6 = 6/7$	122	0.03
$r_7 = 1$	81	0.02

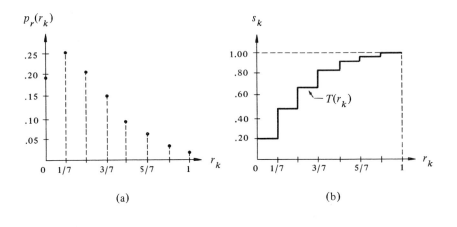

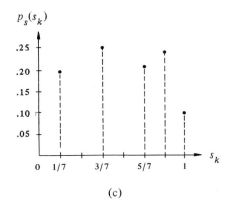

Figure 4.5. Illustration of the histogram-equalization method. (a) Original histogram. (b) Transformation function. (c) Equalized histogram.

The histogram of these gray levels is shown in Fig. 4.5(a).

The transformation function is obtained by using Eq. (4.2-9). For instance,

$$s_0 = T(r_0) = \sum_{j=0}^{0} p_r(r_j)$$

$$= p_r(r_0)$$

$$= 0.19$$

Similarly,

$$s_1 = T(r_1) = \sum_{j=0}^{1} p_r(r_j)$$

$$= p_r(r_0) + p_r(r_1)$$

$$= 0.44$$

and

$$s_2 = 0.65 \qquad s_5 = 0.95$$

$$s_3 = 0.81 \qquad s_6 = 0.98$$

$$s_4 = 0.89 \qquad s_7 = 1.00$$

The transformation function has the staircase form shown in Fig. 4.5(b).

Since only eight equally-spaced levels are allowed in this case, each of the transformed values must be assigned to its closest valid level. Thus we have

$$s_0 \cong 1/7 \qquad s_4 \cong 6/7$$

$$s_1 \cong 3/7 \qquad s_5 \cong 1$$

$$s_2 \cong 5/7 \qquad s_6 \cong 1$$

$$s_3 \cong 6/7 \qquad s_7 \cong 1$$

It is noted that there are only five distinct histogram-equalized gray levels. Redefining the notation to take this into account yields the levels

$$s_0 = 1/7 \qquad s_3 = 6/7$$

$$s_1 = 3/7 \qquad s_4 = 1$$

$$s_2 = 5/7$$

Since $r_0 = 0$ was mapped to $s_0 = 1/7$, there are 790 transformed pixels with this new value. Also, there are 1023 pixels with value $s_1 = 3/7$ and 850 pixels with value $s_2 = 5/7$. However, since both levels r_3 and r_4 were mapped to $s_3 = 6/7$, there are now $656 + 329 = 985$ pixels with this new value. Similarly, there are $245 + 122 + 81 = 448$ pixels with value $s_4 = 1$. Dividing these numbers by $n = 4096$ yields the histogram shown in Fig. 4.5(c). Since a histogram is an approximation to a probability density function, perfectly flat results are seldom obtained when working with discrete levels. $\square$

Example: As a practical illustration of histogram equalization consider the image shown in Fig. 4.6(a) which is a picture of a dollar that, due to a heavy shadow, is barely visible. The narrow range of values occupied by the pixels of this image is evident in the histogram shown in Fig. 4.6(b). The equalized histogram is shown in Fig. 4.6(c) and the processed image in Fig. 4.6(d). While the equalized histogram is, as expected, not perfectly flat throughout the full range of gray levels, considerable improvement over the original image was achieved by the spreading effect of the histogram-equalization technique. Although the final result is not an ideal picture, one should keep in mind the poor quality of the original.   □

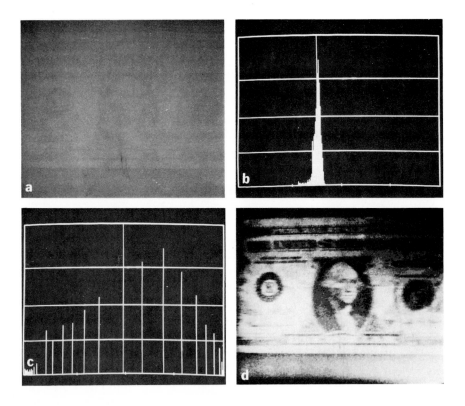

Figure 4.6. Illustration of the histogram-equalization approach. (a) Original image. (b) Original histogram. (c) Equalized histogram. (d) Enhanced image.

4.2.3 Direct Histogram Specification

Although the method discussed in the previous section is quite useful, it does not lend itself to interactive image enhancement applications because the capabilities of this method are limited to the generation of only one result—an approximation to a uniform histogram.

It is sometimes desirable to be able to specify interactively particular histograms capable of highlighting certain gray level ranges in an image. To see how this can be accomplished, let us return for a moment to continuous gray levels, and let $p_r(r)$ and $p_z(z)$ be the original and desired probability density functions, respectively. Suppose that a given image is first histogram equalized using Eq. (4.2-5); that is

$$s = T(r) = \int_0^r p_r(w)\, dw \qquad (4.2\text{-}10)$$

If the desired image were available, its levels could also be equalized by using the transformation function

$$v = G(z) = \int_0^z p_z(w)\, dw \qquad (4.2\text{-}11)$$

The inverse process, $z = G^{-1}(v)$, would then yield the desired levels back. This, of course, is a hypothetical formulation since the z levels are precisely what we are trying to obtain. It is noted, however, that $p_s(s)$ and $p_v(v)$ would be identical uniform densities since the final result of Eq. (4.2-5) is independent of the density inside the integral. Thus, if instead of using v in the inverse process we use the uniform levels s obtained from the original image, the resulting levels, $z = G^{-1}(s)$, would have the desired probability density function. Assuming that $G^{-1}(s)$ is single-valued, the procedure can be summarized as follows:

(1) Equalize the levels of the original image using Eq. (4.2-5).
(2) Specify the desired density function and obtain the transformation function $G(z)$ using Eq. (4.2-11).
(3) Apply the inverse transformation function, $z = G^{-1}(s)$ to the levels obtained in Step (1).

This procedure yields a processed version of the original image where the new gray levels are characterized by the specified density $p_z(z)$.

Although the method of histogram specification involves two transformation functions, $T(r)$ followed by $G^{-1}(s)$, it is a simple matter to combine both enhancement steps into one function which will yield the desired levels starting with the original pixels. From the above discussion, we have that

$$z = G^{-1}(s) \tag{4.2-12}$$

Substitution of Eq. (4.2-5) in Eq. (4.2-12) results in the combined transformation function:

$$z = G^{-1}[T(r)] \tag{4.2-13}$$

which relates r to z. It is noted that, when $G^{-1}[T(r)] = T(r)$, this expression reduces to histogram equalization.

The implication of Eq. (4.2-13) is simply that an image need not be histogram-equalized explicitly. All that is required is that $T(r)$ be determined and combined with the inverse transformation function G^{-1}. The real problem in using the preceding method for continuous variables lies in obtaining the inverse function analytically. In the discrete case this problem is circumvented by the fact that the number of distinct gray levels is usually relatively small and it becomes feasible to calculate and store a mapping for each possible pixel value. The discrete formulation of the procedure parallels Eqs. (4.2-8) through (4.2-10), as illustrated by the following example.

Example: Consider the 64 × 64, 8-level image values used in the second example given above. The histogram of this image is shown again in Fig. 4.7(a) for easy reference. It is desired to transform this histogram so that it

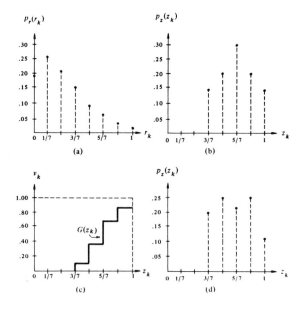

Figure 4.7. Illustration of the histogram-specification method. (a) Original histogram. (b) Specified histogram. (c) Transformation function. (d) Resulting histogram.

Table 4.2

z_k	$p_z(z_k)$
$z_0 = 0$	0.00
$z_1 = 1/7$	0.00
$z_2 = 2/7$	0.00
$z_3 = 3/7$	0.15
$z_4 = 4/7$	0.20
$z_5 = 5/7$	0.30
$z_6 = 6/7$	0.20
$z_7 = 1$	0.15

will have the shape shown in Fig. 4.7(b). The values of the specified histogram are listed in Table 4.2.

The first step in the procedure is to obtain the histogram-equalization mappings. This was done in the second example above with the results shown in Table 4.3.

Table 4.3

$r_j \rightarrow s_k$	n_k	$p_s(s_k)$
$r_0 \rightarrow s_0 = 1/7$	790	0.19
$r_1 \rightarrow s_1 = 3/7$	1023	0.25
$r_2 \rightarrow s_2 = 5/7$	850	0.21
$r_3, r_4 \rightarrow s_3 = 6/7$	985	0.24
$r_5, r_6, r_7 \rightarrow s_4 = 1$	448	0.11

Next, we compute the transformation function using Eq. (4.2-9):

$$v_k = G(z_k) = \sum_{j=0}^{k} p_z(z_j)$$

This yields the values

$$v_0 = G(z_0) = 0.00 \qquad v_4 = G(z_4) = 0.35$$
$$v_1 = G(z_1) = 0.00 \qquad v_5 = G(z_5) = 0.65$$
$$v_2 = G(z_2) = 0.00 \qquad v_6 = G(z_6) = 0.85$$
$$v_3 = G(z_3) = 0.15 \qquad v_7 = G(z_7) = 1.00$$

The transformation function is shown in Fig. 4.7(c).

To obtain the z levels we apply the inverse of the G transformation obtained above to the histogram-equalized levels s_k. Since we are dealing with discrete values, an approximation must usually be made in the inverse mapping. For example, the closest match to $s_0 = 1/7 \approx 0.14$ is $G(z_3) = 0.15$ or, using the inverse, $G^{-1}(0.15) = z_3$. Thus, s_0 is mapped to the level z_3. Using this procedure yields the following mappings:

$$s_0 = 1/7 \rightarrow z_3 = 3/7$$

$$s_1 = 3/7 \rightarrow z_4 = 4/7$$

$$s_2 = 5/7 \rightarrow z_5 = 5/7$$

$$s_3 = 6/7 \rightarrow z_6 = 6/7$$

$$s_4 = 1 \quad \rightarrow z_7 = 1$$

As indicated in Eq. (4.2-13), these results can be combined with those of histogram equalization to yield the following direct mappings:

$$r_0 = 0 \quad \rightarrow z_3 = 3/7 \qquad r_4 = 4/7 \rightarrow z_6 = 6/7$$

$$r_1 = 1/7 \rightarrow z_4 = 4/7 \qquad r_5 = 5/7 \rightarrow z_7 = 1$$

$$r_2 = 2/7 \rightarrow z_5 = 5/7 \qquad r_6 = 6/7 \rightarrow z_7 = 1$$

$$r_3 = 3/7 \rightarrow z_6 = 6/7 \qquad r_7 = 1 \quad \rightarrow z_7 = 1$$

Redistributing the pixels according to these mappings and dividing by $n = 4096$ results in the histogram shown in Fig. 4.7(d). The values are listed in Table 4.4.

It is noted that, although each of the specified levels was filled, the resulting histogram is not particularly close to the desired shape. As in the case of histogram equalization, this error is due to the fact that the transformation is guaranteed to yield exact results only in the continuous case. As the number of levels decreases, the error between the specified and

Table 4.4

z_k	n_k	$p_z(z_k)$
$z_0 = 0$	0	0.00
$z_1 = 1/7$	0	0.00
$z_2 = 2/7$	0	0.00
$z_3 = 3/7$	790	0.19
$z_4 = 4/7$	1023	0.25
$z_5 = 5/7$	850	0.21
$z_6 = 6/7$	985	0.24
$z_7 = 1$	448	0.11

resulting histograms tends to increase. As will be seen below, however, very useful enhancement results can be obtained even with an approximation to a desired histogram.   □

In practice, the inverse transformation from s to z is often not single-valued. This situation arises when there are unfilled levels in the specified histogram (which makes the CDF remain constant over the unfilled intervals), or in the process of rounding off $G^{-1}(s)$ to the nearest allowable level, as was done in the above example. Generally, the easiest solution to this problem is to assign the levels in such a way as to match the given histogram as closely as possible. Consider, for example, the specified histogram shown in Fig. 4.8(a). The original transformation function is shown in Fig. 4.8(b) and the rounded-off function is shown in Fig. 4.8(c). It is noted that a value of $s = 3/7$ can be mapped either to $z = 3/7$ or $z = 4/7$. Using the given histogram to guide the process, however, we see that the latter level was specified as being empty, so $z = 3/7$ is used in the inverse mapping. This procedure can still be used when the ambiguous levels are filled in the specified histogram. The ambiguity arises in this case due to round-off, as illustrated in Figs. 4.8(b) and (c). In this example, the level $s = 1$ can be assigned to either $z = 6/7$ or $z = 1$. However, $z = 6/7$ is chosen for the inverse mapping because the height of the specified histogram is larger at this level. It should be noted that assigning some levels to $z = 6/7$ and some levels to $z = 1$ simply to match the histogram would make little sense since this would require dividing the

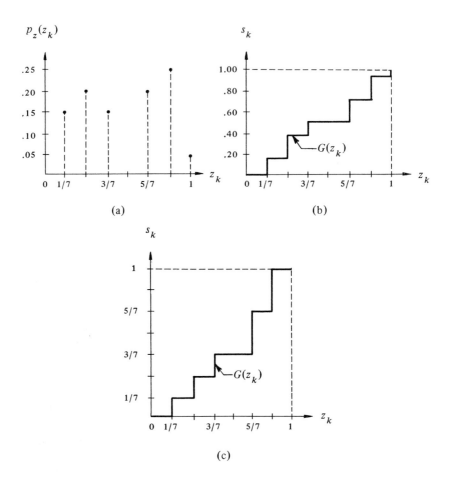

Figure 4.8. A nonunique transformation function.

pixels which have the same level into two groups. This is an artificial manipulation which could distort the appearance of the image.

The principal difficulty in applying the histogram specification
method to image enhancement lies in being able to construct a meaningful
histogram. Two solutions to this problem are as follows. The first is to
specify a particular probability density function (i.e., Gaussian, Rayleigh,
log-normal, etc.) and then form a histogram by digitizing the given function.
The second approach consists of specifying a histogram of arbitrary (but
controllable) shape by, for example, forming a string of connected straight

line segments. After the desired shape has been obtained, the function is digitized and normalized to unit area.

Figure 4.9(a) illustrates this approach where the histogram is formed from line segments which are controlled by the parameters m, h, θ_L, and θ_R. Point m can be chosen anywhere in the interval $[0, 1]$ and point h can assume any nonnegative value. The parameters θ_L and θ_R specify the angle from vertical and can assume values from $0°$ to $90°$. The inflection joint j moves along the line segment that connects the points $(0, 1)$ and $(m, 0)$ as θ_L varies. Similarly, point k moves along the line segment that connects the points $(1, 1)$ and $(m, 0)$ as θ_R varies. The inflection points j and k are determined uniquely by the values of m, h, θ_L, and θ_R and are, therefore, not specifiable. This histogram-specification method can generate a variety of useful histograms. For example, setting $m = 0.5$, $h = 1.0$, and $\theta_L = \theta_R = 0°$, yields a uniform histogram.

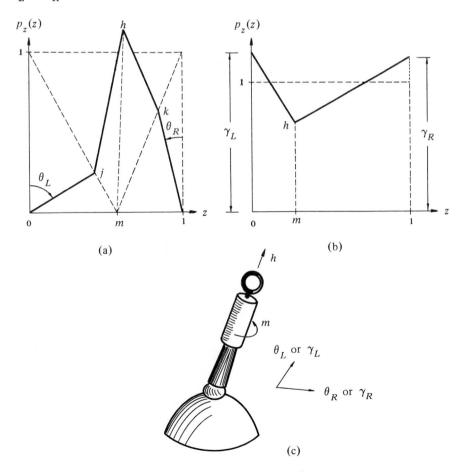

Figure 4.9. Interactive specification of histograms.

A second, somewhat simpler, approach is shown in Fig. 4.9(b). This technique, which is also based on four parameters, is useful for contrast enhancement by emphasizing the ends of the pixel spectrum. Both of these methods depend on a small number of parameters and, therefore, lend themselves well to fast, interactive image enhancement. Figure 4.9(c) shows how the histogram parameters could be specified interactively by means of a joy stick. This approach is particularly attractive for on-line applications where an operator has a continuous visual output and can control the enhancement process to fit a given situation.

Example: As a practical illustration of the direct histogram specification approach, consider Fig. 4.10(a) which shows a semi-dark room viewed from

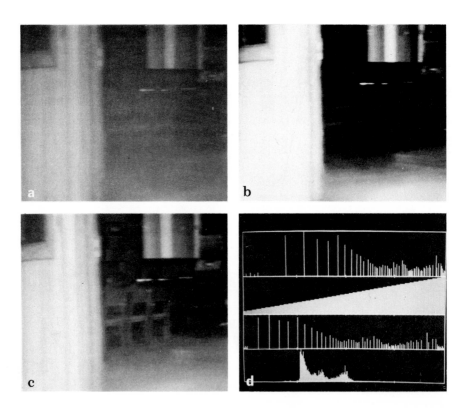

Figure 4.10. Illustration of the histogram specification method. (a) Original image. (b) Histogram-equalized image. (c) Image enhanced by histogram specification. (d) Histograms.

a doorway. Figure 4.10(b) shows the histogram-equalized image, and Fig. 4.10(c) is the result of interactive histogram specification. The histograms are shown in Fig. 4.10(d) which includes, from top to bottom, the original, equalized, specified, and resulting histograms.

It is noted that histogram equalization produced an image whose contrast was somewhat high, while the result shown in Fig. 4.10(c) has a much more balanced appearance. Because of its flexibility, the direct specification method can often yield results which are superieor to histogram equalization. This is particularly true in cases where the image has been degraded and control of the pixel distribution becomes necessary to obtain good enhancement results. An extreme case which illustrates this point is shown in Fig. 4.11. Part (a) of this figure shows a picture of a coin taken under very poor lighting conditions. Figure 4.11(b) is the histogram equalized

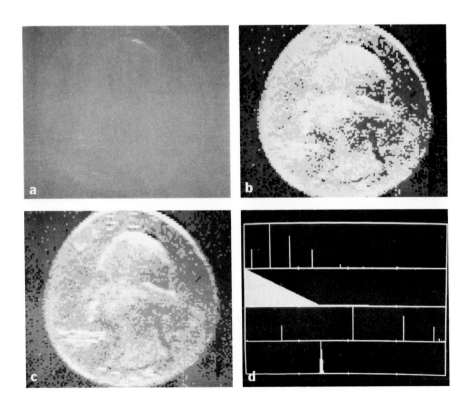

Figure 4.11. Another illustration of the histogram specification method. (a) Original image. (b) Histogram-equalized image. (c) Image enhanced by histogram specification. (d) Histograms.

image and Fig. 4.ll(c) is the result of direct histogram specification. Figure 4.ll(d) contains the histograms, which are shown in the same order as in the Fig. 4.10(d).

Although histogram equalization improved the image considerably, it is evident that a more balanced, clearer result was achieved by the histogram specification method. The reason for the improvement can be easily explained by examining Fig. 4.ll(d). The original histogram is very narrow, indicating that the pixels in the original image have little dynamic range. This is to be expected since the object of interest has such low contrast that it is barely visible. Histogram equalization improved the image by increasing the dynamic range of the pixels, thus enhancing the contrast between the coin and the background.

Since the number of different levels in this case is rather small, it is logical to deduce that having control over the shape of the final histogram should lead to some improvement over histogram equalization. In this example the improvement was realized by the specified histogram shown in Fig. 4.11(d), which achieved an increase in dynamic range while biasing the histogram toward the dark side of the spectrum.   □

4.3 IMAGE SMOOTHING

Smoothing operations are used primarily for diminishing spurious effects that may be present in a digital image as a result of a poor sampling system of transmission channel. In this section we consider smoothing techniques in both the spatial and frequency domains.

4.3.1 Neighborhood Averaging

Neighborhood averaging is a straightforward spatial-domain technique for image smoothing. Given an $N \times N$ image $f(x, y)$, the procedure is to generate a smoothed image $g(x, y)$ whose gray level at every point (x, y) is obtained by averaging the gray-level values of the pixels of f contained in a predefined neighborhood of (x, y). In other words, the smoothed image is obtained by using the relation

$$g(x, y) = \frac{1}{M} \sum_{(n, m) \in S} f(n, m) \qquad (4.3\text{-}1)$$

for $x, y = 0, 1, ..., N - 1$. S is the set of coordinates of points in the neighborhood of (but not including) the point (x, y), and M is the total number of points defined by the coordinates in S. For example, the set of coordinates of the neighbors which are unit distance from a point (x, y) is $S = \{(x, y + 1), (x, y - 1), (x + 1, y), (x - 1, y)\}$.

Figure 4.12 shows one approach for extracting neighborhoods from an image array. The neighborhood of a point is defined in this case by the set of points inside, or on the boundary of, a circle centered about this point in question. In Fig. 4.12(a) a radius equal to Δx (this separation is assumed for convenience to be equal to Δy) was used, and Fig. 4.12(b) was obtained with a radius of $\sqrt{2}\Delta x$. It is noted that this approach yields symmetrical neighborhoods, except on or near the edges of the image. Points in these regions may be replaced by their partial neighborhoods.

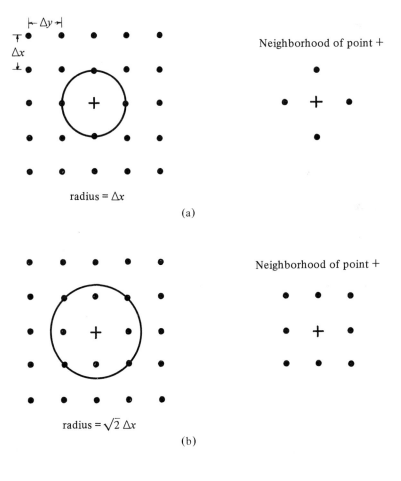

Figure 4.12. Extraction of symmetric neighborhoods from a digital image.

Example: Figure 4.13 illustrates the smoothing effect produced by neigh-borhood averaging. Figure 4.13(a) is a simple four-level image corrupted by noise. Figures 4.13(b) through (f) are the results of processing the noisy image with neighborhoods of radii equal to 1, 2, 4, 8, and 16, respectively. It is noted that the degree of blurring is strongly proportional to the size of radius used. □

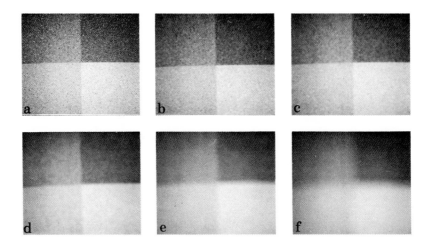

Figure 4.13. Example of neighborhood averaging. (a) Noisy image. (b) through (f), image processed using Eq. (4.3-1) with neighborhood radii equal to 1, 2, 4, 8, and 16, respectively.

For a given radius, the blurring effect produced by neighborhood averaging can be reduced by using a thresholding procedure; that is, instead of using Eq. (4.3-1), we form $g(x, y)$ according to the following criterion:

$$g(x,y) = \begin{cases} \dfrac{1}{M} \sum_{(m,n)\in S} f(m,n) & \text{if } \left| f(x,y) - \dfrac{1}{M} \sum_{(m,n)\in S} f(m,n) \right| > T \\[4mm] f(x,y) & \text{otherwise} \end{cases} \tag{4.3-2}$$

where T is a specified nonnegative threshold. The motivation behind this approach is to reduce blurring by leaving unchanged those points whose difference from their neighborhood values do not exceed the specified value of T.

Example: Figure 4.14 shows the result of processing the image of Fig. 4.13(a) with Eq. (4.3-2). A radius of 4 and $T = 10$ was used in this case. We note by comparing this result with Fig. 4.13(d) that, although equivalent smoothing was achieved by use of a threshold, the boundaries between the four squares are much sharper in Fig. 4.14. □

Figure 4.14. Smoothed image obtained by processing Fig. 4.13(a) with Eq. (4.3-2), using a neighborhood radius of 4 and $T = 10$. Compare with Fig. 4.13(d).

4.3.2 Lowpass Filtering

Edges and other sharp transitions (such as noise) in the gray levels of an image contribute heavily to the high-frequency content of its Fourier transform. It follows, therefore, that blurring can be achieved via the frequency domain by attenuating a specified range of high-frequency components in the transform of a given image.

From Eq. (4.1-2) we have the relation

$$G(u, v) = H(u, v)F(u, v) \tag{4.3-3}$$

where $F(u, v)$ is the transform of the image we wish to smooth. The problem is to select a function $H(u, v)$ which yields $G(u, v)$ by attenuating the high-frequency components of $F(u, v)$. The inverse transform of $G(u, v)$ will then yield the desired smoothed image $g(x, y)$. Since high-frequency components are "filtered out", and information in the low-frequency range is "passed" without attenuation, this method is commonly referred to as *lowpass filtering.* The function $H(u, v)$ is referred to in this context as a *filter transfer function.* Several lowpass filtering approaches are discussed in the following paragraphs. In all cases, the filters are functions which affect corresponding real and imaginary components of the Fourier transform in exactly the same manner. Such filters are referred to as *zero-phase-shift-filters* because they do not alter the phase of the transform.

4.3.2.1 Ideal filter

A two-dimensional ideal lowpass filter (ILPF) is one whose transfer function satisfies the relation

$$H(u, v) = \begin{cases} 1 & \text{if } D(u, v) \leqslant D_0 \\ 0 & \text{if } D(u, v) > D_0 \end{cases} \tag{4.3-4}$$

where D_0 is a specified nonnegative quantity, and $D(u, v)$ is the distance from point (u, v) to the origin of the frequency plane, that is,

$$D(u, v) = \{u^2 + v^2\}^{1/2} \tag{4.3-5}$$

A three-dimensional perspective plot of $H(u, v)$ *vs.* u and v is shown in Fig. 4.15(a). The name *ideal* filter arises from the fact that all frequencies inside a circle of radius D_0 are passed with no attenuation, while all frequencies outside this circle are completely attenuated.

All low-pass filters considered in this chapter are radially symmetric about the origin. For this type of filter it suffices to specify a cross section extending as a function of distance from the origin along a radial line, as shown in Fig. 4.15(b). The complete filter transfer function can then be

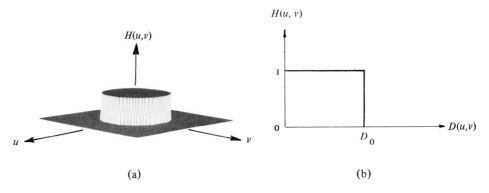

(a)   (b)

Figure 4.15. (a) Perspective plot of an ideal lowpass filter transfer function. (b) Filter cross section.

generated by rotating the cross section $360°$ about the origin. It should also be noted that specification of radially-symmetric filters centered on the $N \times N$ frequency square is based on the assumption that the origin of the Fourier transform has been centered on the square, as discussed in Section 3.3.2.

For an ideal lowpass filter cross section, the point of transition between $H(u, v) = 1$ and $H(u, v) = 0$ is often called the *cut-off frequency*. In the case of Fig. 4.15(b), for example, the cut-off frequency is D_0. As the cross section is rotated about the origin, the point D_0 traces a circle and we obtain a locus of cut-off frequencies all of which are a distance D_0 from the origin. As will be seen below, the cut-off frequency concept is quite useful in specifying the characteristics of a given filter, and it also serves as a common base for comparing the behavior of different types of filters.

The sharp cut-off frequencies of an ideal lowpass filter cannot be realized with electronic components, although they can certainly be simulated in a computer. The effects of using these "nonphysical" filters on a digital image are discussed after the following example.

Example: The 256×256 image shown in Fig. 4.16(a) will be used to illustrate all the lowpass filters discussed in this section. The performance of

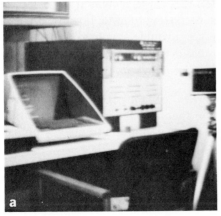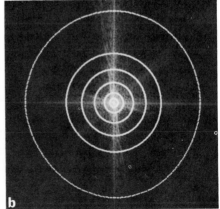

Figure 4.16. (a) A 256×256 image, and (b) its Fourier spectrum. The superimposed circles, which have radii equal to 5, 11, 22, 36, 53, and 98, enclose, respectively, 90, 95, 98, 99, 99.5, and 99.9 percent of the image energy.

these filters will be compared by using the same cut-off frequency loci. One way to establish a set of standard loci is to compute circles which enclose various amounts of the total signal energy, E_T. This quantity is obtained by summing the energy at each point (u, v) for $u, v = 0, 1, ..., N - 1$; that is,

$$E_T = \sum_{u=0}^{N-1} \sum_{v=0}^{N-1} E(u, v)$$

where $E(u, v)$ is given by Eq. (3.1-13). Assuming that the transform has been centered, a circle of radius r with origin at the center of the frequency square encloses β percent of the energy, where

$$\beta = 100 \left[\sum_u \sum_v E(u, v) / E_T \right]$$

and the summation is taken over values of (u, v) which lie inside, or on the boundary of, the circle.

Figure 4.16(b) shows the Fourier transform of Fig. 4.16(a). The super-imposed circles, which have radii of 5, 11, 22, 36, 53, and 98, enclose β percent of the energy for $\beta = 90, 95, 98, 99, 99.5,$ and 99.9, respectively. It is noted that the energy spectrum falls off rather rapidly, with 90% of the total energy being enclosed by the relatively small radius of 5. Since we are dealing with a 256 $\times$ 256 image and the Fourier transform has been centered, a circle of radius $(\sqrt{2})(128)$ would enclose 100% of the energy.

The results of applying ideal lowpass filters with cut-off frequencies at the above radii are shown in Fig. 4.17. Part (a) of this figure is, for all practical purposes, useless. The severe blurring in this image is a clear indication that most of the edge information in the picture is contained within the 10% energy removed by the filter. As the filter radius was increased the degree of blurring was, of course, decreased. It is interesting to note, however, that all the filtered images are characterized by considerable "ringing". This phenomenon, which is explained below, is visible even in Fig. 4.17(f) where only 0.1% of the energy was removed.   □

The blurring and ringing properties of the ILPF can be easily explained by resorting to the convolution theorem. Since the Fourier transforms of the original and blurred images are related in the frequency domain by the equation

$$G(u, v) = H(u, v)F(u, v)$$

it follows from this theorem that the following expression holds in the spatial domain:

$$g(x, y) = h(x, y) * f(x, y)$$

where $h(x, y)$ is the inverse Fourier transform of the filter transfer function $H(u, v)$.

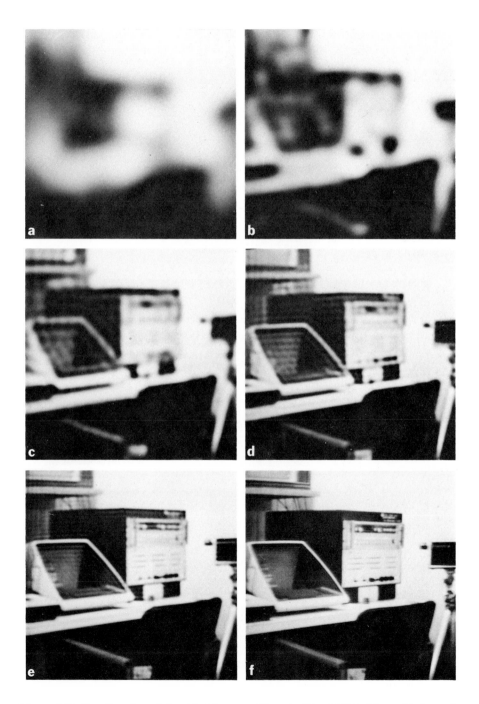

Figure 4.17. Results of applying ideal lowpass filters to Fig. 4.16(a). The radii shown in Fig. 4.16(b) were used.

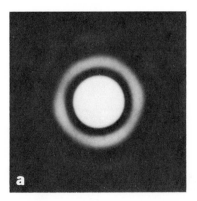

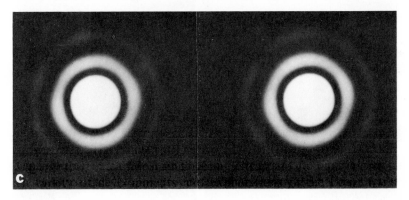

Figure 4.18. Illustration of the blurring process in the spatial domain. (a) Blurring function $h(x, y)$ for an ideal lowpass filter. (b) A Simple image composed of two bright dots. (c) Convolution of $h(x, y)$ and $f(x, y)$.

The key to understanding blurring as a convolution process in the spatial domain lies in the nature of $h(x, y)$. For an ILPF, $h(x, y)$ has the general form shown in Fig. 4.18(a)[†]. Suppose that $f(x, y)$ is a simple image composed of two bright pixels on a black background, as shown in Fig. 4.18 (b). We may view the two bright points as approximations of two impulses whose strength depends on the brightness of these points. Then, the convolution of $h(x, y)$ and $f(x, y)$ is simply a process of "copying" $h(x, y)$ at the location of each impulse, as explained in Section 3.3.8. The result of this operation, which is shown in Fig. 4.18(c), explains how the two original points are blurred as a consequence of convolving $f(x, y)$ with the blurring function $h(x, y)$. These concepts are extended to more complex images by considering each pixel as an impulse with a strength proportional to the gray level of the pixel.

The shape of $h(x, y)$ depends on the radius of the filter function in the frequency domain. By computing the inverse transform of $H(u, v)$ for an ILPF, it can be shown that the radii of the concentric rings in $h(x, y)$ are inversely proportional to the value of D_0 in Eq. (4.3-4). Thus, severe filtering in the frequency domain (i.e., choice of a small D_0) produces a large number of rings in the $N \times N$ region of $h(x, y)$ and, consequently, pronounced ringing in $g(x, y)$. As D_0 increases, the number of rings in a given region decreases, thus reducing the ringing effect in $g(x, y)$. If D_0 is outside the $N \times N$ domain of definition of $F(u, v)$, $h(x, y)$ becomes unity in its corresponding $N \times N$ spatial region and the convolution of $h(x, y)$ and $f(x, y)$ is simply $f(x, y)$. This situation, of course, corresponds to no filtering at all. The spatial domain effect of the filters discussed below can be explained in a similar manner as for the ideal filter.

4.3.2.2 Butterworth Filter

The transfer function of the Butterworth lowpass filter (BLPF) of order n and with cut-off frequency locus at a distance D_0 from the origin is defined by the relation

$$H(u, v) = \frac{1}{1 + \left[D(u, v) / D_0 \right]^{2n}} \qquad (4.3\text{-}6)$$

where $D(u, v)$ is given by Eq. (4.3-5). A perspective plot and cross section of the BLPF function are shown in Fig. 4.19.

Unlike the ILPF, the BLPF transfer function does not have a sharp discontinuity that establishes a clear cut-off between passed and filtered frequencies. For filters with smooth transfer functions it is customary to define a "cut-off" frequency locus at points for which $H(u, v)$ is down to a

---

[†]The reader can verify this by taking the inverse Fourier transform of Eq. (4.3-4).

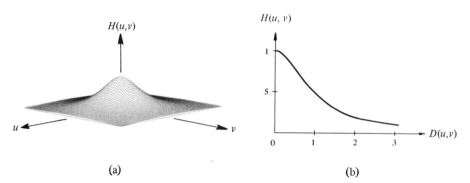

Figure 4.19. (a) A Butterworth lowpass filter. (b) Radial cross sections for $n = 1$.

certain fraction of its maximum value. In the case of Eq. (4.3-6) we see that $H(u, v) = 0.5$ (down 50% from its maximum value of one) when $D(u,v) = D_0$. Another value commonly used is $1/\sqrt{2}$ of the maximum value of $H(u, v)$. For Eq. (4.3-6), the following simple modification yields the desired value when $D(u, v) = D_0$:

$$H(u, v) = \frac{1}{1 + [\sqrt{2} - 1][D(u, v)/D_0]^{2n}}$$

$$= \frac{1}{1 + 0.414[D(u, v)/D_0]^{2n}} \qquad (4.3\text{-}7)$$

Example: Figure 4.20 shows the results of applying BLPFs [Eq. (4.3-7)] to Fig. 4.16(a), with $n = 1$ and D_0 equal to the first five radii shown in Fig. 4.16(b). It is noted that these images are considerably less blurred than the corresponding results obtained with ideal lowpass filters. The reason is that the "tail" in the BLPF passes a fairly high amount of high-frequency information, thus preserving more of the edge content in the picture. It is also of interest to note that no ringing is evident in any of the images processed with the BLPF, a fact attributed to the filter's smooth transition between low and high frequencies.   □

As indicated in Section 4.1, all the filtering results presented in this section were obtained by directly computing the FFT without extending the images to avoid wraparound error. As shown in Fig. 4.20(e), the error is certainly not objectionable since this image is essentially of the same quality as the original, in spite of the fact that convolution with a broad filter was

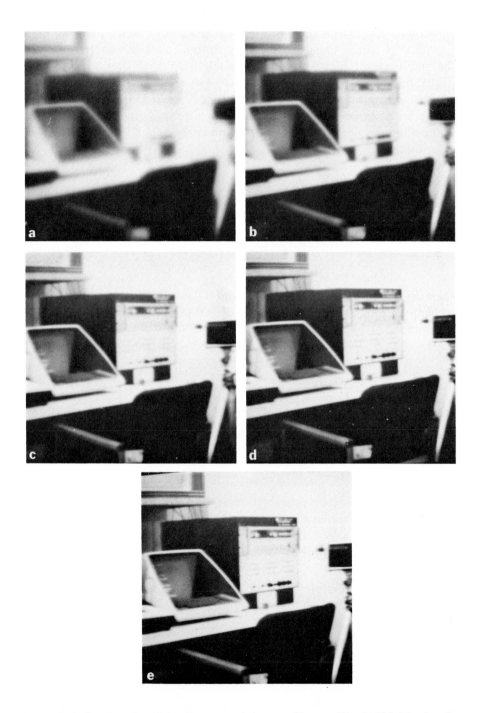

Figure 4.20. Results of applying Butterworth lowpass filters to Fig. 4.16(a). The first five radii shown in Fig. 4.16(b) were used.

carried out. The reason for this is that the spectrum of $f(x, y)$ falls off rapidly, with 90% of the signal energy being contained inside a circle of radius 5. The amplitude of $F(u, v)$, therefore, is relatively small over a large portion of the frequency plane, and these small values attenuate the wraparound error caused by overlaps in the convolution periods. This behavior is typical in practice and often allows us to ignore the error incurred in discrete convolution when the periodic constraints imposed by Eqs. (3.3-33) and (3.3-34) are not satisfied.

4.3.2.3 Exponential Filter

The exponential lowpass filter (ELPF) is another smooth filter commonly used in image processing. The ELPF with cut-off frequency locus at a distance D_0 from the origin has a transfer function given by the relation

$$H(u, v) = e^{-[D(u, v)/D_0]^n} \qquad (4.3\text{-}8)$$

where $D(u, v)$ is given by Eq. (4.3-5), and n controls the rate of decay of the exponential function. A perspective plot and cross section of the ELPF are shown in Fig. 4.21.

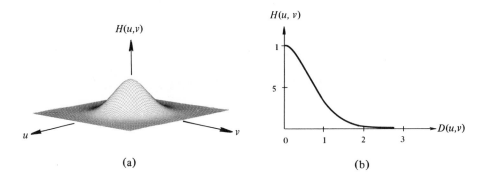

Figure 4.21. (a) An exponential lowpass filter. (b) Radial cross sections for $n = 2$.

When $D(u, v) = D_0$ we have from Eq. (4.3-8) that $H(u, v) = 1/e$. A simple modification given by

$$H(u, v) = e^{[\ln(1/\sqrt{2})][D(u, v)/D_0]^n}$$

$$= e^{-0.347[D(u, v)/D_0]^n} \qquad (4.3\text{-}9)$$

forces $H(u, v)$ to be equal to $1/\sqrt{2}$ of its maximum value at frequencies in the cut-off locus.

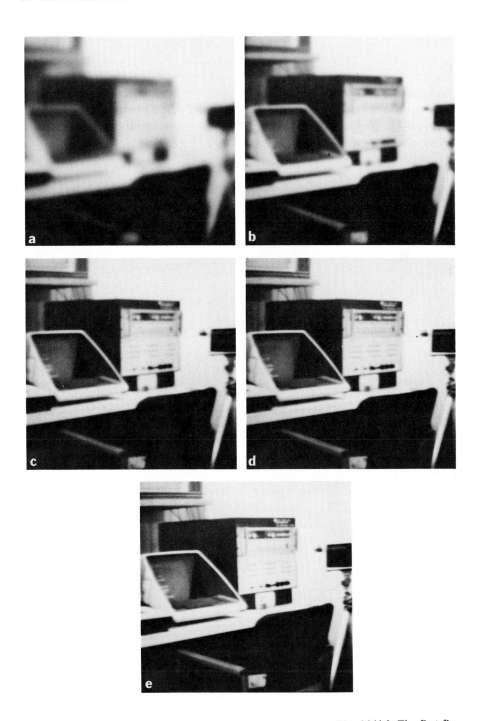

Figure 4.22. Results of applying exponential lowpass filters to Fig. 4.16(a). The first five radii shown in Fig. 4.16(b) were used.

Example: Figure 4.22 shows the results of applying ELPF's [Eq. (4.3-9)] to Fig. 4.16(a), with $n = 2$ and D_0 equal to the first five radii shown in Fig. 4.16(b). Because of their faster roll-off rate, the exponential filters achieved slightly more blurring than the corresponding Butterworth filters. It is also noted that no ringing is evident in any of the images processed with the ELPF. As in the case of the BLPF, this is due to the smooth characteristics of the exponential filter.   □

4.3.2.4 Trapezoidal Filter

A trapezoidal lowpass filter (TLPF) is a compromise between the ILPF and a completely smooth filter. TLPFs can be defined by the relation

$$H(u, v) = \begin{cases} 1 & \text{if } D(u, v) < D_0 \\ \dfrac{1}{[D_0 - D_1]}\,[D(u, v) - D_1] & \text{if } D_0 \leqslant D(u, v) \leqslant D_1 \\ 0 & \text{if } D(u, v) > D_1 \end{cases} \qquad (4.3\text{-}10)$$

where $D(u, v)$ is given by Eq. (4.3-5), D_0 and D_1 are specified, and it is assumed that $D_0 < D_1$. A perspective plot and cross section of a typical TPLF transfer function are shown in Fig. 4.23.

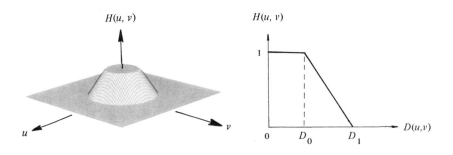

Figure 4.23. Perspective plot and cross section of a typical trapezoidal lowpass filter.

For convenience in implementation, we define the cut-off locus to be at the first breakpoint (D_0) of the transfer function. The second variable, D_1, is arbitrary, as long as it is greater than D_0.

Example: Figure 4.24 shows the results of applying TLPFs to Fig. 4.16(a), with D_0 equal to the first five radii in Fig. 4.16(b). The value of D_1 corresponding to $D_0 = 5$ was 45, and subsequent values of D_1 were chosen to keep the slope of the transfer function nearly constant as D_0 was increased. It is noted that the trapezoidal filtering results have a certain amount of ringing, but it is not as pronounced as in the corresponding ideal filter results. The ringing is caused by the corners of the transfer function at D_0 and D_1. $\square$

All the lowpass filtering results given thus far have been with images of good quality in order to illustrate and compare the effect of the various filters discussed in this section. Figure 4.25 shows two practical applications of lowpass filtering for image smoothing. The image shown in Fig. 4.25(a) was digitized with only 16 gray levels and, as a consequence, exhibits a considerable amount of false contouring. Figure 4.25(b) is the result of smoothing this image with a lowpass Butterworth filter of order 1. Similarly, Fig. 4.25(d) shows the effect of applying a BLPF to the noisy image of Fig. 4.25(c). It is noted from these examples that lowpass filtering is a cosmetic process which reduces spurious effects at the expense of image sharpness.

4.3.3 Averaging of Multiple Images

Consider a noisy image $g(x, y)$ which is formed by the addition of noise $\eta(x, y)$ to an original image $f(x, y)$; that is

$$g(x, y) = f(x, y) + \eta(x, y) \tag{4.3-11}$$

where it is assumed that at every pair of coordinates (x, y) the noise is uncorrelated and has zero average value. The objective of the following procedure is to obtain a smoothed result by adding a given set of noisy images $\{g_i(x, y)\}$.

If the noise satisfies the constraints just stated, it is a simple problem to show (e. g., see Papoulis [1965]) that if an image $\bar{g}(x, y)$ is formed by averaging M different noisy images,

$$\bar{g}(x, y) = \frac{1}{M} \sum_{j=1}^{M} g_i(x, y) \tag{4.3-12}$$

then it follows that

$$E\{\bar{g}(x, y)\} = f(x, y) \tag{4.3-13}$$

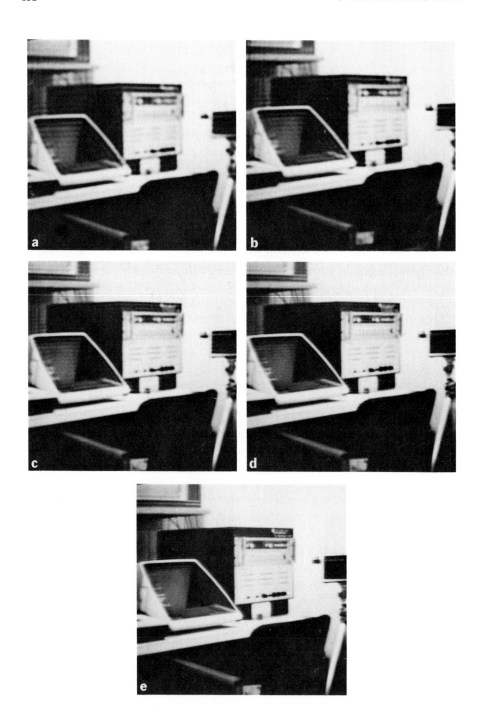

Figure 4.24. Results of applying trapezoidal lowpass filters to Fig. 4.16(a). The first five radii shown in Fig. 4.16(b) were used.

Figure 4.25. Two examples of image smoothing by lowpass filtering.

and

$$\sigma^2_{\bar{g}(x,y)} = \frac{1}{M}\,\sigma^2_{\eta(x,y)} \tag{4.3-14}$$

where $E\{\bar{g}(x, y)\}$ is the expected value of $\bar{g}$, and $\sigma^2_{\bar{g}(x,y)}$ and $\sigma^2_{\eta(x,y)}$ are the variances of $\bar{g}$ and η, all at coordinates (x, y). The standard deviation at any point in the average image is given by

$$\sigma_{\bar{g}(x,\,y)} = \frac{1}{\sqrt{M}}\,\sigma_{\eta(x,\,y)} \tag{4.3-15}$$

Equations (4.3-14) and (4.3-15) indicate that as M increases the variability of the pixel values decreases. Since $E\{\bar{g}(x,\ y)\} = f(x,\ y)$, this means that $\bar{g}(x,\ y)$ will approach $f(x,\ y)$ as the number of noisy images used in the averaging process increases. In practice, the principal difficulty in using this method lies in being able to register the images so that corresponding pixels line-up correctly.

Example: As an illustration of the averaging method, consider the images shown in Fig. 4.26. Part (a) of this figure shows an original image, and part (b) is the same image in which each pixel is corrupted by additive Gaussian noise with zero mean and standard deviation equal to 20. Since negative pixel values were not allowed, any $\bar{g}(x,\ y)$ that was negative as a result of adding noise to $f(x,\ y)$ was replaced by $|\bar{g}(x,\ y)|$. In this sense, the noise can only be considered approximately Gaussian. Registration was not a problem in this case because all noisy images were generated from the same source.

Figures 4.26(c) through (h) show the results of using $M = 2, 5, 10, 25,$ 50, and 100, respectively, in the averaging process given in Eq. (4.3-12). There is little discernible difference between the original noisy image and the sum of two samples. When $M = 5$, however, the reduction in noise is quite apparent. The same is true of the case where $M = 10$, although some noisy regions are still clearly visible. [Compare, for example, the edges of the robe sleeve and leg nearest the right side of Figs. 4.26(a) and (e).] Averaging 25 copies [Fig. 4.26(f)] yielded an image that is almost of the same quality and definition as the original. A little noise is still visible in regions such as the sleeve and leg mentioned above, but it is difficult to tell the two images apart by casual observation. The results for $M = 50$ and $M = 100$ are, for all practical purposes, of the same quality as the original image.   □

4.4 IMAGE SHARPENING

Sharpening techniques are useful primarily as enhancement tools for highlighting edges in an image. Following the same format as in Section 4.3, we present below sharpening methods in both the spatial and frequency domains.

4.4.1 Sharpening by Differentiation

It was noted in Section 4.3 that averaging pixels over a region tends to blur detail in an image. Since averaging is analogous to integration, it is natural to expect that differentiation will have the opposite effect and thus sharpen a given image.

The most commonly used method of differentiation in image processing applications is the *gradient*. Given a function $f(x, y)$, the gradient of f at coordinates (x, y) is defined as the *vector*

$$G[f(x,y)] = \begin{bmatrix} \dfrac{\partial f}{\partial x} \\[2mm] \dfrac{\partial f}{\partial y} \end{bmatrix} \qquad (4.4\text{-}1)$$

Two important properties of the gradient are: (1) the vector $G[f(x, y)]$ points in the direction of the maximum rate of increase of the function $f(x, y)$; and (2) the magnitude of $G[f(x, y)]$, denoted by $G[f(x, y)]$, and given by

$$G[f(x,y)] = \text{mag}[G] = \left[(\partial f/\partial x)^2 + (\partial f/\partial y)^2 \right]^{1/2} \qquad (4.4\text{-}2)$$

equals the maximum rate of increase of $f(x, y)$ per unit distance in the direction of G.

Equation (4.4-2) is the basis for a number of approaches to image differentiation. It is noted that this expression is in the form of a two-dimensional derivative function and that it is always positive. In practice, the scalar function $G[f(x, y)]$ is commonly referred to as the gradient of f. This terminology will be used throughout the following discussion to avoid having to continually refer to $G[f(x, y)]$ as "the magnitude of the gradient." The reader should, however, keep in mind the basic difference between Eqs. (4.4-1) and (4.4-2).

For a digital image, the derivatives in Eq. (4.4-2) are approximated by differences. One typical approximation is given by the relation

$$G[f(x,y)] \cong \left[\left[f(x,y) - f(x+1,y) \right]^2 \right.$$
$$\left. + \left[f(x,y) - f(x,y+1) \right]^2 \right]^{1/2} \qquad (4.4\text{-}3)$$

Similar results are obtained by using absolute values, as follows:

$$G[f(x,y)] \cong |f(x,y) - f(x+1,y)|$$
$$+ |f(x,y) - f(x,y+1)| \qquad (4.4\text{-}4)$$

This formulation is more desirable for a computer implementation of the gradient. It is also easier to program in assembly language if speed of computation is an essential requirement.

The relationship between pixels in Eq. (4.4-3) and (4.4-4) is shown in

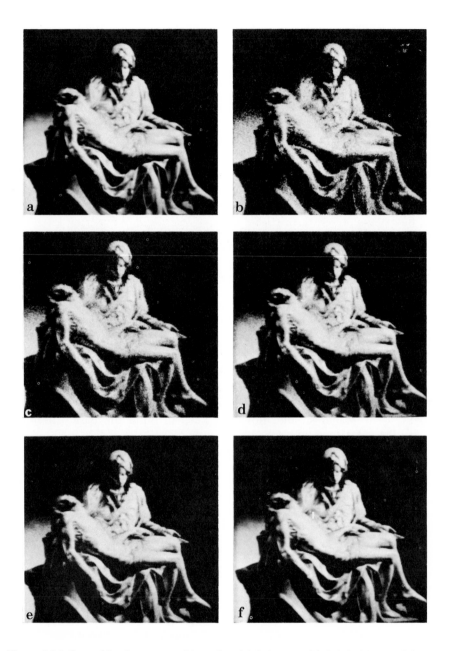

Figure 4.26. Smoothing by superposition of multiple images. (a) Original image. (b) Noisy image. (c) through (h) Results obtained by averaging 2, 5, 10, 25, 50, and 100 copies, respectively.

Figure 4.26. (Continued.)

Fig. 4.27(a). For an $N \times N$ image, it is noted that it is not possible to take the gradient for pixels in the last row ($x = N$) or the last column ($y = N$). If an $N \times N$ gradient image is desired, one procedure that can be followed for pixels in these regions is to duplicate the gradients obtained in the previous row when $x = N$ and the previous column when $y = N$.

The above arrangement for approximating the gradient is by no means unique. Another useful approximation, sometimes called the *Roberts*

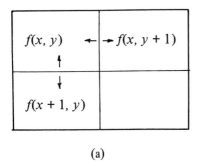

(a)

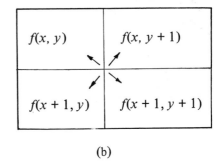

(b)

Figure 4.27. Two procedures for computing a two-dimensional, discrete gradient.

gradient, uses the cross-differences shown in Fig. 4.27(b). This approximation is given by the relation

$$G[f(x,y)] = \Big[[f(x,y) - f(x+1,y+1)]^2$$

$$+ [f(x+1,y) - f(x,y+1)]^2 \Big]^{1/2} \qquad (4.4\text{-}5)$$

or, using absolute values,

$$G[f(x,y)] \cong |f(x,y) - f(x+1,y+1)|$$

$$+ |f(x+1,y) - f(x,y+1)| \qquad (4.4\text{-}6)$$

It is noted that in all the approximations given above the value of the gradient is proportional to the difference in gray level between adjacent pixels.
Thus, as expected, the gradient assumes relatively large values for prominent
edges in an image, and small values in regions that are fairly smooth, being
zero only in regions that have a constant gray level. These properties of the
gradient are illustrated in Fig. 4.28. The digital image shown in Fig. 4.28(a)
is composed of two levels. As shown in Fig. 4.28(b), the gradient operation
[Eq. (4.4-4) was used] reduces all the constant white regions to zero (black),
leaving only the points associated with abrupt changes in gray level (in this
case the edge boundaries and the small spot on upper-right part of the letter
T).

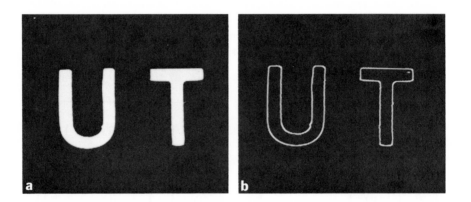

Figure 4.28. (a) A binary image. (b) Result of computing the gradient.

Once a method for approximating the gradient has been selected,
there are numerous ways of using the results for generating a gradient image
g(x, y). The simplest approach is to let the value of *g* at coordinates *(x, y)* be

equal to the gradient of f at that point, that is,

$$g(x, y) = G[f(x, y)] \tag{4.4-7}$$

The principal disadvantage of this method is that all smooth regions in $f(x, y)$ appear dark in $g(x, y)$ because of the relatively small values of the gradient in these regions. One solution to this problem is to form $g(x, y)$ as follows:

$$g(x, y) = \begin{cases} G[f(x, y)] & \text{if } G[f(x, y)] \geqslant T \\ f(x, y) & \text{otherwise} \end{cases} \tag{4.4-8}$$

where T is a nonnegative threshold. By properly selecting T, it is possible to emphasize significant edges without destroying the characteristics of smooth backgrounds. A variation of this approach where edges are set to a specified gray level, L_G, is given by

$$g(x, y) = \begin{cases} L_G & \text{if } G[f(x, y)] \geqslant T \\ f(x, y) & \text{otherwise} \end{cases} \tag{4.4-9}$$

It is sometimes desirable to study the gray-level variation of edges without interference from the background. This can be accomplished by forming the gradient image as follows:

$$g(x, y) = \begin{cases} G[f(x, y)] & \text{if } G[f(x, y)] \geqslant T \\ L_B & \text{otherwise} \end{cases} \tag{4.4-10}$$

where L_B is a specified level for the background.

Finally, if only the location of edges is of interest, the relation

$$g(x, y) = \begin{cases} L_G & \text{if } G[f(x, y)] \geqslant T \\ L_B & \text{otherwise} \end{cases} \tag{4.4-11}$$

gives a binary gradient picture where the edges and background are displayed in any two specified gray levels.

Example: The types of edge enhancement that can be obtained by using Eq. (4.4-4) and Eqs. (4.4-7) through (4.4-11) are illustrated in Fig. 4.29. Part (a) of the figure shows an original image of moderate complexity. Figure 4.29 (b) is the result of using the gradient scheme given by Eq. (4.4-7). It is noted that a considerable amount of small segments appeared in the resulting

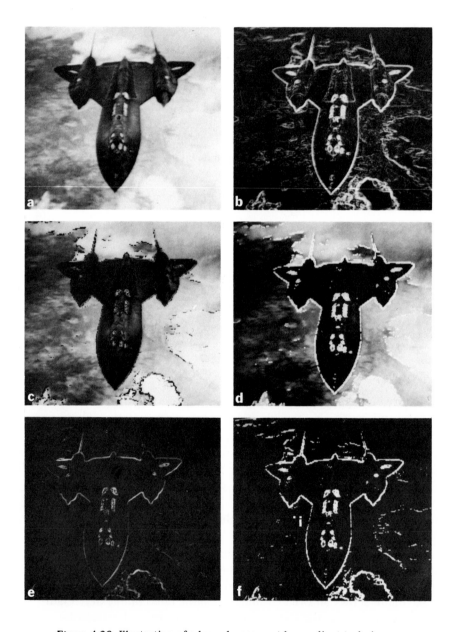

Figure 4.29. Illustration of edge enhancement by gradient techniques.

image, with the strongest intensities taking place around the border of the aircraft. This is an expected result since the magnitude of the gradient is proportional to changes in gray levels and should be more prominent in regions of an image containing distinct edges.

Figure 4.29(c) was obtained by using Eq. (4.4-8) with $T = 25$, which is approximately 10% of the maximum gray-level value in the original image. The gradient values appear a dark shade of gray because they are displayed on the relatively light background of the image. The important points to note in connection with this figure is that only prominent edges are out-lined as a result of using a threshold, and also that the background has not been completely obliterated.

Figure 4.29(d) is the result of using Eq. (4.4-9) with $T = 25$ and $L_G = 255$, the latter being the brightest possible level in the system used to display the results. It is noted that Figs. 4.29(c) and (d) are the same, with the exception that the gradient points exceeding the threshold are much more visible in the latter image.

Figure 4.29(e) was obtained by using Eq. (4.4-10) with the same threshold as above and a background level of $L_B = 0$, which is the darkest possible display level. The principal use of this particular approach is to examine the relative strength of gradient points which exceed the specified threshold. In this case, we see that the outline of the aircraft and the cloud near the bottom of the image are quite prominent in relation to other sections of the picture.

Finally, Fig. 4.29(f) was obtained by using Eq. (4.4-11) with $T = 25$, $L_G = 255$, and $L_B = 0$. This equation is useful for displaying all the gradient points above the specified threshold. □

4.4.2 Highpass Filtering

It was shown in Section 4.3.2 that an image can be blurred by attenu-ating the high-frequency components of its Fourier transform. Since edges and other abrupt changes in gray levels are associated with high-frequency components, image sharpening can be achieved in the frequency domain by a *highpass filtering* process which attenuates the low-frequency components without disturbing high-frequency information in the Fourier transform.

We consider below the high-frequency counterparts of the filters developed in Section 4.3.2. As before, we will only consider zero-phase-shift filters which are radially symmetric and can be completely specified by a cross section extending as a function of distance from the origin.

4.4.2.1 Ideal filter

A two-dimensional ideal highpass filter (IHPF) is one whose transfer function satisfies the relation

$$H(u, v) = \begin{cases} 0 & \text{if } D(u, v) \leqslant D_0 \\ 1 & \text{if } D(u, v) > D_0 \end{cases} \qquad (4.4\text{-}12)$$

where D_0 is the cut-off distance measured from the origin of the frequency plane, and $D(u, v)$ is given by Eq. (4.3-5). A perspective plot and cross section of the IHPF function are shown in Fig. 4.30. It is noted that this filter is just the opposite of the ideal lowpass filter discussed in Section 4.3.2

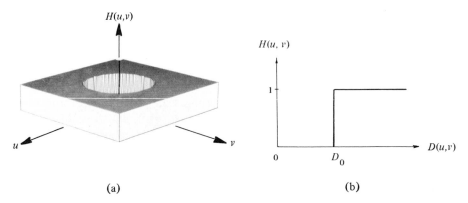

(a)   (b)

Figure 4.30. Perspective plot and radial cross section of ideal highpass filter.

since it completely attenuates all frequency inside a circle of radius D_0 while passing, without attenuation, all frequencies outside the circle. As in the case of the ideal lowpass filter, the IHPF is not physically realizable.

4.4.2.2 Butterworth filter

The transfer function of the Butterworth highpass filter (BHPF) of order n and with cut-off frequency locus as a distance D_0 from the origin is defined by the relation

$$H(u, v) = \frac{1}{1 + \left[D_0/D(u, v) \right]^{2n}} \qquad (4.4\text{-}13)$$

where $D(u, v)$ is given by Eq. (4.3-5). A perspective plot and cross section of the BHPF function are shown in Fig. 4.31.

It is noted that, when $D(u, v) = D_0$, $H(u, v)$ is down to 1/2 of its maximum value. As in the case of the Butterworth lowpass filter, it is common practice to select the cut-off frequency locus at points for which $H(u, v)$ is down to $1/\sqrt{2}$ of its maximum value. Equation (4.4-13) is easily modified to satisfy this constraint by using the following scaling:

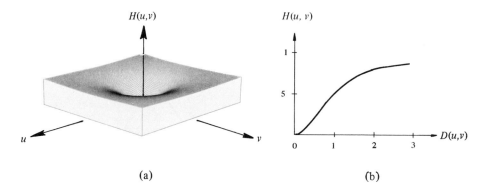

$H(u,v)$

$H(u, v)$

(a)

(b)

Figure 4.31. Perspective plot and radial cross section of Butterworth highpass filter for $n = 1$.

$$H(u, v) = \frac{1}{1 + [\sqrt{2} - 1][D_0/D(u, v)]^{2n}}$$

$$= \frac{1}{1 + 0.414[D_0/D(u, v)]^{2n}} \qquad (4.4\text{-}14)$$

4.4.2.3 Exponential filter

The exponential highpass filter (EHPF) with cut-off frequency locus at a distance D_0 from the origin has a transfer function given by the relation

$$H(u, v) = e^{-[D_0/D(u, v)]^n} \qquad (4.4\text{-}15)$$

where $D(u, v)$ is given by Eq. (4.3-5), and the parameter n controls the rate of increase of $H(u, v)$ as a function of increasing distance from the origin. A perspective plot and cross section of the EHPF transfer function are shown in Fig. 4.32.

When $D(u, v) = D_0$ we have from Eq. (4.4-15) that $H(u, v) = 1/e$. A simple modification given by

$$H(u, v) = e^{[\ln 1/\sqrt{2}][D_0/D(u, v)]^n}$$

$$= e^{-0.347[D_0/D(u, v)]^n} \qquad (4.4\text{-}16)$$

forces $H(u, v)$ to be equal to $1/\sqrt{2}$ of its maximum value at frequencies in the cut-off locus.

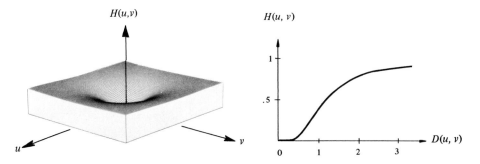

Figure 4.32. Perspective plot and radial cross section of exponential highpass filter.

4.3.2.4 Trapezoidal filter

A trapezoidal highpass filter (THPF) can be defined by the relation

$$H(u, v) = \begin{cases} 0 & \text{if } D(u, v) < D_1 \\[2mm] \dfrac{1}{[D_0 - D_1]}[D(u, v) - D_1] & \text{if } D_1 \leqslant D(u, v) \leqslant D_0 \qquad (4.4\text{-}17) \\[2mm] 1 & \text{if } D(u, v) > D_0 \end{cases}$$

where $D(u, v)$ is given by Eq. (4.3-5), D_0 and D_1 are specified, and it is assumed that $D_0 > D_1$. A cross section and perspective view of a typical THPF transfer function are shown in Fig. 4.33.

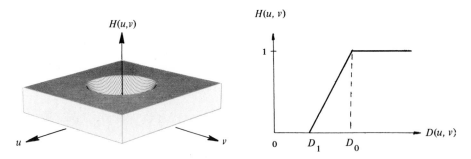

Figure 4.33. Perspective plot and radial cross section of a trapezoidal highpass filter.

For convenience in implementation, we define the cut-off frequency locus to be at D_0 rather than at the radius for which $H(u, v)$ is $1/\sqrt{2}$ of its maximum value. The second variable, D_1, is arbitrary, as long as it is less than D_0.

Example: Figure 4.34(a) shows a chest x-ray which was poorly developed, and Fig. 4.34(b) shows the image after it was processed with a high-pass exponential filter. Only the edges are predominant in this image because the low-frequency components were severely attenuated, thus making different (but smooth) gray-level regions appear essentially the same.

A technique often used to alleviate this problem consists of adding a constant (e.g., 1) to a high-pass filter transfer function in order to preserve the low-frequency components. This, of course, amplifies the high-frequency

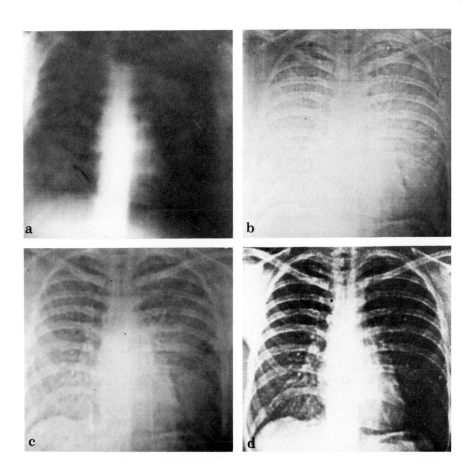

Figure 4.34. Example of highpass filtering. (a) Original image. (b) Image processed with a highpass exponential filter. (c) Result of high-frequency emphasis. (d) High-frequency emphasis and histogram equalization. (From Hall, *et al* [1971].)

components to values which are higher than in the original transform. This technique, called *high-frequency emphasis*, is illustrated in Fig. 4.34(c). It is noted that the image in this case has a little better tonality, particularly in the lower-left part of the photograph.

Although high-frequency emphasis preserves the low-frequency components, the proportionally larger high-frequency terms tend to obscure the result, as shown by the small gain in quality from Fig. 4.34(b) to Fig. 4.34(c) A technique often used to compensate for this problem is to do some post-filtering processing to redistribute the gray levels. Histogram equalization is ideally suited for this because of its simplicity. Figure 4.34(d) shows the significant improvement that can be obtained by histogram-equalizing an image that has been processed by high-frequency emphasis.   □

4.5 ENHANCEMENT BASED ON AN IMAGE MODEL

The illumination-reflectance model introduced in Section 2.2 can be used as the basis for a frequency-domain procedure that is useful for improving the appearance of an image by simultaneous brightness range compression and contrast enhancement. From the discussion in Section 2.2 we have that an image $f(x, y)$ can be expressed in terms of its illumination and reflectance components by means of the relation

$$f(x, y) = i(x, y)r(x, y) \qquad (4.5\text{-}1)$$

Equation (4.5-1) cannot be used directly in order to operate separately on the frequency components of illumination and reflectance because the Fourier transform of the product of two functions is not separable; in other words,

$$\mathcal{F}\{f(x, y)\} \neq \mathcal{F}\{i(x, y)\}\mathcal{F}\{r(x, y)\}$$

Suppose, however, that we let

$$z(x, y) = \ln f(x, y)$$

$$= \ln i(x, y) + \ln r(x, y) \qquad (4.5\text{-}2)$$

Then, it follows that

$$\mathcal{F}\{z(x, y)\} = \mathcal{F}\{\ln f(x, y)\}$$

$$= \mathcal{F}\{\ln i(x, y)\} + \mathcal{F}\{\ln r(x, y)\} \qquad (4.5\text{-}3)$$

or

$$Z(u, v) = I(u, v) + R(u, v) \qquad (4.5\text{-}4)$$

where $I(u, v)$ and $R(u, v)$ are the Fourier transform of ln $[i(x, y)]$ and ln $[r(x, y)]$, respectively.

If we process $Z(u, v)$ by means of a filter function $H(u, v)$, it follows from Eq. (4.1-2) that

$$S(u, v) = H(u, v)Z(u, v)$$

$$= H(u, v)I(u, v) + H(u, v)R(u, v) \qquad (4.5\text{-}5)$$

where $S(u, v)$ is the Fourier transform of the result. In the spatial domain, we have the relation

$$s(x, y) = \mathscr{F}^{-1}\{S(u, v)\}$$

$$= \mathscr{F}^{-1}\{H(u, v)I(u, v)\} + \mathscr{F}^{-1}\{H(u, v)R(u, v)\} \qquad (4.5\text{-}6)$$

By letting

$$i'(x, y) = \mathscr{F}^{-1}\{H(u, v)I(u, v)\} \qquad (4.5\text{-}7)$$

and

$$r'(x, y) = \mathscr{F}^{-1}\{H(u, v)R(u, v)\} \qquad (4.5\text{-}8)$$

we can express Eq. (4.5-6) in the form

$$s(x, y) = i'(x, y) + r'(x, y) \qquad (4.5\text{-}9)$$

Finally, since $z(x, y)$ was formed by taking the logarithm of the original image $f(x, y)$, we now perform the inverse operation to obtain the desired enhanced image $g(x, y)$; that is,

$$g(x, y) = \exp\{s(x, y)\}$$

$$= \exp\{i'(x, y)\} \cdot \exp\{r'(x, y)\}$$

$$= i_0(x, y)r_0(x, v) \qquad (4.5\text{-}10)$$

where

$$i_0(x, y) = \exp\{i'(x, y)\} \qquad (4.5\text{-}11)$$

and

$$r_0(x, y) = \exp\{r'(x, y)\} \qquad (4.5\text{-}12)$$

are the illumination and reflectance components of the output image.

The enhancement approach using the foregoing concepts is summarized in Fig. 4.35. This method is based on a special case of a class of systems known as *homomorphic systems*. In this particular application, the key to the approach is the fact that separation of the illumination and reflectance components is achieved in the form shown in Eq. (4.5-4). It is then

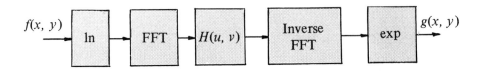

Figure 4.35. Homomorphic filtering approach for image enhancement.

possible for the *homomorphic filter function H(u, v)* to operate on these components, separately, as indicated in Eq. (4.5-5).

The illumination component of an image is generally characterized by slow spatial variations. The reflectance component, on the other hand, tends to vary abruptly, particularly at the junctions of very dissimilar objects. These characteristics lead us to associate the low frequencies of the Fourier transform of the logarithm of an image with illumination, and the high frequencies with reflectance. Although this is a rough approximation, it can be used to advantage in image enhancement.

Illumination is directly responsible for the dynamic range achieved by the pixels in an image. Similarly, contrast is a function of the reflective nature of the objects in the image. A good deal of control can be gained over these components by using a homomorphic filter. This requires specification of a filter function $H(u, v)$ which will affect the low and high frequency components of the Fourier transform in different ways. A cross section of such a function is shown in Fig. 4.36. A complete specification of $H(u, v)$ is obtained by rotating the cross section 360° about the vertical axis. If the parameters γ_L and γ_H are chosen so that $\gamma_L < 1$ and $\gamma_H > 1$, the filter function shown in Fig. 4.36 will tend to decrease the low frequencies and amplify the high frequencies. The net result is simultaneous dynamic range compression and contrast enhancement.

Example: Figure 4.37 is typical of the results that can be obtained with the homomorphic filter function shown in Fig. 4.36. We note in the original image, Fig. 4.37(a), that the details inside the room are obscured by the glare from the outside walls. Figure 4.37(b) shows the result of processing this image by homomorphic filtering with $\lambda_L = 0.5$ and $\lambda_H = 2.0$ in the above filter function. A reduction of dynamic range in the brightness, together with an increase in contrast, brought out the details of objects inside the room and balanced the levels of the outside wall. □

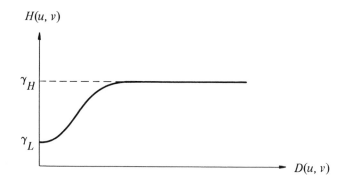

Figure 4.36. Cross section of a circularly symmetric filter function for use in homomorphic filtering. $D(u, v)$ is the distance from the origin.

4.6 PSEUDO-COLOR IMAGE PROCESSING

Attention has been focused thus far on processing techniques for monochrome images. A relatively recent and potentially powerful area of digital image processing is the use of pseudo color for image display and enhancement. The motivation for using color in image processing is provided by the fact that the human eye can discern thousands of color shades and intensities. This is in sharp contrast with the eye's relatively poor performance with gray levels where, as indicated in Section 2.1, only one to two dozen shades of gray are detectable at any one point in an image by the average observer. The reader needs only to turn off the color next time he views a TV set in order to verify the superior performance of the eye when interpreting color *vs.* monochrome information.

The basic idea behind the use of pseudo color should be carefully distinguished from techniques known as false-color processing. The latter are analogous to true color photography, with the exception that they may deal with light bands that are outside the visible spectrum. Infrared photography is an example of this where interest lies not in true color fidelity, but rather on information that is most evident in the infrared spectrum. In pseudo color, the situation is fundamentally different in that processing starts out with a monochrome image. The objective is then to assign a color to each pixel based, for example, on its intensity. The range of techniques for color assignment are limited only by the capabilities of the display system and the ingenuity of the user. As will be seen in the following discussion, even some straightforward techniques for color coding can sometimes bring out information which is often difficult to detect and interpret in a monochrome image.

Figure 4.37. (a) Original image. (b) Image processed by homomorphic filtering to achieve simultaneous dynamic range compression and contrast enhancement. (From Stockham [1972].)

4.6.1 Color Fundamentals

Although the process followed by the human brain in perceiving color is a physio-psychological phenomenon which is not yet fully understood, the physical nature of color can be expressed on a formal basis supported by experimental and theoretical results.

In 1666, Sir Isaac Newton discovered that when a beam of sunlight is passed through a glass prism the emerging beam of light is not white, but consists instead of a continuous spectrum of colors ranging from violet at one end to red at the other. As shown in Plate I,[†] the color spectrum may be divided into six broad regions: violet, blue, green, yellow, orange, and red. When viewed in full color (Plate II) no color in the spectrum ends abruptly, but rather we have a situation where each color blends slowly into the next.

Basically, the colors we perceive in an object are determined by the nature of the light reflected from the object. As illustrated in Plate II, visible light is composed of a relatively narrow band of frequencies in the electromagnetic energy spectrum. A body which reflects light that is relatively balanced in all visible wavelengths appears white to the observer. On the other hand, a body which favors reflectance in a limited range of the visible spectrum will exhibit some shades of color. For example, green objects reflect light with wavelengths primarily in the 500 to 570 nm (10^{-9} meters) range, while absorbing most of the energy at other wavelengths.

The ranges of colors we are used to observing in our normal visual activities are the result of a mixture of light of different wavelengths. It has been determined (Kiver [1965]) that certain wavelengths of red (R), green (G), and blue (B), when combined with each other in various proportions (intensities) will produce a wider range of colors than any other combination of three colors. These colors are thus referred to as the *primary* colors of light, although use of the word "primary" has been widely misinterpreted to mean that red, green, and blue will, in various combinations, produce *all* colors. This is not true since many color shades cannot be obtained by any combination of the three RGB primaries alone.

From Plate II, it is evident that there is no single color which may be called red, green, or blue; instead, transitions of these colors in the spectrum are smooth, blending variations. For the purpose of standardization, the CIE (Commission Internationale de l'Eclairage - the International Commission on Illumination) designated in 1931 the following specific wavelength values to the three primary colors: blue = 435.8 nm, green = 546.1 nm, and red = 700 nm.

The primary colors can be added to produce the *secondary* colors of light—magenta (red plus blue), cyan (green plus blue), and yellow (red plus

---

† Following page 176.

green). Mixing the three primaries, or a secondary with its opposite primary color, in the right intensities produces white light. This is shown in Plate III (a) which also illustrates the three primary colors and their combinations to produce the secondary colors.

It is important to differentiate between the primary colors of light and the primary colors of pigments or colorants. In the latter, a primary color is defined as one that subtracts or absorbs a primary color of light and reflects or transmits the other two. Therefore, the primary colors of pigments are magenta, cyan, and yellow, while the secondary colors are red, green, and blue. These colors are shown in Plate III(b). It is noted that a proper combination of the three pigment primaries, or a secondary with its opposite primary, produces black.

Color television reception is an example of the additive nature of light colors. The interior of many color TV tubes is composed of a large array of triangular dot patterns of electron-sensitive phosphor. When excited, each dot in a triad is capable of producing light in one of the primary colors. The intensity of the red-emitting phosphor dots is modulated by an electron gun inside the tube which generates pulses corresponding to the "red energy" seen by the TV camera. The green and blue phosphor dots in each triad are modulated in the same manner. The effect, viewed on the television receiver, is that the three primary colors from each phosphor triad are "added" together and received by the color-sensitive cones in the eye, and a full color image is perceived. Thirty successive image changes per second in all three colors complete the illusion of a continuous image display on the screen.

The nature of light colors is further illustrated in Plate IV in what is commonly called a *chromaticity diagram*. The colors in this diagram were generated by deriving mathematically and plotting as a chromaticity point the relative percentages of each of the primary colors required to produce all possible colors. Since, as indicated earlier, not all colors can be obtained by use of three single primary colors alone, derivation of the chromaticity diagram required the introduction of fictitious primaries (i. e., primaries that cannot be obtained in practice).

The coordinates in the chromaticity diagram represent the relative fractions of each of the primary colors present in a given color. These coordinates follow the convention: x = red, y = green, z = blue. Since the sum of all three primaries must add to 1, we have the relation

$$x + y + z = 1 \qquad\qquad (4.6\text{-}1)$$

or

$$z = 1 - (x + y) \qquad\qquad (4.6\text{-}2)$$

Only x and y are required to specify the chromaticity diagram since z can be obtained from these coordinates by using Eq. (4.6-2). The point marked

"Green" in Plate IV, for example, has approximately 62% green and 25% red content. It then follows from Eq. (4.6-2) that the composition of blue is approximately 13%.

The positions of the various spectrum colors, from violet at 380 nm to red at 780 nm, are indicated around the boundary of the tongue-shaped chromaticity diagram. These are the "pure" colors shown in the spectrum of Plate II. Any point not actually on the boundary but within the diagram represents some mixture of spectrum colors. The point of equal energy shown in Plate IV corresponds to equal fractions of the three primary colors; it represents the CIE standard for white light. Any point located on the boundary of the chromaticity chart is said to be completely *saturated*. As a point leaves the boundary and approaches the point of equal energy, more white light is added to the color and it becomes less saturated. The saturation at the point of equal energy is zero.

The term *hue* is often used in connection with saturation. Hue represents color such as red, orange, and yellow. It is associated with color wavelength, and when we call a certain color red, orange, or yellow, we are specifying its hue. Thus, hue refers to the basic color as it appears to the observer, while saturation indicates how deep the color is.

The chromaticity diagram is useful for color mixing because a straight line segment joining any two points in the diagram defines all the different color variations that can be obtained by combining these two colors, additively. Consider, for example, a straight line drawn from the Red to the Green points shown in Plate IV. If there is more red light than green light, the exact point representing the new color will be on the line segment, but it will be closer to the red point than to the green point. Similarly, a line drawn from the point of equal energy to any point on the boundary of the chart will define all the shades of a particular spectrum color.

Extension of the above procedure to three colors is straightforward. To determine the range of colors which can be obtained from any three given colors in the chromaticity diagram, we simply draw connecting lines to each of the three color points. The result is a triangle, and any color inside the triangle can be produced by various combinations of the three initial colors. It is noted that a triangle with vertices at any three colors does not enclose the entire color region in Plate IV. This supports graphically the remark made earlier that not all colors can be obtained with three single primaries.

In the following discussion, the intensity of the three primary colors at coordinates (x, y) of an image will be denoted by $I_R(x, y)$, $I_G(x, y)$, and $I_B(x, y)$, respectively. Thus, each pixel in a color image will be considered as an additive combination of these three intensity values at the coordinates of the pixel.

4.6.2 Density Slicing

The technique of *density* (or *intensity*) *slicing* and color coding is one of the simplest examples of pseudo-color image processing. If an image is viewed as a two-dimensional intensity function (see Section 1.2), the method can be interpreted as one of placing planes parallel to the coordinate plane of the image; each plane then "slices" the function in the area of intersection. Figure 4.38 shows a simple example of this where a plane at $f(x, y) = l_i$ is used to slice a function into two levels. The term "density slicing" arises from calling the gray levels densities, a terminology that is commonly associated with this particular method.

It is evident that if a different color is assigned to each side of the plane shown in Fig. 4.38, then any pixel whose gray level is above the plane will be coded with one color, while any pixel below the plane will be coded with the other. Levels that lie on the plane itself may be arbitrarily assigned one of the two colors. The result of this scheme would produce a two-color image whose relative appearance can be controlled by moving the slicing plane up and down the gray-level axis.

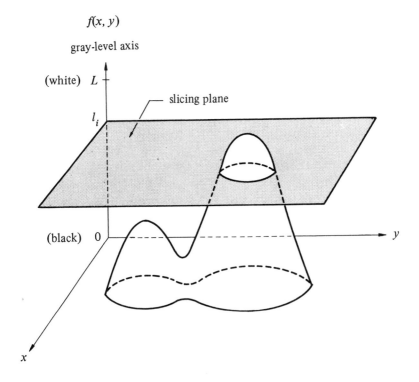

Figure 4.38. Geometrical interpretation of the density-slicing technique.

In general, the technique may be summarized as follows. Suppose that M planes are defined at levels l_1, l_2, ..., l_M, and let l_0 represent black [$f(x, y) = 0$] and l_L represent white [$f(x, y) = L$]. Then, assuming that $0 < M < L$, the M planes partition the gray scale into $M + 1$ regions and color assignments are made according to the relation:

$$f(x, y) = c_k \qquad \text{if } f(x, y) \in R_k \qquad (4.6\text{-}3)$$

where c_k is the color associated with the kth region, R_k, defined by the partitioning planes.

It is important to note that the idea of planes is useful primarily for a geometrical interpretation of the density-slicing technique. An alternative representation is shown in Fig. 4.39, which defines the same mapping as Fig. 4.38. According to the mapping function shown in Fig. 4.39, any input gray level is assigned one of two colors, depending on whether it is above or below the value of l_i. When more levels are used, the mapping function assumes a staircase form. This type of mapping is a special case of the approach discussed in the following section.

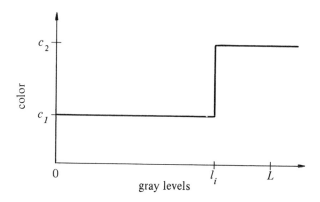

Figure 4.39. An alternate representation of the density-slicing method.

Example: An example of density slicing is shown in Plate V. Part (a) is a monochrome image of the Picker Thyroid Phantom (a radiation test pattern), and Plate V(b) is the result of density slicing this image into eight color regions. It is noted that regions which appear of constant intensity in the monochrome image are really quite variable, as shown by the various colors in the sliced image. The left lobe, for instance, is a dull gray in the monochrome image, and it is difficult to pick out variations in intensity. By contrast, the color image clearly shows eight different regions of constant intensity, one for each of the colors used. ☐

4.6.3 Gray-Level-to-Color Transformations

It is possible to specify other types of transformations that are more general and thus are capable of achieving a wider range of pseudo-color enhancement results than the simple density-slicing technique discussed in the previous section. An approach that is particularly attractive is shown in Fig. 4.40. Basically, the idea underlying this approach is to perform three independent transformations on the gray levels of any input pixel. The three results are then fed separately into the red, green, and blue guns of a color television monitor. This produces a composite image whose color content is modulated by the nature of the transformation function. It should be kept in mind that these are transformations on the gray-level values of an image and that they are not functions of position.

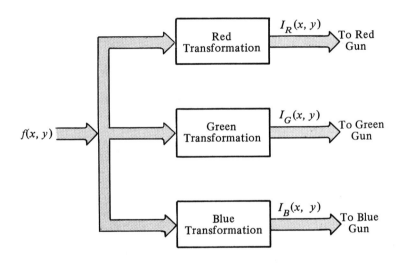

Figure 4.40. Functional block diagram for pseudo-color image processing.

Figure 4.41 shows a set of typical transformation functions. Part (a) of this figure, for example, indicates that the red transformation maps any level lower than $L/2$ into the darkest available shade of red. The intensity of the color input increases linearly between $L/2$ and $3L/4$ and remains constant at the brightest shade of red for gray levels in the range $3L/4$ and L. The other color mappings are interpreted in a similar manner. The three transformation functions are shown together in Fig. 4.41(d). It is noted from

SPECIAL ILLUSTRATIONS

Plates I-VII

ISBN 0-201-02596-7
ISBN 0-201-02597-5 (pbk.)

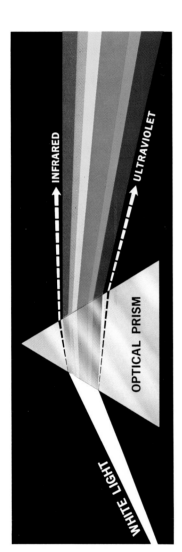

Plate I. Color spectrum seen by passing white light through a prism. (Courtesy of General Electric Co., Lamp Business Division.)

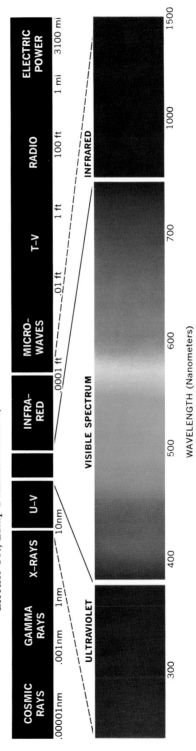

Plate II. A section of the electromagnetic energy spectrum showing the range of wavelengths comprising the visible spectrum. (Courtesy of General Electric Co., Lamp Business Division.)

Illus-iii

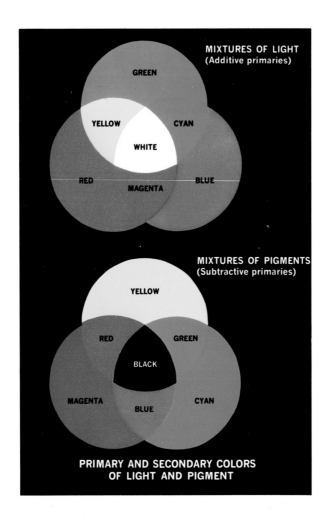

Plate III. Primary and secondary colors of light and pigments. (Courtesy of General Electric Co., Lamp Business Division.)

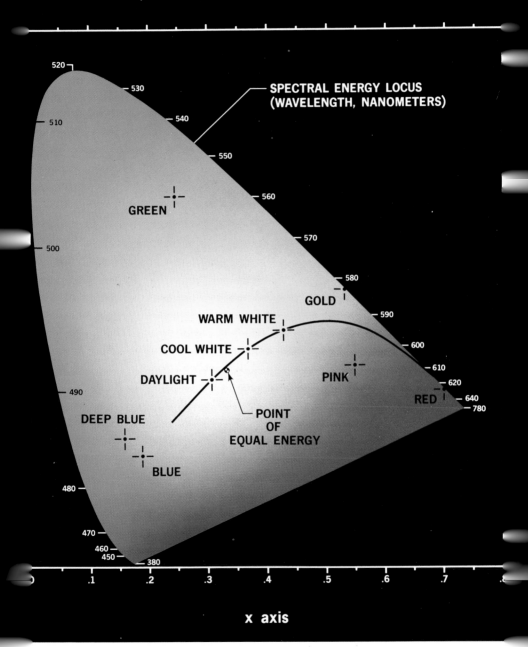

(C. I. E. CHROMATICITY DIAGRAM)

IV. Chromaticity diagram. (Courtesy of General Electric Co., Lamp Business Divi-

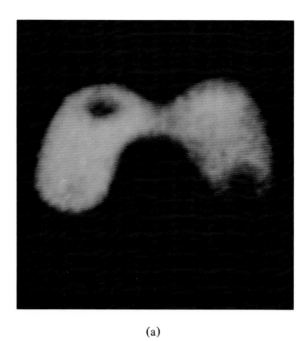

(a)

(b)

Plate V. (a) Monochrome image of the Picker Thyroid Phantom. (b) Result of density slicing into eight color regions. (Courtesy of Dr. J. L. Blankenship, Instrumentation and Controls Division, Oak Ridge National Laboratory.)

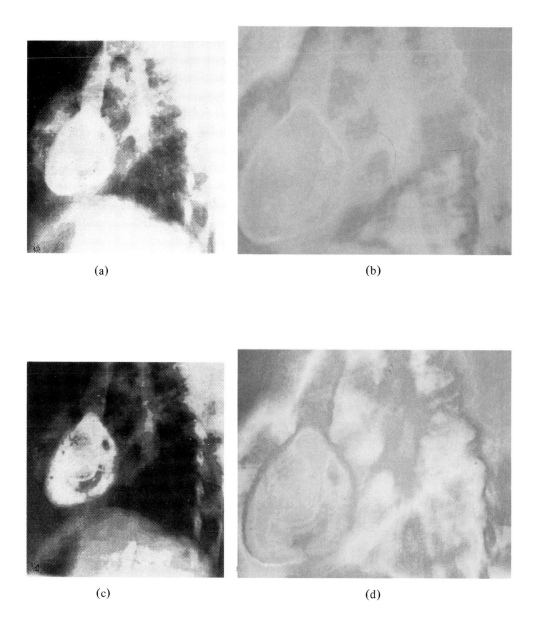

(a)

(b)

(c)

(d)

Plate VI. Pseudo-color enhancement by using the gray-level to color transformations shown in Fig. 4.41. (From Andrews, Tescher, and Kruger [1972].)

(a)

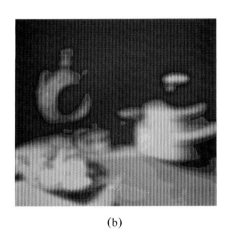

(b)

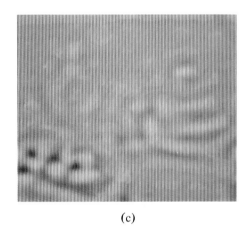

(c)

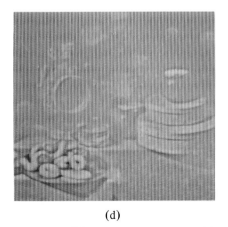

(d)

(e)

Plate VII. (a) Monochrome image. (b) Lowpass filtered. (c) Bandpass filtered. (d) Highpass filtered. (e) Color composite image.

Illus-viii

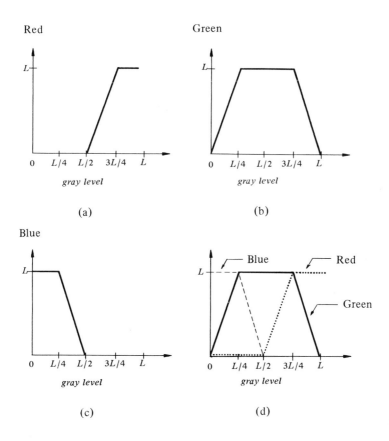

Figure 4.41. A typical set of color-transformation functions.

this figure that the pure primaries occur only at the ends and exact center of the gray scale.

Example: Plate VI(a) shows an x-ray image of a dog's heart, and Plate VI(b) shows the result of color-coding the monochrome image with the transformation functions of Fig. 4.41. Of particular interest are the border of the heart and the dark area to the right of the heart. These areas are difficult to interpret in the original image, but show up quite clearly in the processed image. An even better result is obtained by histogram-equalizing the monochrome image [Plate VI(c)] and then applying the above color-mapping technique. The result is shown in Plate VI(d). ☐

4.6.4 A Filtering Approach

Figure 4.42 shows a color coding scheme that is based on frequency-domain operations. The idea depicted in this figure is the same as the basic filtering approach discussed earlier in this chapter, with the exception that the Fourier transform of an image is modified independently by three filter functions to produce three images that can be fed into the red, green, and blue inputs of a color monitor. Consider, for example, the sequence of steps followed in obtaining the image for the red channel. The Fourier transform of the input image is altered by using a specified filter function. The processed image is then obtained by using the inverse Fourier transform. This can then be followed by additional processing (such as histogram equalization) before the image is fed into the red input of the monitor. Similar comments apply to the other two paths in Fig. 4.42.

The objective of this color processing technique is to color-code regions of an image based on frequency content. A typical filtering approach is to use lowpass, bandpass (or bandreject), and highpass filters to obtain three ranges of frequency components. We have already discussed lowpass and highpass filters. Bandreject and bandpass filters are an extension of these concepts.

A simple approach for generating filters which reject or attenuate frequencies about a circular neighborhood of a point (u_0, v_0) is to perform a translation of coordinates for any of the highpass filters discussed in Section 4.4.2. We illustrate the procedure for the ideal filter.

An ideal bandreject filter (IBRF) which suppresses all frequencies in a neighborhood of radius D_0 about a point (u_0, v_0) is given by the relation

$$H(u, v) = \begin{cases} 0 & \text{if } D(u, v) \leqslant D_0 \\ 1 & \text{if } D(u, v) > D_0 \end{cases} \tag{4.6-4}$$

where

$$D(u, v) = \left\{ (u - u_0)^2 + (v - v_0)^2 \right\}^{1/2} \tag{4.6-5}$$

It is noted that Eq. (4.6-4) is identical in form to Eq. (4.4-12), but the distance function $D(u, v)$ is computed about the point (u_0, v_0) instead of the origin.

Due to the symmetry of the Fourier transform, band rejection that is not about the origin must be carried out in symmetric *pairs* in order to obtain meaningful results. In the case of ideal filter we modify Eq. (4.6-4) as follows:

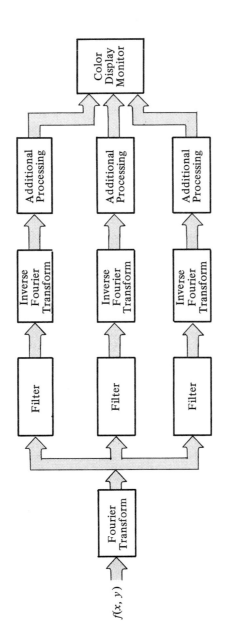

Figure 4.42. A filtering model for pseudo-color image enhancement.

$$D_1(u, v) = \begin{cases} 0 & \text{if } D_1(u, v) \leqslant D_0 \text{ or } D_2(u, v) \leqslant D_0 \\ 1 & \text{otherwise} \end{cases} \qquad (4.6\text{-}6)$$

where

$$D_1(u, v) = \left\{(u - u_0)^2 + (v - v_0)^2\right\}^{1/2} \qquad (4.6\text{-}7)$$

and

$$D_2(u, v) = \left\{(u + u_0)^2 + (v + v_0)^2\right\}^{1/2} \qquad (4.6\text{-}8)$$

The procedure can be extended in a similar manner to four or more regions. The other filters given in Section 4.4.2 can also be applied directly to band rejection by following the technique just described for the ideal filter. Figure 4.43 shows a perspective plot of a typical IBRF transfer function.

The filter discussed above is localized about some point off the origin of the Fourier transform. If it is desired to remove a band of frequencies centered about the origin, we can consider symmetric filters similar to the low and highpass filters discussed earlier. The procedure is illustrated for the ideal and Butterworth filters.

A radially-symmetric ideal bandreject filter which removes a band of frequencies about the origin is given by the relation

$$H(u, v) = \begin{cases} 1 & \text{if } D(u, v) < D_0 - \dfrac{W}{2} \\ 0 & \text{if } D_0 - \dfrac{W}{2} \leqslant D(u, v) \leqslant D_0 + \dfrac{W}{2} \\ 1 & \text{if } D(u, v) > D_0 + \dfrac{W}{2} \end{cases} \qquad (4.6\text{-}9)$$

where W is the width of the band and D_0 is its radial center. As is the case with all radially-symmetric filters, this filter can be completely specified by a cross section.

A radially-symmetric Butterworth bandreject filter (BBRF) of order n has the transfer function

$$H(u, v) = \frac{1}{1 + \left[\dfrac{D(u, v)W}{D^2(u, v) - D_0^2}\right]^{2n}} \qquad (4.6\text{-}10)$$

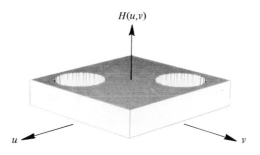

Figure 4.43. Ideal bandreject filter.

where W is defined as the "width" of the band and D_0 is its center.

Bandpass filters pass frequencies in a specified band or region, while attenuating, or completely suppressing, all other frequencies. They are, therefore, exactly the opposite of bandreject filters. It then follows that, if $H_R(u, v)$ is the transfer function of any of the bandreject filters just discussed, the corresponding bandpass function, $H(u, v)$, can be obtained simply by "flipping" $H_R(u, v)$, that is,

$$H(u, v) = -\left[H_R(u, v) - 1 \right]$$
(4.6-11)

Example: Plate VII(a) shows a monochrome image and Plates VII(b) through (d) are the results of using Butterworth filters. The red image is the lowpass filtered result with the cut-off point at a circle enclosing 90% of the image energy (see Section 4.3.2). The blue image is the bandpassed result [Eq. (4.6-10)] with D_0 at the circle of 83% energy and W chosen so that the end of the bandpass region was at the circle of 93% energy. The green image is the highpass filtered result with the cut-off point at the circle enclosing 95% of the image energy. Plate VII(e) is the result of inputing the above three images into the RGB inputs of a color monitor.   □

4.7 CONCLUDING REMARKS

The material presented in this chapter is representative of techniques commonly used in practice for digital image enhancement. It must be kept in mind, however, that this area of image processing is a dynamic field where reports of new techniques and applications are commonplace in the literature. For this reason, the topics included in this chapter were selected mostly for their value as fundamental material which would serve as foundation for further study of this field.

REFERENCES

References for the material in Section 4.2 are Hall *et al* [1971], Hall [1974], Hummel [1974], and Gonzalez and Fittes [1975, 1977]. The neighborhood averaging approach introduced in Section 4.3.1 is based on the discussion of this topic by Rosenfeld [1969]. The lowpass filtering concepts developed in Section 4.3.2 are based on a direct extension of one-dimensional filters where, instead of using a single variable, we use the distance from the origin of the Fourier transform in order to obtain circularly-symmetric filter functions. This is also true of the other filters discussed in this chapter. For a discussion of one-dimensional filters see, for example, the books by Weinberg [1962] and Budak [1974]. For additional details on smoothing by averaging multiple images see the paper by Kohler and Howell [1963].

Early references on image sharpening by differentiation are Goldmark and Hollywood [1951] and Kovasznay and Joseph [1953, 1955]. The Roberts gradient approach was proposed by Roberts [1965]. The articles by Prewitt [1970] and by Frei and Chen [1977] are also of interest.

The material in Section 4.6 is based on the paper by Stockham [1972]. See also the book by Oppenheim and Schafer [1975]. Additional reading for the material in Section 4.6 may be found in Smith [1963], Kiver [1965], Roth [1968], Billingsley *et al* [1970], and Andrews *et al* [1972].

IMAGE RESTORATION

Things which we see are not by themselves what
we see ... It remains completely unknown to us
what the objects may be by themselves and
apart from the receptivity of our senses.
We know nothing but our manner of
perceiving them. . .
Immanuel Kant

As in image enhancement, the ultimate goal of restoration techniques
is to improve a given image in some sense. For the purpose of differentiation,
we consider restoration to be a process which attempts to reconstruct or
recover an image that has been degraded by using some a priori knowledge of
the degradation phenomenon. Thus, restoration techniques are oriented
toward modeling the degradation and applying the inverse process in order
to recover the original image. This usually involves formulating a criterion of
goodness which will yield some optimum estimate of the desired result.
Enhancement techniques, on the other hand, are basically heuristic procedures which are designed to manipulate an image in order to take advantage
of the psychophysical aspects of the human visual system. For example,
contrast stretching is considered an enhancement technique because it is
based primarily on the pleasing aspects it might present to the viewer, while
removal of image blur by applying a deblurring function is considered a
restoration problem.

Early techniques for digital image restoration were derived mostly
from frequency-domain concepts. Attention is focused in this chapter,
however, on a more modern, algebraic approach to the problem. This
approach has the advantage that it allows the derivation of numerous restoration techniques starting from the same basic principles. Although a direct
solution by algebraic methods generally involves the manipulation of large
systems of simultaneous equations, it is shown in the following sections that,
under certain conditions, it is possible to reduce computational complexity
to the same level as that required by traditional frequency-domain restoration techniques.

The material developed in the following sections is strictly introductory. We consider only the restoration problem from the point where a

Rafael C. Gonzalez and Paul Wintz, Digital Image Processing   ISBN 0-201-02596-5; 0-201-02597-3(pbk.)

degraded, *digital* image is given; thus, topics dealing with sensor, digitizer, and display degradations are not considered in this chapter. These subjects, although of importance in the overall treatment of image restoration applications, are outside the mainstream of our present discussion. The references cited at the end of the chapter provide a starting point to the voluminous literature on these and related topics.

5.1 DEGRADATION MODEL

As shown in Fig. 5.1, the degradation process will be modeled in this chapter as an operator (or system), H, which together with an additive noise term, $\eta(x, y)$, operates on an input image $f(x, y)$ to produce a degraded image $g(x, y)$. The digital image restoration problem may be viewed as that of obtaining an approximation to $f(x, y)$, given $g(x, y)$ and a knowledge of the degradation in the form of the operator H. It is assumed that our knowledge about $\eta(x, y)$ is limited to information of a statistical nature.

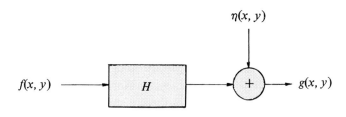

Figure 5.1. A model of the image degradation process.

5.1.1 Some Definitions

The input-output relationship in Fig. 5.1 is given by the expression

$$g(x, y) = Hf(x, y) + \eta(x, y) \tag{5.1-1}$$

For the moment, let us assume that $\eta(x, y) = 0$ so that $g(x, y) = Hf(x, y)$. We define H to be *linear* if

$$H\big[k_1 f_1(x, y) + k_2 f_2(x, y) \big] = k_1 H f_1(x, y) + k_2 H f_2(x, y) \tag{5.1-2}$$

where k_1 and k_2 are constants and $f_1(x, y)$ and $f_2(x, y)$ are any two input images.

If we let $k_1 = k_2 = 1$, Eq. (5.1-2) becomes

$$H[f_1(x, y) + f_2(x, y)] = Hf_1(x, y) + Hf_2(x, y) \qquad (5.1\text{-}3)$$

which is called the property of *additivity*; this property simply says that, if H is a linear operator, the response to a sum of two inputs is equal to the sum of the two responses.

With $f_2(x, y) = 0$, Eq. (5.1-2) becomes

$$H[k_1 f_1(x, y)] = k_1 Hf_1(x, y) \qquad (5.1\text{-}4)$$

This is called the property of *homogeneity*, which says that the response to a constant multiple of any input is equal to the response to that input multiplied by the same constant. Thus, we see that a linear operator possesses both the property of additivity and the property of homogeneity.

An operator having the input-output relationship $g(x, y) = Hf(x, y)$ is said to be *position* (or *space*) *invariant* if

$$Hf(x - \alpha, y - \beta) = g(x - \alpha, y - \beta) \qquad (5.1\text{-}5)$$

for any $f(x, y)$ and any α and β. This definition indicates that the response at any point in the image depends only on the *value* of the input at that point and *not* on the *position* of the point.

5.1.2 Degradation Model for Continuous Functions

From the definition of the impulse function given in Eq. (3.3-46), we can express $f(x, y)$ in the form

$$f(x, y) = \int\limits_{-\infty}^{\infty}\!\!\int f(\alpha, \beta)\delta(x - \alpha, y - \beta)\, d\alpha\, d\beta \qquad (5.1\text{-}6)$$

Then, if $\eta(x, y) = 0$ in Eq. (5.1-1),

$$g(x, y) = Hf(x, y) = H\int\limits_{-\infty}^{\infty}\!\!\int f(\alpha, \beta)\delta(x - \alpha, y - \beta)\, d\alpha\, d\beta \qquad (5.1\text{-}7)$$

If H is a linear operator, and we assume that the additivity property is valid for integrals, then

$$g(x, y) = \int\limits_{-\infty}^{\infty}\!\!\int H[f(\alpha, \beta)\delta(x - \alpha, y - \beta)]\, d\alpha\, d\beta \qquad (5.1\text{-}8)$$

Since $f(\alpha, \beta)$ is independent of x and y, it follows from the homogeneity property that

$$g(x, y) = \int\int_{-\infty}^{\infty} f(\alpha, \beta) H\delta (x - \alpha, y - \beta) \, d\alpha \, d\beta \qquad (5.1\text{-}9)$$

The term

$$h(x, \alpha, y, \beta) = H\delta (x - \alpha, y - \beta) \qquad (5.1\text{-}10)$$

is called the *impulse response* of H. In other words, if $\eta(x, y) = 0$ in Eq. (5.1-1), we see that $h(x, \alpha, y, \beta)$ is the response of H to an impulse at coordinates (α, β). In optics, the impulse becomes a point of light and $h(x, \alpha, y, \beta)$ is commonly referred to in this case as the *point spread function* (PSF).
Substitution of Eq. (5.1-10) into Eq. (5.1-9) yields the expression

$$g(x, y) = \int\int_{-\infty}^{\infty} f(\alpha, \beta) h(x, \alpha, y, \beta) \, d\alpha \, d\beta \qquad (5.1\text{-}11)$$

which is called the *superposition* or *Fredholm integral of the first kind*. This expression is of fundamental importance in linear system theory. It states that if the response of H to an impulse is known, then the response to any input $f(\alpha, \beta)$ can be calculated by means of Eq. (5.1-11). In other words, a linear system H is completely characterized by its impulse response.
If H is position invariant, it follows from Eq. (5.1-5) that

$$H\delta (x - \alpha, y - \beta) = h(x - \alpha, y - \beta) \qquad (5.1\text{-}12)$$

Equation (5.1-11) reduces in this case to

$$g(x, y) = \int\int_{-\infty}^{\infty} f(\alpha, \beta) h(x - \alpha, y - \beta) \, d\alpha \, d\beta \qquad (5.1\text{-}13)$$

which is recognized as the convolution integral defined in Eq. (3.3-30).
In the presence of additive noise the expression describing a linear degradation model becomes

$$g(x, y) = \int\int_{-\infty}^{\infty} f(\alpha, \beta) h(x, \alpha, y, \beta) \, d\alpha \, d\beta + \eta(x, y) \qquad (5.1\text{-}14)$$

If H is position invariant, this expression becomes

$$g(x, y) = \int\int_{-\infty}^{\infty} f(\alpha, \beta) h(x - \alpha, y - \beta) \, d\alpha \, d\beta + \eta(x, y) \qquad (5.1\text{-}15)$$

The noise, of course, is assumed in both cases to be independent of position in the image.

Many types of degradations can be approximated by linear, position-invariant processes. The advantage of this approach is that the extensive tools of linear system theory then become available for the solution of image restoration problems. Nonlinear and space-variant techniques, although more general (and usually more accurate), introduce difficulties which often have no known solution or are very difficult to solve computationally. Attention is focused in this chapter on linear, space-invariant restoration techniques. As will be seen in the following discussion, however, even this simplification can result in computational problems which, if attacked directly, are beyond the capabilities of most present-day computers.

5.1.3 Discrete Formulation

The development of a discrete, space-invariant degradation model is simplified by starting with the one-dimensional case and temporarily neglecting the noise term. Suppose that two functions $f(x)$ and $h(x)$ are sampled uniformly to form arrays of dimension A and B, respectively. In this case, we interpret x as a discrete variable in the ranges $0, 1, 2, ..., A - 1$ for $f(x)$ and $0, 1, 2, ..., B - 1$ for $h(x)$.

The discrete convolution formulation given in Section 3.3.8.1 is based on the assumption that the sampled functions are periodic with a period M. Overlap in the individual periods of the resulting convolution is avoided by choosing $M \geqslant A + B - 1$ and extending the functions with zeros so that their length is equal to M. Letting $f_e(x)$ and $h_e(x)$ represent the extended functions, it follows from Eq. (3.3-29) that their convolution is given by

$$g_e(x) = \sum_{m=0}^{M-1} f_e(m)h_e(x - m) \tag{5.1-16}$$

for $x = 0, 1, 2, ..., M-1$. Since both $f_e(x)$ and $g_e(x)$ are assumed to have a period equal to M, $g_e(x)$ also has this period.

It is easily verified by direct matrix multiplication that Eq. (5.1-16) can be expressed in the form

$$\mathbf{g} = \mathbf{Hf} \tag{5.1-17}$$

where $\mathbf{f}$ and $\mathbf{g}$ are M-dimensional column vectors given by

$$\mathbf{f} = \begin{bmatrix} f_e(0) \\ f_e(1) \\ \vdots \\ f_e(M-1) \end{bmatrix} \tag{5.1-18}$$

and

$$\mathbf{g} = \begin{bmatrix} g_e(0) \\ g_e(1) \\ \vdots \\ g_e(M-1) \end{bmatrix} \qquad (5.1\text{-}19)$$

H is the $M \times M$ matrix

$$\mathbf{H} = \begin{bmatrix} h_e(0) & h_e(-1) & h_e(-2) & \cdots & h_e(-M+1) \\ h_e(1) & h_e(0) & h_e(-1) & \cdots & h_e(-M+2) \\ h_e(2) & h_e(1) & h_e(0) & \cdots & h_e(-M+3) \\ \vdots & & & & \\ h_e(M-1) & h_e(M-2) & h_e(M-3) & \cdots & h_e(0) \end{bmatrix} \qquad (5.1\text{-}20)$$

Because of the periodicity assumption on $h_e(x)$, we have that $h_e(x) = h_e(M + x)$. Using this property, Eq. (5.1-20) may be written in the form

$$\mathbf{H} = \begin{bmatrix} h_e(0) & h_e(M-1) & h_e(M-2) & \cdots & h_e(1) \\ h_e(1) & h_e(0) & h_e(M-1) & \cdots & h_e(2) \\ h_e(2) & h_e(1) & h_e(0) & \cdots & h_e(3) \\ \vdots & & & & \\ h_e(M-1) & h_e(M-2) & h_e(M-3) & \cdots & h_e(0) \end{bmatrix} \qquad (5.1\text{-}21)$$

The structure of this matrix plays a fundamental role throughout the remainder of this chapter. From Eq. (5.1-21) we see that the rows of **H** are related by a *circular shift* to the right; that is, the right-most element in one row is equal to the left-most element in the row immediately below. The shift is called circular because an element shifted out of the right end of a row reappears at the left end. It is also noted in Eq. (5.1-21) that the circularity of **H** is complete in the sense that it extends from the last row back to the first row. A square matrix in which each row is a circular shift of the preceding row, and the first row in a circular shift of the last row, is called a *circulant matrix*. It is important to keep in mind that the circular behavior of **H** is a direct consequence of the assumed periodicity of $h(x)$.

Example: Suppose that $A = 4$ and $B = 3$. We may choose $M = 6$ and then append two zeros to the samples of $f(x)$ and three zeros to samples of $h(x)$. In this case **f** and **g** are 6-dimensional vectors and **H** is the 6×6 matrix

$$
\mathbf{H} = \begin{bmatrix}
h_e(0) & h_e(5) & h_e(4) & \cdots & h_e(1) \\
h_e(1) & h_e(0) & h_e(5) & \cdots & h_e(2) \\
h_e(2) & h_e(1) & h_e(0) & \cdots & h_e(3) \\
\vdots & & & & \\
h_e(5) & h_e(4) & h_e(3) & \cdots & h_e(0)
\end{bmatrix}
$$

However, since $h_e(x) = 0$ for $x = 3, 4, 5$, and $h_e(x) = h(x)$ for $x = 0, 1, 2$, we have

$$
\mathbf{H} = \begin{bmatrix}
h(0) & & & & h(2) & h(1) \\
h(1) & h(0) & & & & h(2) \\
h(2) & h(1) & h(0) & & & \\
& h(2) & h(1) & h(0) & & \\
& & h(2) & h(1) & h(0) & \\
& & & h(2) & h(1) & h(0)
\end{bmatrix}
$$

where all elements not indicated in the matrix are zero. □

Extension of the foregoing discussion to a two-dimensional, discrete degradation model is straightforward. Given two digitized images $f(x, y)$ and $h(x, y)$ of sizes $A \times B$ and $C \times D$, respectively, we form extended images of size $M \times N$ by padding the above functions with zeros. As indicated in Section 3.3.8.1, one procedure for doing this is to let

$$
f_e(x, y) = \begin{cases}
f(x, y) & 0 \leqslant x \leqslant A - 1 \quad \text{and} \quad 0 \leqslant y \leqslant B - 1 \\
0 & A \leqslant x \leqslant M - 1 \quad \text{or} \quad B \leqslant y \leqslant N - 1
\end{cases}
$$

and

$$
h_e(x, y) = \begin{cases}
h(x, y) & 0 \leqslant x \leqslant C - 1 \quad \text{and} \quad 0 \leqslant y \leqslant D - 1 \\
0 & C \leqslant x \leqslant M - 1 \quad \text{or} \quad D \leqslant y \leqslant N - 1
\end{cases}
$$

If we treat the extended functions $f_e(x, y)$ and $h_e(x, y)$ as being periodic in two dimensions, with periods M and N in the x and y directions, respectively, it then follows from Eq. (3.3-35) that the convolution of these two functions is given by the relation

$$
g_e(x, y) = \sum_{m=0}^{M-1} \sum_{n=0}^{N-1} f_e(m, n) h_e(x - m, y - n) \tag{5.1-22}
$$

for $x = 0, 1, 2, ..., M-1$ and $y = 0, 1, 2, ..., N-1$. The convolution function

$g_e(x, y)$ is periodic with the same period of $f_e(x, y)$ and $h_e(x, y)$. Overlap of the individual convolution periods is avoided by choosing $M \geqslant A + C - 1$ and $N \geqslant B + D - 1$. In order to complete the discrete degradation model we add an $M \times N$ extended discrete noise term $\eta_e(x, y)$ to Eq. (5.l-22) so that

$$g_e(x, y) = \sum_{m=0}^{M-1} \sum_{n=0}^{N-1} f_e(m, n) h_e(x - m, y - n) + \eta_e(x, y) \qquad (5.1\text{-}23)$$

for $x = 0, 1, 2, ..., M-1$ and $y = 0, 1, 2, ..., N-1$.

Let $\mathbf{f}$, $\mathbf{g}$, and $\mathbf{n}$ represent MN-dimensional column vectors formed by stacking the rows of the $M \times N$ functions $f_e(x, y)$, $g_e(x, y)$, and $\eta_e(x, y)$. The first N elements of $\mathbf{f}$, for example, are the elements in the first row of $f_e(x, y)$, the next N elements are from the second row, and so forth for all M rows of $f_e(x, y)$. Using this convention, Eq. (5.1-23) can be expressed in the following vector-matrix form:

$$\mathbf{g} = \mathbf{Hf} + \mathbf{n} \qquad (5.1\text{-}24)$$

where $\mathbf{f}$, $\mathbf{g}$, and $\mathbf{n}$ are of dimension $(MN) \times 1$ and $\mathbf{H}$ is of dimension $MN \times MN$. This matrix consists of M^2 partitions, each partition being of size $N \times N$ and ordered according to

$$\mathbf{H} = \begin{bmatrix} \mathbf{H}_0 & \mathbf{H}_{M-1} & \mathbf{H}_{M-2} & \cdots & \mathbf{H}_1 \\ \mathbf{H}_1 & \mathbf{H}_0 & \mathbf{H}_{M-1} & \cdots & \mathbf{H}_2 \\ \mathbf{H}_2 & \mathbf{H}_1 & \mathbf{H}_0 & \cdots & \mathbf{H}_3 \\ \vdots & & & & \\ \mathbf{H}_{M-1} & \mathbf{H}_{M-2} & \mathbf{H}_{M-3} & \cdots & \mathbf{H}_0 \end{bmatrix} \qquad (5.1\text{-}25)$$

Each partition $\mathbf{H}_j$ is constructed from the jth row of the extended function $h_e(x, y)$, as follows:

$$\mathbf{H}_j = \begin{bmatrix} h_e(j, 0) & h_e(j, N-1) & h_e(j, N-2) & \cdots & h_e(j, 1) \\ h_e(j, 1) & h_e(j, 0) & h_e(j, N-1) & \cdots & h_e(j, 2) \\ h_e(j, 2) & h_e(j, 1) & h_e(j, 0) & \cdots & h_e(j, 3) \\ \vdots & & & & \\ h_e(j, N-1) & h_e(j, N-2) & h_e(j, N-3) & \cdots & h_e(j, 0) \end{bmatrix} \qquad (5.1\text{-}26)$$

where, as in Eq. (5.1-21), use was made of the periodicity of $h_e(x, y)$. It is noted that $\mathbf{H}_j$ is a circulant matrix and that the blocks of $\mathbf{H}$ are also subscripted in a circular manner. For these reasons, the matrix $\mathbf{H}$ given in Eq. (5.1-25) is often called a *block-circulant* matrix.

Most of the discussion in the following sections is centered around the discrete degradation model given in Eq. (5.1-24). It is important to keep in mind that this expression was derived under the assumption of a linear, space-invariant degradation process. As indicated earlier, the objective is to estimate the ideal image $f(x, y)$, given $g(x, y)$, and a knowledge of $h(x, y)$ and $\eta(x, y)$. In terms of Eq. (5.1-24), this means that we are interested in estimating $\mathbf{f}$, given $\mathbf{g}$ and some knowledge about $\mathbf{H}$ and $\mathbf{n}$.

Although Eq. (5.1-24) seems deceptively simple, a direct solution of this expression to obtain the elements of $\mathbf{f}$ is a monumental processing task for images of practical size. If, for example, $M = N = 512$, we have that $\mathbf{H}$ is of size $262{,}144 \times 262{,}144$. Thus, to obtain $\mathbf{f}$ directly would require the solution of a system of $262{,}144$ simultaneous linear equations. Fortunately, the complexity of this problem can be reduced considerably by taking advantage of the circulant properties of $\mathbf{H}$, as shown in the following section.

5.2 DIAGONALIZATION OF CIRCULANT AND BLOCK-CIRCULANT MATRICES.

It is shown in this section that solutions which are computationally feasible may be obtained from the model given in Eq. (5.1-24) by diagonalizing the $\mathbf{H}$ matrix. In order to simplify the explanation we begin the discussion by considering circulant matrices; the procedure is then extended to block-circulants.

5.2.1 Circulant Matrices

Consider an $M \times M$ circulant matrix $\mathbf{H}$ of the form

$$\mathbf{H} = \begin{bmatrix} h_e(0) & h_e(M-1) & h_e(M-2) & \cdots & h_e(1) \\ h_e(1) & h_e(0) & h_e(M-1) & \cdots & h_e(2) \\ h_e(2) & h_e(1) & h_e(0) & \cdots & h_e(3) \\ \vdots & & & & \\ h_e(M-1) & h_e(M-2) & h_e(M-3) & \cdots & h_e(0) \end{bmatrix} \tag{5.2-1}$$

Let us define a scalar function $\lambda(k)$ and a vector $\mathbf{w}(k)$ as follows:

$$\lambda(k) = h_e(0) + h_e(M-1) \exp\left[j \frac{2\pi}{M} k \right] + h_e(M-2) \exp\left[j \frac{2\pi}{M} 2k \right]$$
$$+ \cdots + h_e(1) \exp\left[j \frac{2\pi}{M} (M-1)k \right] \tag{5.2-2}$$

where $j = \sqrt{-1}$, and

$$\mathbf{w}(k) = \begin{bmatrix} 1 \\ \exp\left[j \frac{2\pi}{M} k \right] \\ \exp\left[j \frac{2\pi}{M} 2k \right] \\ \vdots \\ \exp\left[j \frac{2\pi}{M} (M-1)k \right] \end{bmatrix} \tag{5.2-3}$$

for $k = 0, 1, 2, ..., M-1$. It can be shown by matrix multiplication that

$$\mathbf{H}\mathbf{w}(k) = \lambda(k)\mathbf{w}(k) \tag{5.2-4}$$

This expression indicates that $\mathbf{w}(k)$ is an eigenvector of the circulant matrix $\mathbf{H}$ and $\lambda(k)$ is its corresponding eigenvalue.

Suppose that we form an $M \times M$ matrix $\mathbf{W}$ by using the M eigenvectors of $\mathbf{H}$ as columns; that is,

$$\mathbf{W} = \begin{bmatrix} \mathbf{w}(0) & \mathbf{w}(1) & \mathbf{w}(2) & \cdots & \mathbf{w}(M-1) \end{bmatrix} \tag{5.2-5}$$

The kith element of $\mathbf{W}$, denoted as $W(k, i)$, is given by

$$W(k, i) = \exp\left[j\frac{2\pi}{M} ki \right] \tag{5.2-6}$$

for $k, i = 0, 1, 2, ..., M-1$. Due to the orthogonality properties of the complex exponential, the inverse matrix, $\mathbf{W}^{-1}$, can be written by inspection; its kith element, symbolized as $W^{-1}(k, i)$, is given by

$$W^{-1}(k, i) = \frac{1}{M} \exp\left[-j\frac{2\pi}{M} ki \right] \tag{5.2-7}$$

It can be verified by using Eqs. (5.2-6) and (5.2-7) that

$$\mathbf{W}\mathbf{W}^{-1} = \mathbf{W}^{-1}\mathbf{W} = \mathbf{I} \tag{5.2-8}$$

where $\mathbf{I}$ is the $M \times M$ identity matrix.

The importance of the existence of the inverse matrix $\mathbf{W}^{-1}$ is that it guarantees that the columns of $\mathbf{W}$ (i. e., the eigenvectors of $\mathbf{H}$) are *linearly independent*. It then follows from elementary matrix theory (Noble [1969]) that $\mathbf{H}$ may be expressed in the form

$$\mathbf{H} = \mathbf{W}\mathbf{D}\mathbf{W}^{-1} \tag{5.2-9}$$

or, using Eq. (5.2-8),

$$\mathbf{D} = \mathbf{W}^{-1}\mathbf{H}\mathbf{W} \tag{5.2-10}$$

where $\mathbf{D}$ is a diagonal matrix whose elements $D(k, k)$ are the eigenvalues of $\mathbf{H}$, that is,

$$D(k, k) = \lambda(k) \tag{5.2-11}$$

Equation (5.2-10) indicates that $\mathbf{H}$ is diagonalized by using $\mathbf{W}^{-1}$ and $\mathbf{W}$ in the order indicated.

5.2.2 Block-Circulant Matrices

The transformation matrix for diagonalizing block circulants is constructed as follows. Let

$$w_M(i, m) = \exp\left[j \frac{2\pi}{M} im \right] \tag{5.2-12}$$

and

$$w_N(k, n) = \exp\left[j \frac{2\pi}{N} kn \right] \tag{5.2-13}$$

Based on this notation, we define a matrix $\mathbf{W}$ that is of size $MN \times MN$, and contains M^2 partitions of size $N \times N$. The imth partition of $\mathbf{W}$ is defined as

$$\mathbf{W}(i, m) = w_M(i, m)\mathbf{W}_N \tag{5.2-14}$$

for $i, m = 0, 1, 2, ..., M-1$. $\mathbf{W}_N$ is an $N \times N$ matrix with elements

$$\mathbf{W}_N(k, n) = w_N(k, n) \tag{5.2-15}$$

for $k, n = 0, 1, 2, ..., N-1$.

The inverse matrix $\mathbf{W}^{-1}$ is also of size $MN \times MN$ with M^2 partitions of size $N \times N$. The imth partition of $\mathbf{W}^{-1}$, symbolized as $\mathbf{W}^{-1}(i, m)$, is defined as

$$\mathbf{W}^{-1}(i, m) = \frac{1}{M} w_M^{-1}(i, m)\mathbf{W}_N^{-1} \tag{5.2-16}$$

where $w_M^{-1}(i, m)$ is given by

$$w_M^{-1}(i, m) = \exp\left[-j \frac{2\pi}{M} im \right] \tag{5.2-17}$$

for $i, m = 0, 1, 2, ..., M-1$. The matrix $\mathbf{W}_N^{-1}$ has elements

$$\mathbf{W}_N^{-1}(k, n) = \frac{1}{N} w_N^{-1}(k, n) \tag{5.2-18}$$

where

$$w_N^{-1}(k, n) = \exp\left[-j \frac{2\pi}{N} kn \right] \tag{5.2-19}$$

for $k, n = 0, 1, 2, ..., N-1$. It can be verified by direct substitution of the above elements of $\mathbf{W}$ and $\mathbf{W}^{-1}$ that

$$\mathbf{W}\mathbf{W}^{-1} = \mathbf{W}^{-1}\mathbf{W} = \mathbf{I} \tag{5.2-20}$$

where $\mathbf{I}$ is the $MN \times MN$ identity matrix.

By making use of the results in the previous section it can be shown (Hunt [1973]) that if $\mathbf{H}$ is a block circulant matrix, it can be written as

$$\mathbf{H} = \mathbf{W}\mathbf{D}\mathbf{W}^{-1} \tag{5.2-21}$$

or

$$\mathbf{D} = \mathbf{W}^{-1}\mathbf{H}\mathbf{W} \tag{5.2-22}$$

where $\mathbf{D}$ is a diagonal matrix whose elements $D(k, k)$ are related to the discrete Fourier transform of the extended function $h_e(x, y)$ given in Section 5.1.3. It can also be shown that the transpose of $\mathbf{H}$, denoted by $\mathbf{H}'$, is given by

$$\mathbf{H}' = \mathbf{WD^*W^{-1}} \qquad (5.2\text{-}23)$$

where $\mathbf{D}^*$ is the complex conjugate of $\mathbf{D}$.

5.2.3 Effect of Diagonalization on the Degradation Model

Since the matrix $\mathbf{H}$ in the discrete, one-dimensional model of Eq. (5.1-17) is circulant, it may be expressed in the form of Eq. (5.2-9). Equation (5.1-17) then becomes

$$\mathbf{g} = \mathbf{WDW^{-1}f} \qquad (5.2\text{-}24)$$

Rearranging this equation we have

$$\mathbf{W^{-1}g} = \mathbf{DW^{-1}f} \qquad (5.2\text{-}25)$$

The product $\mathbf{W^{-1}f}$ is an M-dimensional column vector. From Eq. (5.2-16) and the definition of $\mathbf{f}$ given in Section 5.1.3, we have that the kth element of the product $\mathbf{W^{-1}f}$, which we denote by $F(k)$, is given by

$$F(k) = \frac{1}{M} \sum_{i=0}^{M-1} f_e(i) \exp\left[-j\frac{2\pi}{M} ki \right] \qquad (5.2\text{-}26)$$

for $k = 0, 1, 2, ..., M-1$. This expression is recognized as the discrete Fourier transform of the extended sequence $f_e(x)$. In other words, multiplication of $\mathbf{f}$ by $\mathbf{W^{-1}}$ yields a vector whose elements are the Fourier transform of the elements of $\mathbf{f}$. Similarly, $\mathbf{W^{-1}g}$ yields the Fourier transform of the elements of $\mathbf{g}$, denoted by $G(k)$, $k = 0, 1, 2, ..., M-1$.

Next, we examine the matrix $\mathbf{D}$ in Eq. (5.2-25). From the discussion in Section 5.2.1 we know that the main-diagonal elements of $\mathbf{D}$ are the eigenvalues of the circulant matrix $\mathbf{H}$. The eigenvalues are given in Eq. (5.2-2) which, using the fact that

$$\exp\left[j\frac{2\pi}{M}(M-i)k \right] = \exp\left[-j\frac{2\pi}{M} ik \right] \qquad (5.2\text{-}27)$$

may be written in the form

$$\lambda(k) = h_e(0) + h_e(1) \exp\left[-j\frac{2\pi}{M} k \right] + h_e(2) \exp\left[-j\frac{2\pi}{M} 2k \right]$$

$$+ \cdots + h_e(M-1) \exp\left[-j\frac{2\pi}{M}(M-1) \right] \qquad (5.2\text{-}28)$$

From Eqs. (5.2-11) and (5.2-28) we have

$$D(k, k) = \lambda(k) = \sum_{i=0}^{M-1} h_e(i) \exp\left(-j\frac{2\pi}{M} ki \right) \qquad (5.2\text{-}29)$$

for $k = 0, 1, 2, ..., M-1$. The right side of this equation is recognized as $MH(k)$, where $H(k)$ is the discrete Fourier transform of the extended sequence $h_e(x)$. Thus,

$$D(k, k) = MH(k). \qquad (5.2\text{-}30)$$

We can combine these transforms into one result. Since $\mathbf{D}$ is a diagonal matrix, the product of $\mathbf{D}$ with any vector multiplies each element of that vector by a single diagonal element of $\mathbf{D}$. Consequently, the matrix formulation given in Eq. (5.2-25) can be reduced to a term-by-term product of one-dimensional Fourier transform sequences. In other words,

$$G(k) = MH(k)F(k) \qquad (5.2\text{-}31)$$

for $k = 0, 1, 2, ..., M-1$, where $G(k)$ are the elements of the vector $\mathbf{W}^{-1}\mathbf{g}$ and $MH(k)F(k)$ the elements of vector $\mathbf{DW}^{-1}\mathbf{f}$. The right side of Eq. (5.2-31) is recognized as the convolution of $f_e(x)$ and $h_e(x)$ in the frequency domain (see Section 3.3.8.1). From a computational point of view, this result implies considerable simplification because $G(k)$, $H(k)$, and $F(k)$ are M-sample discrete transforms which can be obtained by using a fast Fourier transform algorithm.

A procedure similar to the above development yields equivalent results for the two-dimensional degradation model. Multiplying both sides of Eq. (5.1-24) by $\mathbf{W}^{-1}$, and using Eqs. (5.2-20) and (5.2-21), yields

$$\mathbf{W}^{-1}\mathbf{g} = \mathbf{DW}^{-1}\mathbf{f} + \mathbf{W}^{-1}\mathbf{n} \qquad (5.2\text{-}32)$$

where $\mathbf{W}^{-1}$ is an $MN \times MN$ matrix whose elements are given in Eq. (5.2-16), $\mathbf{D}$ is an $MN \times MN$ diagonal matrix, $\mathbf{H}$ is the $MN \times MN$ block-circulant matrix defined in Eq. (5.1-25), and $\mathbf{f}$ and $\mathbf{g}$ are vectors of dimension MN formed by stacking the rows of the extended images $f_e(x, y)$ and $g_e(x, y)$.

The left side of Eq. (5.2-32) is a vector of dimension $MN \times 1$. Let us denote its elements by $G(0, 0)$, $G(0, 1)$, ..., $G(0, N-1)$; $G(1, 0)$, $G(1, 1)$, ..., $G(1, N-1)$; ...; $G(M-1, 0)$, $G(M-1, 1)$, ..., $G(M-1, N-1)$. It can be shown (Hunt [1973]) that

$$G(u, v) = \frac{1}{MN} \sum_{x=0}^{M-1} \sum_{y=0}^{N-1} g_e(x, y) \exp\left[-j2\pi(ux/M + vy/N)\right]$$

$$(5.2\text{-}33)$$

for $u = 0, 1, 2, ..., M-1$, and $v = 0, 1, 2, ..., N-1$. This expression is recognized as the two-dimensional Fourier transform of $g_e(x, y)$. In other words, the elements of $\mathbf{W}^{-1}\mathbf{g}$ correspond to the stacked rows of the Fourier transform matrix with elements $G(u, v)$, for $u = 0, 1, 2, ..., M-1$, and $v = 0, 1, 2, ..., N-1$. Similarly, we have that the vectors $\mathbf{W}^{-1}\mathbf{f}$ and $\mathbf{W}^{-1}\mathbf{n}$ are MN-dimensional and contain elements $F(u, v)$ and $N(u, v)$, where

$$F(u, v) = \frac{1}{MN} \sum_{x=0}^{M-1} \sum_{y=0}^{N-1} f_e(x, y) \exp\left[-j2\pi(ux/M + vy/N)\right] \quad (5.2\text{-}34)$$

and

$$N(u, v) = \frac{1}{MN} \sum_{x=0}^{M-1} \sum_{y=0}^{N-1} \eta_e(x, y) \exp\left[-j2\pi(ux/M + vy/N)\right] \quad (5.2\text{-}35)$$

for u = 0, 1, 2, ..., $M-1$ and v = 0, 1, 2, ..., $N-1$.

Finally, we have that the elements of the diagonal matrix **D** are related to the Fourier transform of the extended impulse response function $h_e(x, y)$; that is,

$$H(u, v) = \frac{1}{MN} \sum_{x=0}^{M-1} \sum_{y=0}^{N-1} h_e(x, y) \exp\left[-j2\pi(ux/M + vy/N)\right] \quad (5.2\text{-}36)$$

for u = 0, 1, 2, ..., $M-1$, and v = 0, 1, 2, ..., $N-1$. The MN diagonal elements of **D** are formed as follows. The first N elements are $H(0, 0)$, $H(0, 1)$, ..., $H(0, N-1)$; the next $H(1, 0)$, $H(1, 1)$, ..., $H(1, N-1)$; and so forth, with the last N diagonal elements being $H(M-1, 0)$, $H(M-1, 1)$, ..., $H(M-1, N-1)$. (The off-diagonal elements are, of course, zero). The entire matrix formed from the above elements is then multiplied by MN to obtain **D**. A more concise way of expressing this construction is the following:

$$D(k, i) = \begin{cases} MNH\left(\left[\dfrac{k}{n}\right], k \bmod N\right) & \text{if } i = k \\ \\ 0 & \text{if } i \neq k \end{cases} \quad (5.2\text{-}37)$$

where $[p]$ is used to denote the greatest integer not exceeding p, and $k \bmod N$ is the remainder obtained by dividing k by N.

By using Eqs. (5.2-33) through (5.2-36) it is not difficult to show that the individual elements of Eq. (5.2-32) are related by the expression

$$G(u, v) = MNH(u, v)F(u, v) + N(u, v) \quad (5.2\text{-}38)$$

for u = 0, 1, 2, ..., $M-1$, and v = 0, 1, 2, ..., $N-1$.

Since the term MN is simply a scale factor, it will be convenient for notational purposes to absorb it in $H(u, v)$. With this assumption, Eqs. (5.2-37) and (5.2-38) may be expressed as

$$D(k, i) = \begin{cases} H\left(\left[\dfrac{k}{N}\right], k \bmod N\right) & \text{if } i = k \\ \\ 0 & \text{if } i \neq k \end{cases} \quad (5.2\text{-}39)$$

for $k, i = 0, 1, 2, ..., MN-1$, and

$$G(u, v) = H(u, v)F(u, v) + N(u, v) \qquad (5.2\text{-}40)$$

for $u = 0, 1, 2, ..., M-1$, and $v = 0, 1, 2, ..., N-1$, where it is understood that $H(u, v)$ is now scaled by the factor MN.

The significance of Eq. (5.2-38) or (5.2-40) is that the large system of equations implicit in the model given in Eq. (5.1-24) can be reduced to computation of a few discrete Fourier transforms of size $M \times N$. For $M = N = 512$, for example, this is a simple problem if we use an FFT algorithm. As mentioned earlier, however, the problem becomes almost an infeasible computation task if approached directly from the model given in Eq. (5.1-24).

The model given in Eq. (5.1-24) will be used in the following sections as the basis for deriving several image restoration approaches. The results, given in matrix form, will then be simplified by using the concepts derived in this section. The reader should keep in mind that the simplifications achieved above are the result of assuming that: (1) the degradation is a linear, space-invariant process, and (2) all images are treated as extended, periodic functions.

5.3 ALGEBRAIC APPROACH TO RESTORATION

As indicated in Section 5.1.3, the objective of image restoration is to estimate an original image $\mathbf{f}$, given a degraded image $\mathbf{g}$ and some knowledge or assumption about $\mathbf{H}$ and $\mathbf{n}$. By assuming that these quantities are related according to the model given in Eq. (5.1-24), it is possible to formulate a class of image restoration problems in a unified linear algebraic framework.

Central to the algebraic approach is the concept of seeking an estimate of $\mathbf{f}$, denoted by $\hat{\mathbf{f}}$, which minimizes a predefined criterion of goodness. Because of their simplicity, we focus attention in this chapter on least-squares criterion functions. As will be seen in the following sections, this choice has the added advantage of yielding a central approach for the derivation of several well-known restoration methods. These methods are the result of considering either an unconstrained or a constrained approach to the least-squares restoration problem.

5.3.1 Unconstrained Restoration

From Eq. (5.1-24), the noise term in the degradation model is given by

$$\mathbf{n} = \mathbf{g} - \mathbf{Hf} \qquad (5.3\text{-}1)$$

In the absence of any knowledge about $\mathbf{n}$, a meaningful criterion function is to seek an $\hat{\mathbf{f}}$ such that $\mathbf{H}\hat{\mathbf{f}}$ approximates $\mathbf{g}$ in a least-squares sense by assuming that the norm of the noise term is as small as possible. In other words, we wish to find an $\hat{\mathbf{f}}$ such that

$$\|\mathbf{n}\|^2 = \|\mathbf{g} - \mathbf{H}\hat{\mathbf{f}}\|^2 \tag{5.3-2}$$

is minimum, where, by definition, $\|\mathbf{n}\|^2 = \mathbf{n}'\mathbf{n}$, and $\|\mathbf{g} - \mathbf{H}\hat{\mathbf{f}}\|^2 = (\mathbf{g} - \mathbf{H}\hat{\mathbf{f}})'$ $(\mathbf{g} - \mathbf{H}\hat{\mathbf{f}})$ are the squared norms of $\mathbf{n}$ and $(\mathbf{g} - \mathbf{H}\hat{\mathbf{f}})$, respectively. From Eq. (5.3-2), we may equivalently view this problem as one of minimizing the criterion function

$$J(\hat{\mathbf{f}}) = \|\mathbf{g} - \mathbf{H}\hat{\mathbf{f}}\|^2 \tag{5.3-3}$$

with respect to $\hat{\mathbf{f}}$. Aside from the requirement that it minimize Eq. (5.3-3), $\hat{\mathbf{f}}$ is not constrained in any other way.

Minimization of Eq. (5.3-3) is straightforward. We simply differentiate J with respect to $\hat{\mathbf{f}}$, and set the result equal to zero, that is,

$$\frac{\partial J(\hat{\mathbf{f}})}{\partial \hat{\mathbf{f}}} = 0 = -2\mathbf{H}'(\mathbf{g} - \mathbf{H}\hat{\mathbf{f}}) \tag{5.3-4}$$

Solving Eq. (5.3-4) for $\hat{\mathbf{f}}$ yields

$$\hat{\mathbf{f}} = (\mathbf{H}'\mathbf{H})^{-1}\mathbf{H}'\mathbf{g} \tag{5.3-5}$$

By letting $M = N$ so that $\mathbf{H}$ is a square matrix, and assuming that $\mathbf{H}^{-1}$ exists, Eq. (5.3-5) reduces to

$$\hat{\mathbf{f}} = \mathbf{H}^{-1}(\mathbf{H}')^{-1}\mathbf{H}'\mathbf{g}$$

$$= \mathbf{H}^{-1}\mathbf{g} \tag{5.3-6}$$

5.3.2 Constrained Restoration

Let $\mathbf{Q}$ be a linear operator on $\mathbf{f}$. In this section we consider the least-squares restoration problem as one of minimizing functions of the form $\|\mathbf{Q}\hat{\mathbf{f}}\|^2$, subject to the constraint $\|\mathbf{g} - \mathbf{H}\hat{\mathbf{f}}\|^2 = \|\mathbf{n}\|^2$. This approach introduces considerable flexibility in the restoration process because it yields different solutions for different choices of $\mathbf{Q}$. It is noted that the constraint imposed on a solution is consistent with the model of Eq. (5.1-24).

The addition of an equality constraint in the minimization problem can be handled without difficulty by using the method of Lagrange multipliers (Elsgolc [1961]). The procedure is to express the constraint in the form $\alpha(\|\mathbf{g} - \mathbf{H}\hat{\mathbf{f}}\|^2 - \|\mathbf{n}\|^2)$ and then append it to the function $\|\mathbf{Q}\hat{\mathbf{f}}\|^2$. In other words, we seek an $\hat{\mathbf{f}}$ which minimizes the criterion function

$$J(\hat{\mathbf{f}}) = \|\mathbf{Q}\hat{\mathbf{f}}\|^2 + \alpha(\|\mathbf{g} - \mathbf{H}\hat{\mathbf{f}}\|^2 - \|\mathbf{n}\|^2) \tag{5.3-7}$$

where α is a constant called the *Lagrange multiplier*. Once the constraint has been appended, minimization is carried out in the usual way.

Differentiating Eq. (5.3-7) with respect to $\hat{\mathbf{f}}$ and setting the result equal to zero yields

$$\frac{\partial J(\hat{\mathbf{f}})}{\partial \hat{\mathbf{f}}} = 0 = 2\mathbf{Q}'\mathbf{Q}\hat{\mathbf{f}} - 2\alpha\mathbf{H}'(\mathbf{g} - \mathbf{H}\hat{\mathbf{f}}) \qquad (5.3\text{-}8)$$

The solution is obtained by solving Eq. (5.3-8) for $\hat{\mathbf{f}}$; that is,

$$\hat{\mathbf{f}} = (\mathbf{H}'\mathbf{H} + \gamma\mathbf{Q}'\mathbf{Q})^{-1}\mathbf{H}'\mathbf{g} \qquad (5.3\text{-}9)$$

where $\gamma = 1/\alpha$. This quantity must be adjusted so that the constraint is satisfied, a problem which is considered later in this chapter. Equations (5.3-6) and (5.3-9) are the basis for all the restoration procedures discussed in the following sections. It is shown in Section 5.4, for example, that Eq. (5.3-6) leads to the traditional inverse-filter restoration method. Similarly, the general formulation given in Eq. (5.3-9) can be used to derive results such as the classical Wiener filter, as well as other restoration techniques, simply by selecting an appropriate transformation matrix $\mathbf{Q}$ and using the simplifications derived in Section 5.2.

5.4 INVERSE FILTERING

5.4.1 Formulation

We begin the derivation of image restoration techniques by considering the unconstrained result given in Eq. (5.3-6) which, assuming that $M = N$ and using Eq. (5.2-21), may be expressed in the form

$$\hat{\mathbf{f}} = \mathbf{H}^{-1}\mathbf{g}$$

$$= (\mathbf{W}\mathbf{D}\mathbf{W}^{-1})^{-1}\mathbf{g}$$

$$= \mathbf{W}\mathbf{D}^{-1}\mathbf{W}^{-1}\mathbf{g} \qquad (5.4\text{-}1)$$

Premultiplying both sides of Eq. (5.4-1) by $\mathbf{W}^{-1}$ yields

$$\mathbf{W}^{-1}\hat{\mathbf{f}} = \mathbf{D}^{-1}\mathbf{W}^{-1}\mathbf{g} \qquad (5.4\text{-}2)$$

It then follows from the discussion in Section 5.2.3 that the elements composing Eq. (5.4-2) may be written in the form

$$\hat{F}(u, v) = \frac{G(u, v)}{H(u, v)} \qquad (5.4\text{-}3)$$

for $u, v = 0, 1, 2, ..., N-1$. According to Eq. (5.2-39), $H(u, v)$ is assumed to be scaled by N^2 and use has been made of the fact that $\mathbf{D}$ is a diagonal matrix, thus allowing a straightforward procedure for obtaining $\mathbf{D}^{-1}$.

The image restoration approach given by Eq. (5.4-3) is commonly referred to as the *inverse filter* method. This terminology arises from considering $H(u, v)$ as a "filter" function which multiplies $F(u, v)$ to produce the transform of the degraded image $g_e(x, y)$. The division of $G(u, v)$ by $H(u, v)$ indicated in Eq. (5.4-3) then constitutes an inverse filtering operation in this context. The restored image is, of course, obtained by using the relation

$$\hat{f}(x, y) = \mathscr{F}^{-1}\left[\hat{F}(u, v)\right]$$

$$= \mathscr{F}^{-1}\left[G(u, v)/H(u, v)\right] \tag{5.4-4}$$

for $x, y = 0, 1, 2, ..., N-1$. This procedure is normally implemented by means of an FFT algorithm.

Equation (5.4-4) points out that computational difficulties will be encountered in the restoration process if $H(u, v)$ vanishes or becomes very small in any region of interest in the uv-plane. If the zeros of $H(u, v)$ are located at a few known points in the uv-plane, they can generally be neglected when computing $\hat{F}(u, v)$ without noticeably affecting the restored result.

A more serious difficulty arises in the presence of noise. Substitution of Eq. (5.2-40) into Eq. (5.4-3) yields

$$\hat{F}(u, v) = F(u, v) + \frac{N(u, v)}{H(u, v)} \tag{5.4-5}$$

This expression clearly indicates that if $H(u, v)$ is zero or becomes very small, the term $N(u, v)/H(u, v)$ could dominate the restoration result $\mathscr{F}^{-1}[\hat{F}(u, v)]$. In practice one often finds that $H(u, v)$ drops off rapidly as a function of distance from the origin of the uv-plane. The noise term, on the other hand, usually falls off at a much slower rate. In situations like this, reasonable results can often be obtained by carrying out the restoration in a limited neighborhood about the origin in order to avoid small values of $H(u, v)$. An example of this approach is given below.

Example: Figure 5.2(a) shows a point image $f(x, y)$ and Fig. 5.2(b) is a degraded image $g(x, y)$ obtained by blurring $f(x, y)$. If we consider the point source to be an approximation to a unit impulse function, then it follows that

$$G(u, v) = H(u, v)F(u, v)$$

$$\approx H(u, v)$$

since $\mathscr{F}[\delta(x - x_0, y - y_0)] = 1$. This expression indicates that the transfer function $H(u, v)$ can be approximated by the Fourier transform of the degraded image. The procedure of blurring a known function to obtain an

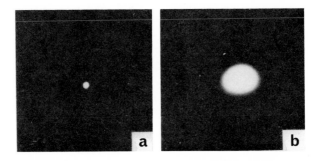

Figure 5.2. Blurring of a point source to obtain $H(u, v)$.

approximation to $H(u, v)$ is a useful one in practice because it can often be used in a trial-and-error mode to restore images for which the blurring function $H(u, v)$ is not known a priori.

The result of applying the same blurring function as above to the ideal image shown in Fig. 5.3(a) is shown in Fig. 5.3(b). The restored image shown in Fig. 5.3(c) was obtained by using Eq. (5.4-4) for values of u and v near enough to the origin of the uv-plane to avoid excessively small values of $H(u, v)$. The result of carrying out the restoration for a larger neighborhood is shown in Fig. 5.3(d). These results clearly point out the difficulties introduced by a vanishing function $H(u, v)$. ☐

Before leaving this section, it is of interest to note that if $H(u, v)$, $G(u, v)$, and $N(u, v)$ are all known, an exact inverse filtering expression can be obtained directly from Eq. (5.2-40), that is,

$$F(u, v) = \frac{G(u, v)}{H(u, v)} - \frac{N(u, v)}{H(u, v)} \tag{5.4-6}$$

The problem with this formulation, of course, is that the noise is seldom known well enough to allow computation of $N(u, v)$.

5.4.2 Removal of Blur Caused by Uniform Linear Motion

There are practical applications in which $H(u, v)$ can be obtained analytically, but the solution has zero values in the frequency range of interest. An example of the difficulties caused by a vanishing $H(u, v)$ was given in the

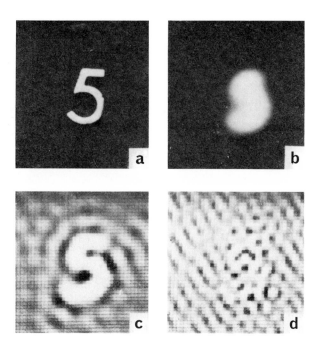

Figure 5.3. Example of image restoration by inverse filtering (a) Original image $f(x, y)$. (b) Degraded (blurred) image $g(x, y)$. (c) Result of restoration by considering a neighborhood about the origin of the uv-plane which does not include excessively small values of $H(u, v)$. (d) Result of using a larger neighborhood in which this condition does not hold. (From McGlamery [1967].)

previous section. In the following discussion we consider the problem of restoring an image that has been blurred by uniform linear motion. We have singled out this problem because of its practical implications, and also because it lends itself well to an analytical formulation. Solution of the uniform blurring case also demonstrates how zeros of $H(u, v)$ can be handled computationally. These considerations are important because they often arise in practice in other contexts of image restoration by inverse filtering.

Suppose that an image $f(x, y)$ undergoes planar motion, and let $x_0(t)$ and $y_0(t)$ be the time-varying components of motion in the x and y directions, respectively. The total exposure at any point of the recording medium (e. g., film) is obtained in this case by integrating the instantaneous exposure over the time interval during which the shutter is open. To isolate the effect of image motion, it is assumed that shutter opening and closing takes place instantaneously, and that the optical imaging process is perfect. Then, if T is the duration of the exposure, we have

$$g(x, y) = \int_0^T f[x - x_0(t), y - y_0(t)] \, dt \tag{5.4-7}$$

where $g(x, y)$ is the blurred image.

From Eq. (3.1-9), the Fourier transform of Eq. (5.4-7) is given by

$$G(u, v) = \int\!\!\int_{-\infty}^{\infty} g(x, y) \exp[-j2\pi(ux + vy)] \, dx \, dy$$

$$= \int\!\!\int_{-\infty}^{\infty} \left[\int_0^T f[x - x_0(t), y - y_0(t)] \, dt \right] \exp[-j2\pi(ux + vy)] \, dx \, dy \tag{5.4-8}$$

If we assume that the order of integration can be reversed, then Eq. (5.4-8) can be expressed in the form

$$G(u, v) = \int_0^T \left[\int\!\!\int_{-\infty}^{\infty} f[x - x_0(t), y - y_0(t)] \exp[-j2\pi(ux + vy)] \, dx \, dy \right] dt \tag{5.4-9}$$

The term inside the outer brackets is recognized as the Fourier transform of the displaced function $f[x - x_0(t), y - y_0(t)]$. By using Eq. (3.3-7b) we then have the relation

$$G(u, v) = \int_0^T F(u, v) \exp[-j2\pi(ux_0(t) + vy_0(t))] \, dt$$

$$= F(u, v) \int_0^T \exp[-j2\pi(ux_0(t) + vy_0(t))] \, dt \tag{5.4-10}$$

where the last step follows from the fact that $F(u, v)$ is independent of t.

By defining

$$H(u, v) = \int_0^T \exp[-j2\pi(ux_0(t) + vy_0(t))] \, dt \tag{5.4-11}$$

Eq. (5.4-10) may be expressed in the familiar form

$$G(u, v) = H(u, v)F(u, v) \tag{5.4-12}$$

If the nature of the motion variables $x(t)$ and $y(t)$ is known, the transfer function $H(u, v)$ can be obtained directly from Eq. (5.4-11). As an illustration, suppose that the image in question undergoes uniform linear motion in the x-direction only, at a rate given by $x_0(t) = at/T$. When $t = T$, we see that the image has been displaced by a total distance a. With $y_0(t) = 0$, Eq. (5.4-11) yields

$$H(u, v) = \int_0^T \exp\left[-j2\pi u x_0(t)\right] dt$$

$$= \int_0^T \exp\left[-j2\pi u at/T\right] dt$$

$$= \frac{T}{\pi u a} \sin(\pi u a)\, e^{-j\pi u a} \qquad (5.4\text{-}13)$$

It is evident that H vanishes at values of u given by $u = n/a$, where n is an integer.

When $f(x, y)$ is zero (or known) outside an interval $0 \leqslant x \leqslant L$, it is possible to avoid the problem presented by Eq. (5.4-13) and reconstruct the image completely from a knowledge of $g(x, y)$ in this interval. Since y is time invariant, let us suppress this variable temporarily and write Eq. (5.4-7) as

$$g(x) = \int_0^T f\left[x - x_0(t)\right] dt$$

$$= \int_0^T f(x - at/T)\, dt \qquad 0 \leqslant x \leqslant L \qquad (5.4\text{-}14)$$

Substituting $\tau = x - at/T$ in this expression and ignoring a scale factor yields

$$g(x) = \int_{x-a}^x f(\tau)\, d\tau \qquad 0 \leqslant x \leqslant L \qquad (5.4\text{-}15)$$

Then, by differentiation,

$$g'(x) = f(x) - f(x - a) \qquad 0 \leqslant x \leqslant L \qquad (5.4\text{-}16)$$

or

$$f(x) = g'(x) + f(x - a) \qquad 0 \leqslant x \leqslant L \qquad (5.4\text{-}17)$$

It will be convenient in the following development to assume that $L = Ka$, where K is an integer. Then the variable x may be expressed in the form

$$x = z + ma \qquad (5.4\text{-}18)$$

where z assumes values in the interval $[0, a]$ and m is the integral part of (x/a). For example, if $a = 2$ and $x = 3.5$, then $m = 1$ (the integral part of $3.5/2$), and $z = 1.5$. Clearly, $z + ma = 3.5$, as required. It is also noted that, for $L = Ka$, the index m can assume any of the integer values 0, 1, ..., $K-1$. For instance, when $x = L$, we have that $z = a$ and $m = K-1$.

Substitution of Eq. (5.4-18) into Eq. (5.4-17) yields the expression

$$f(z + ma) = g'(z + ma) + f[z + (m-1)a] \qquad (5.4\text{-}19)$$

Next, let us denote by $\phi(z)$ the portion of the scene that moves into the range $0 \leqslant z < a$ during exposure, that is,

$$\phi(z) = f(z - a) \qquad 0 \leqslant z < a \qquad (5.4\text{-}20)$$

Equation (5.4-19) can be solved recursively in terms of $\phi(z)$. Thus, for $m = 0$, we have

$$f(z) = g'(z) + f(z - a)$$

$$= g'(z) + \phi(z) \qquad (5.4\text{-}21)$$

For $m = 1$, Eq. (5.4-19) becomes

$$f(z + a) = g'(z + a) + f(z) \qquad (5.4\text{-}22)$$

Substitution of Eq. (5.4-21) into Eq. (5.4-22) yields

$$f(z + a) = g'(z + a) + g'(z) + \phi(z) \qquad (5.4\text{-}23)$$

In the next step we let $m = 2$. This results in the expression

$$f(z + 2a) = g'(z + 2a) + f(z + a) \qquad (5.4\text{-}24)$$

or, substituting Eq. (5.4-23) for $f(z + a)$,

$$f(z + 2a) = g'(z + 2a) + g'(z + a) + g'(z) + \phi(z) \qquad (5.4\text{-}25)$$

It is evident that continuing with this procedure will yield the result

$$f(z + ma) = \sum_{k=0}^{m} g'(z + ka) + \phi(z) \qquad (5.4\text{-}26)$$

However, since $x = z + ma$, Eq. (5.4-26) may be expressed in the form

$$f(x) = \sum_{k=0}^{m} g'(x - ka) + \phi(x - ma) \qquad 0 \leqslant x \leqslant L \qquad (5.4\text{-}27)$$

Since $g(x)$ is known, the problem is reduced to that of estimating $\phi(x)$.

One way to estimate this function directly from the blurred image is as follows. First we note that, as x varies from 0 to L, m ranges from 0 to $K-1$. Since the argument of ϕ is $(x - ma)$, which is always in the range $0 \leqslant x - ma < a$, it follows that ϕ is repeated K times during the evaluation of $f(x)$ for $0 \leqslant x \leqslant L$. Next, we define

$$\tilde{f}(x) = \sum_{k=0}^{m} g'(x - ka) \tag{5.4-28}$$

and rewrite Eq. (5.4-27) as

$$\phi(x - ma) = f(x) - \tilde{f}(x) \tag{5.4-29}$$

If we evaluate the left and right sides of Eq. (5.4-29) for $ka \leqslant x < (k + 1)a$, and add the results for $k = 0, 1, ..., K-1$, we obtain

$$K\phi(x) = \sum_{k=0}^{K-1} f(x + ka) - \sum_{k=0}^{K-1} \tilde{f}(x + ka) \qquad 0 \leqslant x < a \tag{5.4-30}$$

where $m = 0$, since $0 \leqslant x < a$. Dividing through by K yields

$$\phi(x) = \frac{1}{K} \sum_{k=0}^{K-1} f(x + ka) - \frac{1}{K} \sum_{k=0}^{K-1} \tilde{f}(x + ka) \tag{5.4-31}$$

The first sum on the right side of this expression is, of course, unknown. However, it is evident that for large values of K it approaches the average value of f. Thus, this sum may be taken as a constant A and we have the approximation

$$\phi(x) \approx A - \frac{1}{K} \sum_{k=0}^{K-1} \tilde{f}(x + ka) \qquad 0 \leqslant x < a \tag{5.4-32}$$

or

$$\phi(x - ma) \approx A - \frac{1}{K} \sum_{k=0}^{K-1} \tilde{f}\left[x + (k - m)a\right] \qquad 0 \leqslant x \leqslant L \tag{5.4-33}$$

Substitution of Eq. (5.4-28) for $\tilde{f}$ yields

$$\phi(x - ma) \approx A - \frac{1}{K} \sum_{k=0}^{K-1} \sum_{k=0}^{m} g'(x - ma)$$

$$\approx A - mg'(x - ma) \tag{5.4-34}$$

From Eq. (5.4-29) we then have the final result

$$f(x) \approx A - mg'(x - ma) + \sum_{k=0}^{m} g'(x - ka) \qquad 0 \leqslant x \leqslant L \tag{5.4-35}$$

or, reintroducing the suppressed variable y,

$$f(x, y) \approx A - mg'(x - ma, y) + \sum_{k=0}^{m} g'(x - ka, y) \qquad 0 \leqslant x, y \leqslant L$$

$$(5.4\text{-}36)$$

where, as before, it is assumed that $f(x, y)$ is a square image. An expression identical in form to Eq. (5.4-36) would give the reconstruction of an image that moves only in the y direction during exposure. The above concepts can also be used to derive a deblurring expression which takes into account simultaneous uniform motion in both directions.

Example: The image shown in Fig. 5.4(a) was blurred by uniform linear motion in one direction during exposure, with the total distance traveled being approximately equal to one-eighth the width of the photograph. Figure 5.4(b) is the deblurred result obtained by using Eq. (5.4-36). It is noted that the error in the approximation given by this equation is certainly not intolerable.   □

Figure 5.4. (a) Image Blurred by uniform linear motion. (b) Image restored by using Eq. (5.4-36). (From Sondhi [1972].)

5.5 LEAST-SQUARES (WIENER) FILTER

Let $\mathbf{R}_f$ and $\mathbf{R}_n$ be the correlation matrices of $\mathbf{f}$ and $\mathbf{n}$, defined respectively by the relations

$$\mathbf{R_f} = E\{\mathbf{ff'}\} \tag{5.5-1}$$

and

$$\mathbf{R_n} = E\{\mathbf{nn'}\} \tag{5.5-2}$$

where $E\{\cdot\}$ denotes the expected value operation. The ijth element of $\mathbf{R_f}$ is given by $E\{f_i f_j\}$, which is the correlation between the ith and the jth elements of $\mathbf{f}$. Similarly, the ijth element of $\mathbf{R_n}$ gives the correlation between the two corresponding elements in $\mathbf{n}$. Since the elements of $\mathbf{f}$ and $\mathbf{n}$ are real, $E\{f_i f_j\} = E\{f_j f_i\}$, and $E\{n_i n_j\} = E\{n_j n_i\}$, it follows that $\mathbf{R_f}$ and $\mathbf{R_n}$ are real symmetric matrices. For most image functions the correlation between pixels (i. e., elements of $\mathbf{f}$ or $\mathbf{n}$) does not extend beyond a distance of 20 to 30 points in the image (see Section 6.4), so a typical correlation matrix will have a band of nonzero elements about the main diagonal and zeros in the right-upper and left-lower corner regions. Assuming that the correlation between any two pixels is a function of the distance between the pixels and not their position, it can be shown (Andrews and Hunt [1977]) that $\mathbf{R_f}$ and $\mathbf{R_n}$ can be made to approximate block-circulant matrices and, therefore, can be diagonalized by the matrix W using the procedure described in Section 5.2.2. Using $\mathbf{A}$ and $\mathbf{B}$ to denote matrices we then have

$$\mathbf{R_f} = \mathbf{WAW}^{-1} \tag{5.5-3}$$

and

$$\mathbf{R_n} = \mathbf{WBW}^{-1} \tag{5.5-4}$$

Just like the elements of the diagonal matrix $\mathbf{D}$ in the relation $\mathbf{H} = \mathbf{HDW}^{-1}$ were shown to correspond to the Fourier transform of the block elements of $\mathbf{H}$, it can be shown that the elements of $\mathbf{A}$ and $\mathbf{B}$ are the transforms of the correlation elements in $\mathbf{R_f}$ and $\mathbf{R_n}$, respectively. The Fourier transform of these correlations is defined as the *spectral density* of $f_e(x, y)$ and $\eta_e(x, y)$, respectively, and will be denoted in the following discussion by $S_f(u, v)$ and $S_\eta(u, v)$.

By defining

$$\mathbf{Q'Q} = \mathbf{R_f^{-1}R_n} \tag{5.5-5}$$

and substituting this expression in Eq. (5.3-9) we obtain

$$\mathbf{\hat{f}} = \left(\mathbf{H'H} + \gamma \mathbf{R_f^{-1}R_n}\right)^{-1}\mathbf{H'g} \tag{5.5-6}$$

Using Eqs. (5.2-1), (5.2-23), (5.5-3), and (5.5-4) yields

$$\mathbf{\hat{f}} = \left(\mathbf{WD^{*}DW}^{-1} + \gamma \mathbf{WA}^{-1}\mathbf{BW}^{-1}\right)^{-1}\mathbf{WD^{*}W}^{-1}\mathbf{g} \tag{5.5-7}$$

After multiplication of both sides of $\mathbf{W}^{-1}$ and some matrix manipulations this equation reduces to the form

$$\mathbf{W}^{-1}\hat{\mathbf{f}} = (\mathbf{D}^*\mathbf{D} + \gamma\mathbf{A}^{-1}\mathbf{B})^{-1}\mathbf{D}^*\mathbf{W}^{-1}\mathbf{g} \qquad (5.5\text{-}8)$$

Keeping in mind the meaning of the elements of $\mathbf{A}$ and $\mathbf{B}$, recognizing that the matrices inside the parentheses are diagonal, and making use of the concepts developed in Section 5.2.3, allows us to write the elements of Eq. (5.5-8) in the form

$$\hat{F}(u, v) = \left[\frac{H^*(u, v)}{|H(u, v)|^2 + \gamma\left[S_\eta(u, v)/S_f(u, v) \right]} \right] G(u, v)$$

$$= \left[\frac{1}{H(u, v)} \frac{|H(u, v)|^2}{|H(u, v)|^2 + \gamma\left[S_\eta(u, v)/S_f(u, v) \right]} \right] G(u, v) \qquad (5.5\text{-}9)$$

for $u, v = 0, 1, 2, ..., N-1$, where $|H(u, v)|^2 = H^*(u, v)H(u, v)$ and we have assumed that $M = N$.

When $\gamma = 1$, the term inside the outer brackets in Eq. (5.5-9) reduces to the so-called *Wiener filter*. If γ is variable we refer to this expression as the *parametric Wiener filter*. In the absence of noise, $S_\eta(u, v) = 0$ and either form of the Wiener filter reduces to the ideal inverse filter discussed in the previous section. It is important to note that, by setting $\gamma = 1$, we can no longer say in general that the use of Eq. (5.5-9) yields an optimum solution in the sense defined in Section 5.3.2 because, as pointed out in that section, γ must be adjusted to satisfy the constraint $\|\mathbf{g} - \mathbf{H}\hat{\mathbf{f}}\|^2 = \|\mathbf{n}\|^2$. It can be shown, however, that the solution obtained with $\gamma = 1$ is optimum in the sense that it minimizes the quantity $E\{[f(x, y) - \hat{f}(x, y)]^2\}$. Clearly, this is a statistical criterion in which f and $\hat{f}$ are treated as random variables.

When $S_\eta(u, v)$ and $S_f(u, v)$ are not known (a problem often encountered in practice) it is sometimes useful to approximate Eq. (5.5-9) by the relation

$$\hat{F}(u, v) \approx \left[\frac{1}{H(u, v)} \frac{|H(u, v)|^2}{|H(u, v)|^2 + K} \right] G(u, v) \qquad (5.5\text{-}10)$$

where K is a constant. An example of results obtained with this equation is given below. The problem of selecting the optimum γ for image restoration is discussed in some detail in the following section.

Example: The first column in Fig. 5.5 shows three pictures of a domino corrupted by linear motion (at $-45°$ with respect to the horizontal) and noise whose variance at any point in the image was proportional to the brightness of the point. The three images were generated by varying the constant of proportionality so that the ratios of maximum brightness to

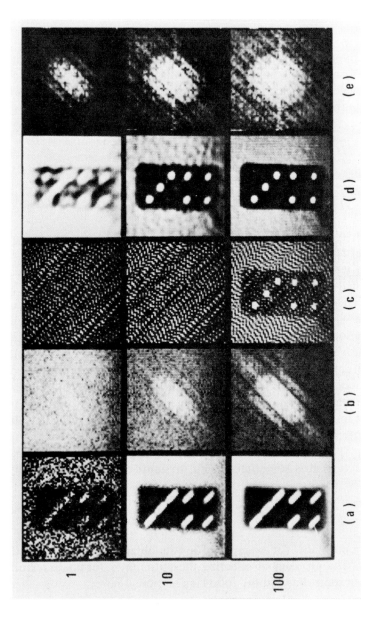

Figure 5.5. Example of image restoration by inverse and Wiener filters. (a) Degraded images, and (b) their Fourier spectra. (c) Images restored by inverse filtering. (d) Images restored by Wiener filtering. (e) Fourier spectra of images in (d). (From Harris [1968].)

noise amplitude were 1, 10, and 100, respectively, as shown on the left side of Fig. 5.5. The Fourier spectra of the degraded images are shown in Fig. 5.5 (b).

Since the effects of linear, uniform motion can be expressed analytically, an equation describing $H(u, v)$ can be obtained without difficulty, as shown in Section 5.4.2. Figure 5.5(c) was obtained by direct inverse filtering following the procedure described in Section 5.4.1. The results are dominated by noise, but as shown in the third image, the inverse filter was successful in removing the degradation (i. e., blur) due to motion. By contrast, Fig. 5.5(d) shows the results obtained using Eq. (5.5-10) with $K = 2\sigma^2$, where σ^2 is the noise variance. The improvements over the direct inverse filtering approach are obvious, particularly for the third image. The Fourier spectra of the restored images are shown in Fig. 5.5(e).   □

5.6 CONSTRAINED LEAST-SQUARES RESTORATION

The least-squares approach derived in the previous section is a statistical procedure since the criterion for optimality is based on the correlation matrices of the image and noise functions. This implies that the results obtained by using a Wiener filter are optimum in an average sense. The restoration procedure developed in this section, on the other hand, is optimum for *each* given image and only requires knowledge of the noise mean and variance. In the course of the following development we also consider the problem of adjusting γ so that the constraint leading to Eq. (5.3-9) is satisfied.

As indicated in Section 5.3.2, the restoration solution obtained by using Eq. (5.3-9) depends on the choice of the matrix **Q**. Due to ill-conditioning, this equation will sometimes yield solutions which are obscured by large oscillating values. It is of interest, therefore, to investigate the feasibility of choosing **Q** such that these adverse effects are minimized. One possibility, suggested by Phillips [1962], is to formulate a criterion of optimality based on a measure of smoothness such as, for example, minimizing some function of the second derivative. In order to see how this criterion can be expressed in a form that is compatible with Eq. (5.3-9), let us first consider the one-dimensional case.

Given a discrete function $f(x)$, $x = 0, 1, 2, ...$, we may approximate its second derivative at a point x by the expression

$$\frac{\partial^2 f(x)}{\partial x^2} \approx f(x + 1) - 2f(x) + f(x - 1) \qquad (5.6\text{-}1)$$

A criterion based on this expression then might be to minimize $[\partial^2 f/\partial x^2]^2$ over x; that is,

$$\text{minimize} \left\{ \sum_x \left[f(x+1) - 2f(x) + f(x-1) \right]^2 \right\} \qquad (5.6\text{-}2)$$

or, in matrix notation,

$$\text{minimize} \left\{ \mathbf{f}'\mathbf{C}'\mathbf{C}\mathbf{f} \right\} \qquad (5.6\text{-}3)$$

where

$$\mathbf{C} = \begin{bmatrix}
1 & & & & & & \\
-2 & 1 & & & & & \\
1 & -2 & 1 & & & & \\
 & 1 & -2 & 1 & & & \\
 & & & \ddots & & & \\
 & & & & 1 & -2 & 1 \\
 & & & & & 1 & -2 \\
 & & & & & & 1
\end{bmatrix} \qquad (5.6\text{-}4)$$

is a smoothing matrix, and $\mathbf{f}$ is a vector whose elements are the samples of $f(x)$.

In the two-dimensional case we consider a direct extension of Eq. (5.6-1). In this case the criterion is to

$$\text{minimize} \left[\frac{\partial^2 f(x,y)}{\partial x^2} + \frac{\partial^2 f(x,y)}{\partial y^2} \right]^2 \qquad (5.6\text{-}5)$$

where the derivative function is approximated by the expression

$$\frac{\partial^2 f}{\partial x^2} + \frac{\partial^2 f}{\partial y^2} \approx f(x+1,y) - 2f(x,y) + f(x-1,y)$$

$$+ f(x,y+1) - 2f(x,y) + f(x,y-1)$$

$$\approx f(x+1,y) + f(x-1,y) + f(x,y+1) + f(x,y-1)$$

$$-4f(x,y) \qquad (5.6\text{-}6)$$

The derivative function given in Eq. (5.6-5) is recognized as the Laplacian operator discussed in Section 3.3.7.

Equation (5.5-6) can be implemented directly in a computer. However, the same operation can be carried out by convolving $f(x, y)$ with the operator

$$p(x, y) = \begin{bmatrix} 0 & 1 & 0 \\ 1 & -4 & 1 \\ 0 & 1 & 0 \end{bmatrix} \tag{5.6-7}$$

As indicated in Section 5.1.3, wraparound error in the discrete convolution process is avoided by extending $f(x, y)$ and $p(x, y)$. We have already considered the formation of $f_e(x, y)$. We form $p_e(x, y)$ in the same manner, that is,

$$p_e(x, y) = \begin{cases} p(x, y) & 0 \leqslant x \leqslant 2 \quad \text{and} \quad 0 \leqslant y \leqslant 2 \\ 0 & 3 \leqslant x \leqslant M - 1 \quad \text{or} \quad 3 \leqslant y \leqslant N - 1 \end{cases}$$

If $f(x, y)$ is of size $A \times B$, we choose $M \geqslant A + 3 - 1$ and $N \geqslant B + 3 - 1$ since $p(x, y)$ is of size 3×3.

The convolution of the extended functions is then

$$g_e(x, y) = \sum_{m=0}^{M-1} \sum_{n=0}^{N-1} f_e(m, n) p_e(x - m, y - n) \tag{5.6-8}$$

which agrees with Eq. (5.1-23).

Following an argument similar to the one given in Section 5.1.3 we may express the above smoothness criterion in matrix form. First we construct a block-circulant matrix of the form

$$\mathbf{C} = \begin{bmatrix} \mathbf{C}_0 & \mathbf{C}_{M-1} & \mathbf{C}_{M-2} & \cdots & \mathbf{C}_1 \\ \mathbf{C}_1 & \mathbf{C}_0 & \mathbf{C}_{M-1} & \cdots & \mathbf{C}_2 \\ \mathbf{C}_2 & \mathbf{C}_1 & \mathbf{C}_0 & \cdots & \mathbf{C}_3 \\ \vdots & & & & \\ \mathbf{C}_{M-1} & \mathbf{C}_{M-2} & \mathbf{C}_{M-3} & \cdots & \mathbf{C}_0 \end{bmatrix} \tag{5.6-9}$$

where each submatrix $\mathbf{C}_j$ is an $N \times N$ circulant constructed from the jth row of $p_e(x, y)$, that is,

$$\mathbf{C}_j = \begin{bmatrix} p_e(j, 0) & p_e(j, N-1) & \cdots & p_e(j, 1) \\ p_e(j, 1) & p_e(j, 0) & \cdots & p_e(j, 2) \\ \vdots & & & \\ p_e(j, N-1) & p_e(j, N-2) & \cdots & p_e(j, 0) \end{bmatrix} \tag{5.6-10}$$

Since $\mathbf{C}$ is block circulant, it is diagonalized by the matrix $\mathbf{W}$ defined in Section 5.2.2. In other words,

$$E = W^{-1}CW \qquad (5.6\text{-}11)$$

where E is a diagonal matrix whose elements are, as in Eq. (5.2-29), given by

$$E(k, i) = \begin{cases} P\left[\dfrac{k}{n} \right], k \bmod N & \text{if } i = k \\ 0 & \text{if } i \neq k \end{cases} \qquad (5.6\text{-}12)$$

In this case $P(u, v)$ is the two-dimensional Fourier transform of $p_e(x, y)$. As indicated in connection with Eqs. (5.2-37) and (5.2-39), it is assumed that Eq. (5.6-12) has been scaled by the factor MN.

Since the convolution operation described above is equivalent to implementing Eq. (5.6-6), we may express the smoothness criterion of Eq. (5.6-5) in the same form as Eq. (5.6-3); that is,

$$\text{minimize } \{f'C'Cf\} \qquad (5.6\text{-}13)$$

where f is an MN-dimensional vector and C is of size $MN \times MN$. By letting $Q = C$, this criterion may be expressed in the form

$$\text{minimize } \|Qf\|^2 \qquad (5.6\text{-}14)$$

which is in the same form as that used in Section 5.3.2. In fact, if we require that the constraint $\|g - H\hat{f}\|^2 = \|n\|^2$ be satisfied, the optimal solution is given by Eq. (5.3-9) with $Q = C$, that is,

$$\hat{f} = (H'H + \gamma C'C)^{-1}H'g \qquad (5.6\text{-}15)$$

By using Eqs. (5.2-21), (5.2-23), and (5.6-7), Eq. (5.6-15) may be expressed in the form

$$\hat{f} = (WD^*DW^{-1} + \gamma WE^*EW^{-1})^{-1}WD^*W^{-1}g \qquad (5.6\text{-}16)$$

After multiplication of both sides by W^{-1} and some matrix manipulations this equation reduces to

$$W^{-1}\hat{f} = (D^*D + \gamma E^*E)^{-1}D^*W^{-1}g \qquad (5.6\text{-}17)$$

By keeping in mind that the elements inside the parentheses are diagonal, and making use of the concepts developed in Section 5.2.3, we can express the elements of Eq. (5.6-17) in the form

$$\hat{F}(u, v) = \left[\frac{H^*(u, v)}{|H(u, v)|^2 + \gamma |p(u, v)|^2} \right] G(u, v) \qquad (5.6\text{-}18)$$

for $u, v = 0, 1, 2, ..., N-1$, where $|H(u, v)|^2 = H^*(u, v)H(u, v)$, and we have assumed that $M = N$. It is noted that Eq. (5.6-18) resembles the parametric Wiener filter derived in the previous section. This principal difference between Eqs. (5.5-9) and (5.6-18) is that the latter does not require explicit knowledge of statistical parameters other than, as will be seen below, an estimate of the noise mean and variance.

The general formulation given in Eq. (5.3-9) requires that γ be adjusted to satisfy the constraint $\|g - Hf\|^2 = \|n\|^2$. Thus, the solution given in Eq. (5.6-18) can be optimal only when γ satisfies this condition. An iterative procedure for estimating this parameter is as follows.

Define a residual vector r as

$$r = g - H\hat{f} \tag{5.6-19}$$

Substituting Eq. (5.6-15) for $\hat{f}$ yields

$$r = g - H(H'H + \gamma C'C)^{-1}H'g \tag{5.6-20}$$

This expression indicates that r is a function of γ. In fact, it can be shown (Hunt [1973]) that

$$\phi(\gamma) = r'r$$

$$= \|r\|^2 \tag{5.6-21}$$

is a monotonically increasing function of γ. What we wish to do is adjust γ so that

$$\|r\|^2 = \|n\|^2 + a \tag{5.6-22}$$

where a is an accuracy factor. Clearly, if $\|r\|^2 = \|n\|^2$ the constraint $\|g - H\hat{f}\|^2 = \|n\|^2$ will be strictly satisfied, in view of Eq. (5.6-19).

Since $\phi(\gamma)$ is monotonic, finding a γ which satisfies Eq. (5.6-17) is not a difficult problem. One simple approach is to: (1) Specify an initial value of γ. (2) Compute $\hat{f}$ and $\|r\|^2$. (3) Stop if Eq. (5.6-22) is satisfied; otherwise return to step 2 after increasing γ if $\|r\|^2 < \|n\|^2 - a$ or decreasing γ if $\|r\|^2 > \|n\|^2 + a$. Other procedures such as a Newton-Raphson algorithm can be used to improve speed of convergence.

Implementation of the above concepts require some knowledge about $\|n\|^2$. The variance of $\eta_e(x, y)$ is given by

$$\sigma_\eta^2 = E\left\{ \left[\eta_e(x, y) - \bar{\eta}_e \right]^2 \right\}$$

$$= E\left\{ \eta_e^2(x, y) - \bar{\eta}_e^2 \right\} \tag{5.6-23}$$

where

$$\bar{\eta}_e = \frac{1}{(M-1)(N-1)} \sum_x \sum_y \eta_e(x, y) \tag{5.6-24}$$

is the mean value of $\eta_e(x, y)$. If we approximate the expected value of $\eta_e^2(x, y)$ by a sample average, Eq. (5.6-23) becomes

$$\sigma_\eta^2 = \frac{1}{(M-1)(N-1)} \sum_x \sum_y \eta_e^2(x, y) - \bar{\eta}_e^2 \qquad (5.6\text{-}25)$$

The summation term simply indicates squaring and adding all values in the array $\eta_e(x, y)$, $x = 0, 1, 2, ..., M-1$, and $y = 0, 1, 2, ..., N-1$. This we recognize simply as the product $\mathbf{n'n}$ which, by definition, is equal to $\|\mathbf{n}\|^2$. Thus, Eq. (5.6-23) reduces to

$$\sigma_\eta^2 = \frac{\|\mathbf{n}\|^2}{(M-1)(N-1)} - \bar{\eta}_e^2 \qquad (5.6\text{-}26)$$

or

$$\|\mathbf{n}\|^2 = (M-1)(N-1)\left[\sigma_\eta^2 + \bar{\eta}_e^2\right] \qquad (5.6\text{-}27)$$

The importance of this equation is that it allows us to establish a value for the constraint in terms of the noise mean and variance, quantities which, if not known, can often be approximated or measured in practice.

The constrained least-squares restoration procedure can be summarized as follows:

Step 1. Choose an initial value of γ, and obtain an estimate of $\|\mathbf{n}\|^2$ using Eq. (5.6-27).

Step 2. Compute $\hat{F}(u, v)$ using Eq. (5.6-18). Obtain $\hat{\mathbf{f}}$ by taking the inverse Fourier transform of $F(u, v)$.

Step 3. Form the residual vector $\mathbf{r}$ according to Eq. (5.6-19), and compute $\phi(\gamma) = \|\mathbf{r}\|^2$.

Step 4. Increment or decrement γ.
a) $\phi(\gamma) < \|\mathbf{n}\|^2 - a$. Increment γ according to the algorithm given above or other appropriate method (such as a Newton-Raphson procedure).
b) $\phi(\gamma) > \|\mathbf{n}\|^2 + a$. Decrement γ according to an appropriate algorithm.

Step 5. Return to Step 2 and continue unless Step 6 is true.

Step 6. $\phi(\gamma) = \|\mathbf{n}\|^2 \pm a$, where a determines the accuracy with which the constraint is satisfied. Stop the estimation procedure, with $\hat{\mathbf{f}}$ for the present value of γ being the restored image.

Example: Figure 5.6(b) was obtained by convolving the Gaussian-shaped point-spread function

$$h(x, y) = \exp\left[-\left(\frac{x^2 + y^2}{2400}\right)^2\right]$$

with the original image shown in Fig. 5.6(a), and adding noise drawn from a uniform distribution in the interval [0, 0.5]. Figure 5.6(c) is the result of using the above algorithm with $\gamma = 0$ (inverse filter). The ill-conditioned nature of the solution is evident by the dominance of the noise on the

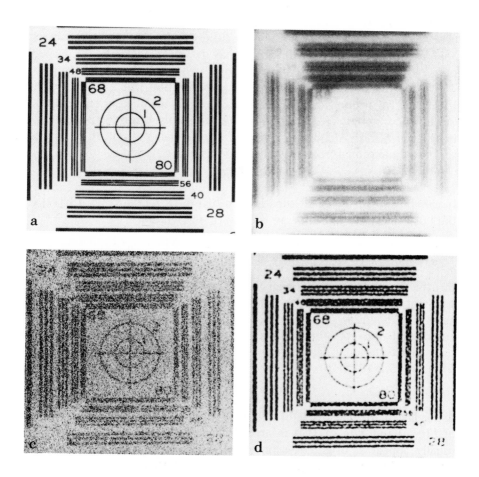

Figure 5.6. (a) Original image. (b) Image blurred and corrupted by additive noise. (c) Image restored by inverse filtering. (d) Image restored by the method of constrained least squares. (From Hunt[1973].)

restored image. Figure 5.6(d) was obtained by allowing the algorithm to seek a γ which would satisfy the constraint. The variance and mean of the uniform density in the interval [0, 0.5] were used to estimate $\|\mathbf{n}\|^2$ and the accuracy factor a was chosen so that $a = 0.025\|\mathbf{n}\|^2$. The improvement of the constrained solution over direct inverse filtering is quite visible. ☐

5.7 INTERACTIVE RESTORATION

Thus far, attention has been focused on a strictly analytical approach to restoration. In many applications, it is practical to take advantage of human intuition, coupled with the versatility of a digital computer, to restore images in an interactive mode. In this case, the observer has control over the restoration process and, by "tuning" the available parameters, is able to obtain a final result which may be quite adequate for a specific purpose. We conclude this chapter with two examples of this approach.

One of the simplest cases of image corruption which lends itself well to interactive restoration is the occurrence of a two-dimensional sinusoidal interference pattern (often called *coherent noise*) superimposed on an image. Let $\eta(x, y)$, denote a sinusoidal interference pattern of amplitude A and two-dimensional frequency components (u_0, v_0), that is,

$$\eta(x, y) = A \sin(u_0 x + v_0 y) \tag{5.7-1}$$

It can be shown by direct substitution of Eq. (5.7-1) into Eq. (3.1-9) that the Fourier transform of $\eta(x, y)$ is given by the relation

$$N(u, v) = \frac{-jA}{2} \left[\delta(u - u_0/2\pi, v - v_0/2\pi) - \delta(u + u_0/2\pi, v + v_0/2\pi) \right] \tag{5.7-2}$$

In other words, the Fourier transform of a two-dimensional sine function is a pair of impulses of strength $-A/2$ and $A/2$, located respectively at coordinates $(u_0/2\pi, v_0/2\pi)$ and $(-u_0/2\pi, -v_0/2\pi)$ of the frequency plane. It is also noted that the transform has only imaginary components in this case.

Since the only degradation being considered is additive noise, we have from Eq. (5.2-40) that

$$G(u, v) = F(u, v) + N(u, v) \tag{5.7-3}$$

A display of the magnitude of $G(u, v)$ will contain the magnitude of the sum of $F(u, v)$ and $N(u, v)$. If A is large enough, the two impulses of $N(u, v)$ will usually appear as bright dots on the display, especially if they are located relatively far from the origin so that the contribution of the components of $F(u, v)$ is small.

If $\eta(x, y)$ were completely known, the original image could, of course, be recovered by subtracting the interference from $g(x, y)$. Since this is seldom the case, a useful approach is to identify visually the location of impulse components in the frequency domain and use a bandreject filter (see Section 4.6.4) at these locations.

Example: The image shown in Fig. 5.7(a) was corrupted by a sinusoidal pattern of the form shown in Eq. (5.7-1). The Fourier spectrum of this image, shown in Fig. 5.7(b), clearly exhibits a pair of symmetric impulses

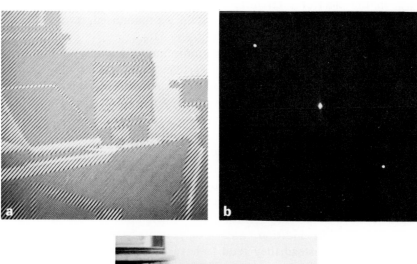

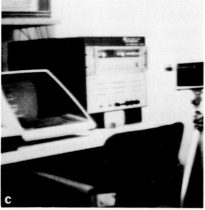

Figure 5.7. Example of sinusoidal interference removal. (a) Corrupted image. (b) Fourier spectrum showing impulses due to sinusoidal pattern. (c) Image restored by using a band-reject filter with a radius of one.

due to the sinusoidal interference. Figure 5.7(c) was obtained by manually placing (from a computer console) two bandreject filters of radius one at the location of the impulses and then taking the inverse Fourier transform of the result. The restored image is seen to be, for all practical purposes, free of interference.   □

The presence of a single, clearly-defined interference pattern such as the one just illustrated seldom occurs in practice. A notable example is found in images that have been derived from electro-optical scanners, such as those commonly used in space missions. A common problem of these sensors is interference caused by coupling and amplification of low-level signals in the electronic circuitry. The result is that images reconstructed from the scanner output tend to contain a pronounced, two-dimensional periodic structure superimposed on the scene data.

An example of this type of periodic image degradation is shown in Fig. 5.8(a), which is a picture of the Martian terrain taken by the Mariner 6 spacecraft. The interference pattern is quite similar to the one shown in Fig. 5.7(a), but the former pattern is considerably more subtle and, consequently, harder to detect in the frequency plane.

Figure 5.8(b) shows the Fourier spectrum of the image in question. The star-like components were caused by the interference, and it is noted that several pairs of components are present, indicating that the pattern was composed of more than just one sinusoidal component. When several interference components are present, the method discussed above is not always acceptable because it may remove too much image information in the filtering process. In addition, these components generally are not single-frequency bursts. Instead, they tend to have broad skirts that carry information about the interference pattern. These skirts are not always easily detectable from the normal transform background.

A procedure that has found acceptance in processing space-related scenes consists of first isolating the principal contributions of the interference pattern and then subtracting a variable, weighted portion of the pattern from the corrupted image. Although the procedure is developed below in the context of a specific application, the basic approach is quite general and can be applied to other enhancement tasks where multiple periodic interference is a problem.

The first step is to extract the principal frequency components of the interference pattern. This can be done by placing a bandpass filter $H(u, v)$ at the location of each spike (see Section 4.6.4). If $H(u, v)$ is constructed to pass only components associated with the interference pattern, it follows that the Fourier transform of the pattern is given by the relation

$$P(u, v) = H(u, v)G(u, v) \qquad (5.7\text{-}4)$$

where $G(u, v)$ is the Fourier transform of the corrupted image $g(x, y)$ and, for $N \times N$ digitization, u and v assume values in the range 0, 1, ..., $N-1$.

It is important to note that the formation of $H(u, v)$ requires a large degree of judgement as to what is or is not an interference spike. For this reason, the bandpass filter is generally constructed interactively by observing the spectrum of $G(u, v)$ on a display. Once a particular filter has been

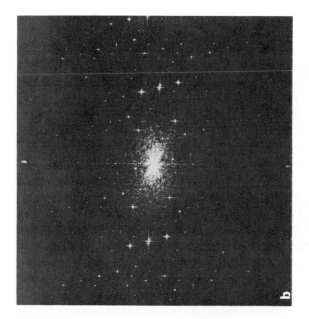

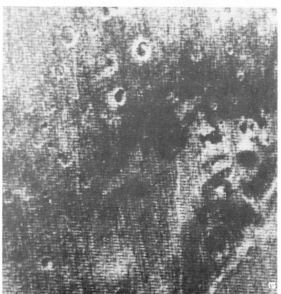

Figure 5.8. (a) Picture of the Martian terrain taken by Mariner 6. (b) Fourier spectrum. Note the periodic interference in the image and the corresponding spikes in the spectrum. (Courtesy of NASA, Jet propulsion Laboratory.)

selected, the corresponding pattern in the spatial domain is obtained from the expression

$$p(x, y) = \mathcal{F}^{-1}\{H(u, v)G(u, v)\} \tag{5.7-5}$$

Since the corrupted image is formed by the addition of $f(x, y)$ and the interference, it would be a simple matter if $p(x, y)$ were completely known to subtract the pattern from $g(x, y)$ to obtain $f(x, y)$. The problem, of course, is that the above filtering procedure usually yields only an approximation to the true pattern. In order to minimize the effects of components not present in the estimate of $p(x, y)$, we instead subtract from $g(x, y)$ a weighted portion of $p(x, y)$ to obtain an estimate of $f(x, y)$; that is,

$$\hat{f}(x, y) = g(x, y) - w(x, y)p(x, y) \tag{5.7-6}$$

where $\hat{f}(x, y)$ is the estimate of $f(x, y)$ and $w(x, y)$ is to be determined. The function $w(x, y)$ is called a *weighting* or *modulation* function, and the objective of the procedure is to select this function so that the result is optimized in some meaningful way. One approach is to select $w(x, y)$ so that the variance of $\hat{f}(x, y)$ is minimized over a specified neighborhood of every point (x, y).

Consider a neighborhood of size $(2X + 1)$ by $(2Y + 1)$ about a point (x, y). The "local" variance of $\hat{f}(x, y)$ at coordinates (x, y) is given by

$$\sigma^2(x, y) = \frac{1}{(2X + 1)(2Y + 1)} \sum_{m=-X}^{X} \sum_{n=-Y}^{Y} \left\{ \hat{f}(x + m, y + n) - \bar{\hat{f}}(x, y) \right\}^2 \tag{5.7-7}$$

where $\bar{\hat{f}}(x, y)$ is the average value of $\hat{f}(x, y)$ in the neighborhood, that is,

$$\bar{\hat{f}}(x, y) = \frac{1}{(2X + 1)(2Y + 1)} \sum_{m=-X}^{X} \sum_{n=-Y}^{Y} \hat{f}(x + m, y + n) \tag{5.7-8}$$

Points on or near the edge of the image can be treated by considering partial neighborhoods.

Substitution of Eq. (5.7-6) into Eq. (5.7-7) yields

$$\sigma^2(x, y) = \frac{1}{(2X + 1)(2Y + 1)} \sum_{m=-X}^{X} \sum_{n=-Y}^{Y} \left\{ \left[\bar{g}(x + m, y + n) \right. \right.$$

$$\left. \left. - w(x + m, y + n)p(x + m, y + n) \right] - \left[\bar{g}(x, y) - \overline{w(x, y)p(x, y)} \right] \right\}^2 \tag{5.7-9}$$

By making the assumption that $w(x, y)$ remains essentially constant over the neighborhood, we obtain the approximations

$$w(x + m, y + n) = w(x, y) \tag{5.7-10}$$

for $-X \leqslant m \leqslant X$ and $-Y \leqslant n \leqslant Y$; also

$$\overline{w(x, y)p(x, y)} = w(x, y)\bar{p}(x, y) \qquad (5.7\text{-}11)$$

in the neighborhood. With these approximations, Eq. (5.7-9) becomes

$$\sigma^2(x, y) = \frac{1}{(2X + 1)(2Y + 1)} \sum_{m=-X}^{X} \sum_{n=-Y}^{Y} \{[g(x + m, y + n)$$

$$- w(x, y)p(x + m, y + n)] - [\bar{g}(x, y) - w(x, y)\bar{p}(x, y)]\}^2 \qquad (5.7\text{-}12)$$

To minimize $\sigma^2(x, y)$ we solve

$$\frac{\partial\sigma^2(x, y)}{\partial w(x, y)} = 0 \qquad (5.7\text{-}13)$$

for $w(x, y)$. The result is

$$w(x, y) = \frac{\overline{g(x, y)p(x, y)} - \bar{g}(x, y)\bar{p}(x, y)}{\overline{p^2}(x, y) - \bar{p}^2(x, y)} \qquad (5.7\text{-}14)$$

To obtain the restored image $\hat{f}(x, y)$ we compute $w(x, y)$ from Eq. (5.7-14) and then make use of Eq. (5.7-6). It is important to note that, since $w(x, y)$ is assumed to be constant in a neighborhood, it is not necessary to compute this function for every value of x and y in the image. Instead, $w(x, y)$ is computed for *one* point in each nonoverlapping neighborhood (preferably the center point) and then used to process all the image points contained in that neighborhood.

Example: Figures 5.9 through 5.11 show the result of applying the above technique to the image shown in Fig. 5.8(a). In this case $N = 512$ and a neighborhood with $X = Y = 15$ was selected. Figure 5.9 is the Fourier spectrum of the corrupted image, but the origin was not shifted to the center of the frequency plane. Figure 5.10(a) shows the spectrum of $P(u, v)$ where only the noise spikes are present, and Fig. 5.10(b) is the interference pattern $p(x, y)$ obtained by taking the inverse Fourier transform of $P(u, v)$. Note the similarity between this pattern and the structure of the noise present in Fig. 5.8(a). Finally, Fig. 5.11 shows the processed image obtained by using Eq. (5.7-6). The periodic interference has, for all practical purposes, been removed, leaving only spotty noise that is not periodic. This noise can be processed by other methods such as neighborhood averaging or lowpass filtering. ◻

Figure 5.9. Fourier spectrum (without shifting) of the image shown in Fig. 5.8(a). (Courtesy of NASA, Jet Propulsion Laboratory.)

5.8 CONCLUDING REMARKS

The principal concepts developed in this chapter are a formulation of the image restoration problem in the framework of linear algebra, and the subsequent simplification of algebraic solutions based on the properties of circulant and block-circulant matrices.

The image restoration techniques derived in the previous sections are all based on a least-squares criterion of optimality. The reader is reminded that the use of the word "optimum" in this context refers strictly to a mathematical concept, and not to optimum response of the human visual system. In fact, our present lack of knowledge about visual perception precludes a general formulation of the image restoration problem which takes into account observer preferences and capabilities. In view of these

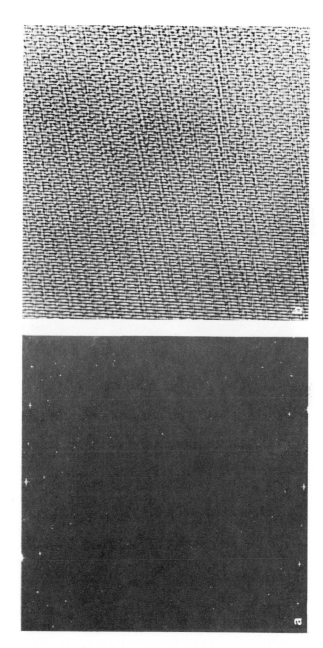

Figure 5.10. (a) Fourier spectrum of $P(u, v)$. (b) Corresponding interference pattern $p(x, y)$. (Courtesy of NASA, Jet Propulsion Laboratory.)

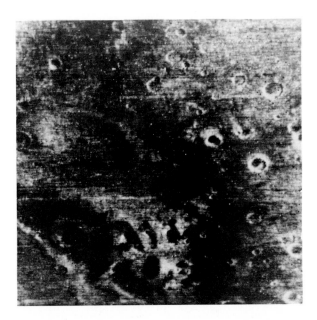

Figure 5.ll. Processed image. (Courtesy of NASA, Jet Propulsion Laboratory.)

limitations, the advantage of the procedure followed in this chapter is the development of a basic approach from which a set of previously known (but not unified) results can be derived. Thus, the power of the algebraic approach is evident in the simplicity by which methods such as the Wiener and constrained least-squares filters can be obtained starting from the same basic relation.

The key points leading to all the results in this chapter are based on the assumption of linear, space-invariant degradations. This assumption leads immediately to the convolution integral, whose discrete formulation can be expressed in terms of the basic degradation model given in Eq. (5.1-24). The assumed periodicity of the input functions further simplified the problem by producing circulant and block-circulant matrices. In terms of implementation, these matrices allow all the derived restoration techniques to be carried out in the frequency domain by means of a two-dimensional FFT algorithm, thus greatly reducing the computational complexity posed by the original matrix formulation of the degradation process.

REFERENCES

The definitions given in Section 5.1 were adapted from Schwarz and Friedland [1965], and a background for most of the basic matrix operations used in this chapter can be found in Deutsch [1965], Noble [1969], and Bellman [1970]. The development of the discrete degradation model in terms of circulant and block-circulant matrices is based on two papers by Hunt [1971, 1973]. These papers and the book by Bellman [1970] also consider the diagonalization properties discussed in Section 5.2. Additional information on the material of Section 5.3, as well as the algebraic derivation of the various restoration techniques used in this chapter, may be found in Andrews and Hunt [1977]. This book, devoted entirely to the topic of image restoration, treats in detail other restoration techniques in addition to the ones developed here.

The inverse filtering approach has been considered by numerous investigators. References for the material in Section 5.4 are McGlamery [1967], Sondhi [1972], Cutrona and Hall [1968], and Slepian [1967]. Additional references on the least-squares restoration approach discussed in Section 5.5 are Helstrom [1967], Slepian [1967], Harris [1968], Rino [1969], Horner [1969], and Rosenfeld and Kak [1976]. It is of interest to compare the classical derivations in these references with the algebraic approach given in Section 5.5. The material in Section 5.6 is based on a paper by Hunt [1973].

Some other references related to the topics discussed in this chapter are Slepian and Pollak [1961], Phillips [1962], Twomey [1963], Shack [1964], Lohman and Paris [1965], Harris [1966], Meuller and Reynolds [1967], Blackman [1968], Huang [1968], Rushforth and Harris [1968], MacAdam [1970], Falconer [1970], Som [1971], Frieden [1972, 1974], Habibi [1972], Sawchuck [1972], Robbins and Huang [1972], Andrews [1974], Jain and Angel [1974], and Anderson and Netravaly [1976].

6

IMAGE ENCODING

But if I'm content with a little,
Enough is as good as a feast.
Isaac Bickerstaffe

As discussed in Section 2.3, digital representations of images usually require a very large number of bits. In many applications, it is important to consider techniques for representing an image, or the information contained in the image, with fewer bits. In the terminology of information theory this is referred to as *source encoding*.

Applications of source encoding in the field of image processing generally fall into one of three categories: (1) image data compression, (2) image transmission, and (3) feature extraction. The methods discussed in this chapter are applicable to any of these three categories. It is important to note, however, that these techniques are very much problem-oriented. In other words, while the final objective of encoding is data reduction, the choice of one encoding technique over another is dictated by the problem at hand. For example, data compression applications are motivated by the need to reduce storage requirements. In this particular problem, it is usually important to employ encoding techniques which allow perfect reconstruction (by means of a *decoder*) of the data from their coded form. Encoder-decoder pairs which incur zero error are referred to as *information preserving*.

In image transmission applications, such as the transmission of space-probe pictures for human interpretation, interest lies in techniques which achieve maximum reduction in the quantity of data to be transmitted, subject to the constraint that a reasonable amount of fidelity be preserved. In this case, emphasis is placed on reducing the amount of data that must be transmitted and the encoding technique need not be information-preserving, as long as the resulting images are acceptable for visual or machine analysis.

Feature extraction applications are used primarily for pattern recognition by computer. In this case, the most important consideration is the choice of encoding techniques which will reduce the data subject to the

Rafael C. Gonzalez and Paul Wintz, Digital Image Processing   ISBN 0-201-02596-5; 0-201-02597-3(pbk.)

constraint that enough information be preserved to allow a machine to differentiate between items of interest in an image. Consider, for example, the problem of classifying by machine different types of agricultural crops in a satellite image. Two types of features are important in this case: those that differentiate between vegetation and nonvegetation; and those which can be used to differentiate between types of vegetation. Other features such as those related to the difference between a road and a river need not be taken into account in selecting an encoding procedure for this particular problem.

6.1 FIDELITY CRITERIA

6.1.1 Objective Fidelity Criteria

In some image transmission systems some errors in the reconstructed image can be tolerated. In this case a fidelity criterion can be used as a measure of system quality. Examples of objective fidelity criteria are the root-mean-square (rms) error between the input image and output image, and the rms signal-to-noise ratio of the output image. Suppose that the input image consists of the $N \times N$ array of pixels $f(x, y)$, $x, y = 0, 1, ..., N - 1$. As discussed in Section 2.3, each pixel is an m-bit binary word corresponding to one of the 2^m possible gray level values. The encoder reduces the data bulk from $N \times N \times m$ bits to a fewer number of bits. The decoder processes these bits to reconstruct the output picture consisting of the $N \times N$ array of picture elements $g(x, y)$, $x, y = 0, 1, ..., N - 1$, where each pixel is also an m-bit binary word corresponding to one of the 2^m possible gray level values.

For any value of x and y in the range $0, 1, ..., N - 1$, the error between an input pixel and the corresponding output pixel is

$$e(x, y) = g(x, y) - f(x, y) \tag{6.1-1}$$

The squared error averaged over the image array is

$$\overline{e^2} = \frac{1}{N^2} \sum_{x=0}^{N-1} \sum_{y=0}^{N-1} e^2(x, y)$$

$$= \frac{1}{N^2} \sum_{x=0}^{N-1} \sum_{y=0}^{N-1} \left[g(x, y) - f(x, y) \right]^2 \tag{6.1-2}$$

and the rms error is defined as

$$e_{rms} = \left[\overline{e^2} \right]^{1/2} \tag{6.1-3}$$

We can also consider the difference between the output and input images to be "noise", so that each output signal (pixel) consists of an input signal (the corresponding input pixel) plus noise (the error), that is,

$$g(x, y) = f(x, y) + e(x, y) \qquad (6.1\text{-}4)$$

The *mean-square signal-to-noise ratio* of the output image is defined as the average of $g^2(x, y)$ divided by the average of $e^2(x, y)$ over the image array. In other words,

$$(\text{SNR})_{\text{ms}} = \sum_{x=0}^{N-1} \sum_{y=0}^{N-1} g^2(x, y) \Big/ \sum_{x=0}^{N-1} \sum_{y=0}^{N-1} e^2(x, y) \qquad (6.1\text{-}5)$$

The rms value of (SNR) is then given by

$$(\text{SNR})_{\text{rms}} = \left[\sum_{x=0}^{N-1} \sum_{y=0}^{N-1} g^2(x, y) \Big/ \sum_{x=0}^{N-1} \sum_{y=0}^{N-1} [g(x, y) - f(x, y)]^2 \right]^{1/2} \qquad (6.1\text{-}6)$$

where the variable term in the denominator is the noise expressed in terms of the input and output images.

An alternate definition of signal-to-noise ratio is the square root of the peak value of $g(x, y)$ squared (assuming the minimum value is zero) and the rms noise; that is,

$$(\text{SNR})_{\text{p}} = \left\{ [\text{peak value of } g(x, y)]^2 / e_{\text{rms}} \right\}^{1/2} \qquad (6.1\text{-}7)$$

where e_{rms} is given by Eq. (6.1-3). The peak value of $g(x, y)$ is the total dynamic range of the output image. Hence, $(\text{SNR})_{\text{rms}}$ and $(\text{SNR})_{\text{p}}$ differ by a scale constant equal to the ratio of maximum signal level to the average signal level.

6.1.2 Subjective Fidelity Criteria

When the output images are to be viewed by people, as in the case of broadcast television, it is more appropriate to use a subjective fidelity criterion corresponding to how good the images look to human observers. The human visual system has peculiar characteristics so that two pictures having the same amount of rms error may appear to have drastically different visual qualities. As indicated in Section 2.1, an important characteristic of the human visual system is its logarithmic sensitivity to light intensity so that errors in dark areas of an image are much more noticeable than errors in light areas. The human visual system is also sensitive to abrupt spatial changes in gray level so that errors on or near the edges are more bothersome than errors in background texture. The subjective quality of an image can be evaluated by showing the image to a number of observers and averaging their evaluations. One possibility is to use an absolute scale such as the one used by Panel 6 of the Television Allocations Study Organization (Frendendall and Behrend [1960]:

1) *Excellent.* — The image of extremely high quality, as good as you could desire.

2) *Fine.* — The image is of high quality, providing enjoyable viewing. Interference is not objectionable.

3) *Passable.* — The image is of acceptable quality. Interference is not objectionable.

4) *Marginal.* — The image is of poor quality and you wish you could improve it. Interference is somewhat objectionable.

5) *Inferior.* — The image is very poor but you could watch it. Objectionable interference is definitely present.

6) *Unusable.* — The image is so bad that you could not watch it.

Another possibility is to use the pair-comparison method where observers are shown images two at a time and asked to express a preference. Both methods have advantages and disadvantages. By averaging the results of many observers (20 observers are usually adequate) the first method results in an absolute number between 1 and 6 for each image but some observers may allow the scale to drift during the course of looking at a sequence of images. The second method avoids this difficulty but yields only a rank ordering of the images.

6.2 THE ENCODING PROCESS

Encoders can be modelled as a sequence of three operations, as illustrated in Fig. 6.1, where images are expressed in vector form (see Section 3.5.5). The mapping operation maps the input data from the pixel domain

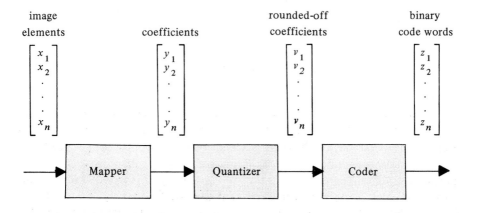

Figure 6.1. An encoder model.

into another domain where the quantizer and coder can be used more efficiently in the sense that fewer bits are required to code the mapped data
than would be required to code the original input data. The quantizer
rounds off each mapped datum to one of a smaller number of possible values
so that fewer code words with fewer bits are required. The coder assigns a
code word to each quantizer output.

6.2.1 The Mapping

The mapping operation maps the input set of numbers (pixels) into
another set of numbers. The basic procedure is best explained by means of
some simple examples.

In *run length* encoding, the sequence of image elements along a scan
line (row) $x_1, x_2, ..., x_N$ is mapped into a sequence of pairs $(g_1, l_1), (g_2, l_2),$
..., (g_k, l_k) where g_i denotes the gray level and l_i the run length of the ith
run, as illustrated in Fig. 6.2. For pictures such as weather maps, significantly fewer bits may be required to encode the run length sequence than the
image element sequence. This mapping is *reversible* because the sequence of
image elements can be reconstructed from the sequence of runs.

Another mapping of utility in image encoding is the linear transformation

$$
\begin{bmatrix} y_1 \\ y_2 \\ \vdots \\ y_n \end{bmatrix} = \begin{bmatrix} a_{11} & a_{12} & \cdots & a_{1n} \\ a_{21} & a_{22} & \cdots & a_{2n} \\ \vdots & \vdots & \ddots & \vdots \\ a_{n1} & a_{n2} & \cdots & a_{nn} \end{bmatrix} \begin{bmatrix} x_1 \\ x_2 \\ \vdots \\ x_n \end{bmatrix}
\tag{6.2-1}
$$

or

$$
\mathbf{y} = \mathbf{A}\mathbf{x} \tag{6.2-2}
$$

This transformation may or may not be reversible, depending on the choice
of $\mathbf{A}$. In this case the vector of pixels $\mathbf{x}$ is transformed into a vector of coefficients $\mathbf{y}$. For some sets of vectors $\mathbf{x}$ and some transformations $\mathbf{A}$ fewer bits
are required to encode the n coefficients of $\mathbf{y}$ than the n pixels of $\mathbf{x}$. In
particular, if the elements $x_1, x_2, ..., x_n$ are highly correlated and the transformation matrix $\mathbf{A}$ is chosen such that the coefficients $y_1, y_2, ..., y_n$ are
less correlated, then the y_i's can be individually coded more efficiently than
the x_i's.

A *difference mapping* is obtained if we use the matrix

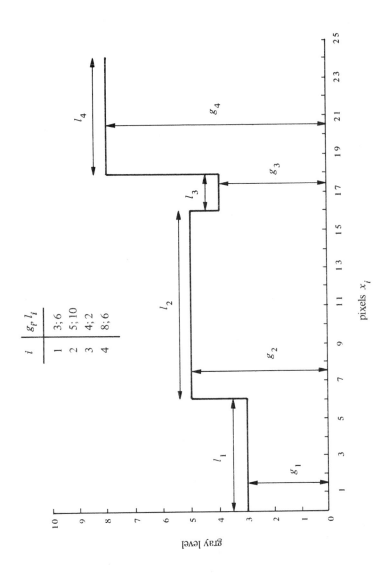

Figure 6.2. Example of run length mapping.

$$\mathbf{A} = \begin{bmatrix} 1 & 0 & 0 & 0 & 0 & 0 \\ 1 & -1 & 0 & 0 & 0 & 0 \\ 0 & 1 & -1 & 0 & 0 & 0 \\ 0 & 0 & 1 & -1 & 0 & 0 \\ 0 & 0 & 0 & 1 & -1 & 0 \\ 0 & 0 & 0 & 0 & 1 & -1 \end{bmatrix} \qquad (6.2\text{-}3)$$

in Eq. (6.2-2). The first element of y is $y_1 = x_1$. However, all subsequent coefficients are given by $y_i = x_{i-1} - x_i$. If the gray levels of adjacent pixels are similar, then the differences $y_i = x_{i-1} - x_i$, will, on the average, be smaller than the gray levels so that it should require fewer bits to code them. This mapping is also reversible.

The above examples are typical of mapping procedures used in image encoding. Some additional techniques are developed later in this chapter in the context of specific encoding applications.

6.2.2 The Quantizer

Consider the number of possible values for each of the coefficients y_i resulting from the linear transformation given by Eq. (6.2-1). Each coefficient is a linear combination of n pixels, that is,

$$y_i = a_{i1}x_1 + a_{i2}x_2 + \cdots + a_{in}x_n \qquad (6.2\text{-}4)$$

If each element x_j in the sum can have any of 2^m different values then each $a_{ij}x_j$ term can also have any of 2^m different values. The sum of n such terms could have any of $(2^m)^n = 2^{mn}$ different values. Consequently, a natural binary representation would require mn-bit code words to assign a unique word to each of the possible 2^{mn} values of y_i. Since only m-bit words would be required to code any x_j, and our objective is to use fewer bits to code the y_i, we must round off the y_i to a fewer number of allowed levels.

A quantizer is a device whose output can have only a limited number of possible values. Each input is forced to one of the allowable output values. One way to accomplish this is to divide the input range into a number of bins, as illustrated in Fig. 6.3. If an input falls into the kth bin, then the output is the value w_k associated with that bin. One possibility is to make w_k correspond to the center of the kth bin so that each input is rounded off to the center of the bin into which it falls. A *uniform quantizer* is one in which all bin widths are equal. *Nonuniform quantizers* allow different bins to have different widths.

The quantizer operation is nonreversible because, given the output value, we cannot in general determine the input value. Let y represent any input value from the vector $\mathbf{y}$ shown in Fig. 6.1, and let v be the corresponding output of the quantizer. The quantization error is the difference

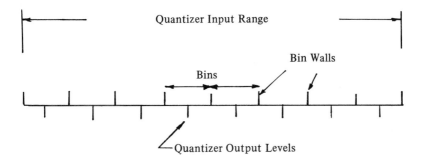

Figure 6.3. Input-output relantionship.

between the quantizer output and input, that is,

$$e_q = v - y \tag{6.2-5}$$

Clearly, the minimum error is zero, which happens only when the input equals one of the allowed output values. The maximum error for a uniform quantizer of bin width Δ is half the bin width; that is, $e_{q\,max} = \Delta/2$.

The rms error is the square root of the squared error averaged over all possible values of the input y. If y is equally likely to be any value within the bin, then the rms error is

$$e_{qrms} = \int_{v-\Delta/2}^{v+\Delta/2} (v - y)^2 \, dy \tag{6.2-6}$$

If all the quantizer bins have the same width, Δ, and if for each bin the input values y are equally likely to be any value within that bin, then the error e_{qrms} is the same for all bins so that the quantizer rms erros is given by Eq. (6.2-6), even for the inputs y that are more likely to fall in some bins than others.

If the rms value of the input y is given by

$$y_{rms} = \sqrt{\int y^2 \, dy} \tag{6.2-7}$$

then the signal-to-quantization distortion (signal-to-quantization noise) ratio is given by

$$Q_{SNR} = y_{rms}/e_{qrms} \tag{6.2-8}$$

If all values of y within the bins are not equally likely then the squared error $(v - y)^2$ must be weighted by the probability density function $p(y)$

$$e_{\text{qrms}} = \int_{v-\Delta/2}^{v+\Delta/2} (v-y)^2 p(y) \, dy \qquad (6.2\text{-}9)$$

and the total quantization error is a weighted average of these terms. In other words, the error term for each bin must be weighted by the probability

$$\int_{v-\Delta/2}^{v+\Delta/2} p(y) dy$$

that y fell in that bin.

For some mappings greater efficiency can be achieved by using a different quantizer and/or code for each of the different coefficients produced by the mapping. For example, the linear transformation defined by Eq. (6.2-1) results in coefficients of unequal variances. Since the dynamic ranges of the coefficients can differ greatly, we can use more quantizer bins and, therefore, more bits to code the coefficients with the larger variances and fewer quantizer bins and fewer bits to code the coefficients with the smaller variances. The quantizers could be uniform or nonuniform, and the codes could be equal length or unequal length. This concept is discussed in greater detail in Section 6.4.2.

6.2.3 The Coder

The inputs to the coder are the n elements of the vector **v** in Fig. 6.1. Suppose that each element v_i can assume one of M values (levels) w_1, w_2, ..., w_M. For each input v_i, the coder outputs a binary word whose value depends on the value w_k of the input. The coder input-output relationship is one-to-one in that a unique code word c_k is assigned to each possible input value w_k. Hence, the process is reversible because given a code word c_k we know w_k. The coder does not introduce any error into the encoding process. If the coder must handle M possible input values, then designing the coder amounts to choosing M unique binary code words and assigning one of them to each input.

An *equal-length code* is a set of code words each of which has the same number of bits, along with a rule for assigning code words to quantizer output levels. One example of an equal-length code is the natural binary code. One possible assignment rule for the natural code is to order the code words according to their binary value. For example, suppose that there are eight possible coder input values (quantizer output levels) ordered w_1, w_2, ..., w_8; then the natural code is c_1 = 000, c_2 = 001, ..., c_8 = 111, as illustrated in Table 6.1. There are 8 possible assignments of the eight code words to the eight inputs. The *reflected binary* or *Gray code*, also illustrated in

Table 6.1. Some typical codes.

Input	Natural code	Gray code	B_1-code	B_2-code	S_2-code
w_1	000	111	C0	C00	00
w_2	001	110	C1	C01	01
w_3	010	100	C0C0	C10	10
w_4	011	101	C0C1	C11	1100
w_5	100	001	C1C0	C00C00	1101
w_6	101	000	C1C1	C00C01	1110
w_7	110	010	C0C0C0	C00C10	111100
w_8	111	011	C0C0C1	C00C11	111101

Table 6.1, has the property that any two adjacent code words in the set differ in only one bit position.

A *uniquely decodable code* is a code with the property that a sequence of code words can be decoded in only one way. The code $c_1 = 0$, $c_2 = 1$, $c_3 = 01$, $c_4 = 10$ is not unique because the sequence of bits 0011 could be decoded as $c_1 c_1 c_2 c_2$ or as $c_1 c_3 c_2$. All the codes presented in Table 6.1 are uniquely decodable.

An *instantaneous code* is one that can be decoded instantaneously. That is, if we look at the sequence of incoming bits bit-by-bit we know the value of the input when we come to the end of a code word. We do not have to look ahead at any future incoming bits in order to decode the bit stream. All of the codes in Table 6.1 are instantaneous except for the B-codes which require that we look one bit ahead to decode. The B-codes are discussed in Section 6.2.3.3

We would like to design the coder to use as few bits as possible. Since there are 2^M unique equal length code words of length $b = \log_2 M$ bits, the b-bit natural code can handle up to 2^M possible input levels and the number of bits output for each input is b. This is an optimum code only when all input levels w_1, w_2, ..., w_M are equally likely. When some input levels occur more often than others, greater efficiency can be achieved by using an unequal length code and assigning the shortest code words to the most likely inputs and longer code words to the least likely inputs.

Given the coder input probabilities, it is of interest to determine the minimum number of bits required to code these inputs and generate a code that would achieve this minimum. In order to reach this objective we must first master the concept of entropy.

6.2.3.1 Entropy

Suppose we have a set of M random variables α_1, α_2, ..., α_M with probabilities $p_1 = p(\alpha_1)$, $p_2 = p(\alpha_2)$, ..., $p_M = p(\alpha_M)$. Then the entropy, in bits, is defined as

$$H = - \sum_{k=1}^{M} p_k \log_2 p_k \qquad (6.2\text{-}10)$$

Suppose there are $M = 8$ random variables and that they are equally likely, i. e., $p_1 = p_2 = ... = p_8 = 1/8$. Then the entropy is

$$H = - \sum_{k=1}^{8} \frac{1}{8} \log_2 \frac{1}{8}$$

$$= 3$$

On the other hand, if $p_1 = 1$, $p_2 = p_3 = ... = p_8 = 0$ then the entropy is

$$H = 0$$

In general, the entropy for M random variables can range from 0 to $\log_2 M$.

Entropy is a measure of the degree of randomness of the set of random variables. The least random case is when one of the random variables has probability 1 so that the outcome is known in advance and $H = 0$. The most random case is when all events are equally likely. In this case $p_1 = p_2 = ... = p_M = 1/M$ and $H = \log_2 M$. This concept is similar to the entropy concept in thermodynamics.

In our coding applications, entropy represents the amount of information associated with the set of coder input values and gives a lower bound on the average number of bits required to code those inputs. If the set of coder input levels is w_1, w_2, ..., w_M with probabilities $p_1, p_2, ..., p_M$, then we are guaranteed that it is not possible to code them using less than

$$H = - \sum_{k=1}^{M} p_k \log_2 p_k$$

bits on the average. Therefore, the entropy concept provides a performance criterion against which we can measure any particular code. That is, if we design a code with code words c_1, c_2, ..., c_M with word lengths $\beta_1, \beta_2, ...,$ β_M, the average number of bits required by the coder is

$$R = \sum_{k=1}^{M} \beta_k p_k \qquad (6.2\text{-}11)$$

If R is close to H the coder is near optimum; if it is significantly different from H it is not.

The entropy defined in Eq. (6.2-10) is the first order entropy. It takes into account only the relative probabilities of the M possible input values

w_1, w_2, ..., w_M. If successive inputs are independent, then the first order entropy is also a bound on the average number of bits per input required to code a sequence of inputs. If successive inputs are not independent then the entropy associated with a sequence of inputs is less per input than for an individual input. The second order entropy is defined as

$$H_2 = - \sum_{i=1}^{M} \sum_{j=1}^{M} p(w_i, w_j) \log_2 p(w_i, w_j) \qquad (6.2\text{-}12)$$

where $p(w_i, w_j)$ is the joint probability density function of the two random variables w_i and w_j. It represents a lower bound on the number of bits required to code a sequence of inputs if we code them two at a time (e.g., we input two successive quantizer levels and output a single code word). Similarly, we can define a third order entropy

$$H_3 = - \sum_{i=1}^{M} \sum_{j=1}^{M} \sum_{k=1}^{M} p(w_i, w_j, w_k) \log_2 p(w_i, w_j, w_k) \qquad (6.2\text{-}13)$$

which is a bound on the number of bits required to code the inputs three at a time. It can be shown that $H_1 \geqslant H_2 \geqslant \dots$. We do not pursue these higher order entropies for two reasons. First, the amount of computation required to obtain higher order probabilities is prohibitive in practice and second, the purpose of the mapping operation in the encoder is to transform the input picture elements which are usually highly dependent into a set of coefficients which are much less dependent so that they can be efficiently coded one at a time.

6.2.3.2 Huffman Code

A *compact code* is a code with an average word length less than or equal to the average length of all other uniquely decodable codes for the same set of input probabilities, that is, it is a minimum length code. Given a set of input probabilities we can generate a compact code using an algorithm due to Huffman [1952]. A Huffman code can be constructed by first ordering the input probabilities according to their magnitudes, as illustrated in Fig. 6.4 for six input values. The two smallest probabilities are combined by addition to form a new set of probabilities. The new set of probabilities, which has one fewer probability than the original set, is again ordered according to magnitude. Equal probabilities can be ordered in any way, (e.g., the 0.1 obtained by combining input probabilities 0.06 and 0.04 could be placed in any three of the bottom step 1 entries). When we get down to two probabilities we stop, as in step 4. Code words are generated by starting at the last step and working backwards. We start by assigning a 0 to one of the last two combined probabilities, and a 1 to the other, as illustrated in Fig. 6.5, where we have placed a 0 to the left of the 0.6 in step 4 and a 1 to the left of the 0.04. We now proceed backwards to step 3, decomposing

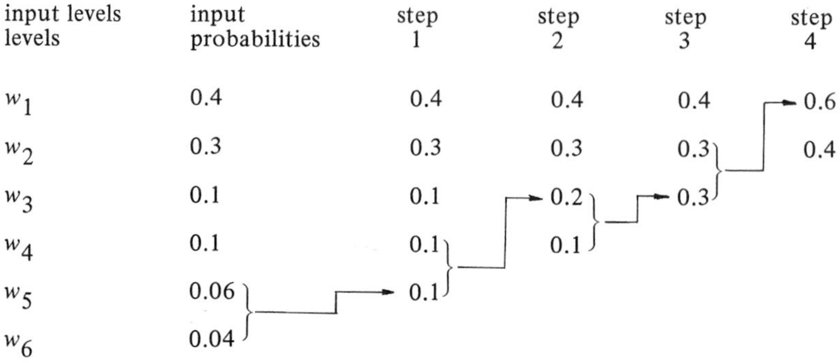

input levels levels	input probabilities	step 1	step 2	step 3	step 4
w_1	0.4	0.4	0.4	0.4	0.6
w_2	0.3	0.3	0.3	0.3	0.4
w_3	0.1	0.1	0.2	0.3	
w_4	0.1	0.1	0.1		
w_5	0.06	0.1			
w_6	0.04				

Figure 6.4. Construction of a Huffman code.

probabilities and generating code words as we go. For example, the 0.6 in step 4 is decomposed back into the two 0.3 probabilities in step 3. The 0 associated with the 0.6 remains the first bit of each of its decomposed code words and the 1 associated with the 0.4 remains the first bit of the 0.04 in step 3. A second bit, a 0 and 1, respectively, is appended to each of the code words associated with their reconstructed probalilities to obtain the code words in step 3. The same procedure is repeated to go back to step 2, and again to step 1, and finally to the input probabilities, at which point we have a code word assigned to each input level w_i. It can be proved that the procedure outlined above generates a compact code.

For the input probabilities listed in Fig. 6.4 the entropy is

$$H = (-.4) \log(.4) - (.3) \log(.3) - (.1) \log(.1) - (.1) \log (.1)$$

$$- (.06) \log(.06) - (.04) \log(.04)$$

$$= 2.14 \text{ bits}$$

The average word length of the Huffman code for this example is

$$R = 1(.4) + 2(.3) + 3(.1) + 4(.1) + 5(.06) + 5(.04)$$

$$= 2.20 \text{ bits}$$

6.2.3.3 B-Codes

In some applications the probabilities of the coder inputs obey a power law; that is, the probabilities of the M coder inputs are of the form

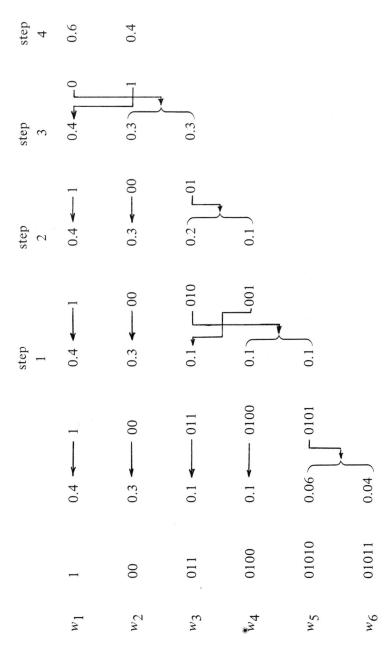

Figure 6.5. Construction of Huffman code words.

$$p_k = k^{-\gamma} \tag{6.2-14}$$

for $k = 1, 2, ..., M$, and some positive constant γ. For example, the distribution of run lengths for many types of graphics (e.g, typewritten text) is approximately exponential. The B-codes are nearly optimal for data that obey Eq. (6.2-14).

The B_1-code is presented in Table 6.1. Half of the bits in each code word are "continuation" bits labelled C and the other half are "information" bits. The information bits use a natural code which increases in length, as illustrated in Table 6.1. The continuation bit is, of course, either a 0 or a 1, but it can be determined by either of two rules: for 2-level data where each pixel is white or black, the continuation bit can be set equal to the gray level, say, $C = 0$ for black and $C = 1$ for white. The other possibility is to let it alternate with each code word since its only purpose is to signify how long a code word is. For example, the sequence of code words for the sequence of inputs w_1, w_8, w_5 could be $\underline{0}0\ \underline{1}0\ \underline{1}0\ \underline{1}1\ \underline{0}1\ \underline{0}0$ or $\underline{1}0\ \underline{0}0\ \underline{0}0\ \underline{0}1\ \underline{1}1\ \underline{1}0$ where we have underlined the continuation bits. A change in the continuation bits signifies the start of a new code word. Note that the code is not instantaneous because the decoder must look ahead to the next continuation bit in order to determine whether or not the present code word has ended.

Implementation of the B_1-code is much simpler than for the Huffman code. For example, for coding run lengths the coder for the information bits is simply an up-counter that counts up by one for each new datum until the end of the run is reached. At the end of each run the counter is reset to zero and the continuation bit is flipped. Similarly, decoding can be accomplished by presetting a down-counter with information bits and letting it count down until the continuation bit changes state. Higher order B-codes can also be reconstructed. A B_n-code uses n information bits for each continuation bit, as illustrated in Table 6.1 for $n = 2$.

For the set of input probabilities listed in Fig. 6.4 the average length of the B_1 code is

$$R = 2(.4) + 2(.3) + 4(.1) + 4(.1) + 4(.06) + 4(.04)$$

$$= 2.6$$

and the average length of the B_2 code is

$$R = 3(.4) + 3(.3) + 3(.1) + 3(.1) + 6(.06) + 6(.04)$$

$$= 3.3$$

6.2.3.4 Shift Codes

Another unequal length code that is easy to implement and is reasonably effecient for inputs with monotonically decreasing probabilities is the S_n-code. For example, the S_2 code uses 2-bit code words so that we have a total of four distinct 2-bit code words c_1, c_2, c_3, and c_4. Three of these code words c_1, c_2, c_3 are assigned to the first three input values w_1, w_2, w_3 and the remaining code word is used to signify that the input is outside this range. When this event occurs the first three code words are shifted by 3 and assigned to the inputs w_4, w_5, w_6. If the input is still outside this range the shift code word is used again and the three code words are shifted to w_7, w_8, w_9, and so on. As an example, suppose that an input falls on the range w_1 to w_3, say w_2; then the coder outputs code word c_2. Suppose the input is w_5; then the coder outputs code word c_4 (to indicate shift) followed by code word c_2. For input w_9 the shift code word c_4 is used twice followed by code word c_3. This code is illustrated in Table 6.1, where $c_1 = 00$, $c_2 = 01, c_3 = 10$, and $c_4 = 11$.

For the set of input probabilities listed in Fig. 6.4 the average length of the S_3 code is:

$$R = 2(.4) + 2(.3) + 2(.1) + 4(.1) + 4(.06) + 4(.04)$$

$$= 2.4$$

These results are summarized in Table 6.2, along with the results for the previous examples.

Table 6.2.

Inputs	Probabilities	Huffman Code	B_1 Code	B_2 Code	S_2 Code	Natural Code
w_1	.4	1	C0	C00	00	000
w_2	.3	00	C1	C01	01	001
w_3	.1	011	C0C0	C10	10	010
w_4	.1	0100	C0C1	C11	1100	011
w_5	.06	01010	C1C0	C00C00	1101	100
w_6	.04	01011	C1C1	C00C01	1110	101
Entropy	2.14					
Average Code Word Length		2.2	2.6	3.3	2.4	3.0

The general encoding process described in Sections 6.2.1 through 6.2.3 is illustrated in the context of image processing in Sections 6.3 and 6.4, where several examples of error-free encoding and encoding relative to a fidelity criterion are given. Although these techniques are developed in the context of specific applications to clarify the presentation, the approaches shown are quite general and can be applied to a much larger class of problems than those considered in the following discussion.

6.3 ERROR-FREE ENCODING

As indicated at the beginning of this chapter, it is of interest in some applications to compress the amount of data in an image, subject to the constraint that the encoding process be reversible in the sense that an exact replica of the original image must be reconstructible from its encoded form. In this section we consider three examples of error-free encoding.

6.3.1 Example 1. Differential Encoding for Storage of LANDSAT Imagery

One frame of LANDSAT[†] imagery consists of four digital images. Each image is of the same scene, but taken through a different spectral window. Two of the spectral windows are in the visible region of the spectrum (corresponding more or less to the green and the red regions of the visible spectrum) and two are in the infrared region. An example of a LANDSAT frame is presented in Fig. 6.6. The white line sloping down to the right in the bottom two pictures is an interstate highway; the small white puffs in the top parts of the bottom two pictures are clouds. Figure 6.7 is another LANDSAT image of the same location but taken on a different day. These scenes are 100 × 100 nautical miles. Each image is represented by a 2340 × 3380 array of pixels. Each pixel is a 7-bit binary word corresponding to one of 128 gray levels, with 0 corresponding to black and 127 corresponding to white.

LANDSAT images are stored on magnetic tape. The number of bits required to store one frame is $(2340)(3234)(7)(4) = 211,000,000$. LANDSAT collects 30 frames per day so that the number of bits to be archived is 6,000,000,000 per day or approximately 2.2×10^{11} bits per year. If the data bulk (number of bits) could be reduced by a factor of 2, the number of data tapes (and the number of warehouses required to store them) could be reduced by the same factor. In order to avoid the possibility of destroying information that may be useful to future users of the data, it

---

[†] LANDSAT is an abbreviation for *Land Satellite*, a name given by NASA to satellites designed to monitor the surface of the Earth.

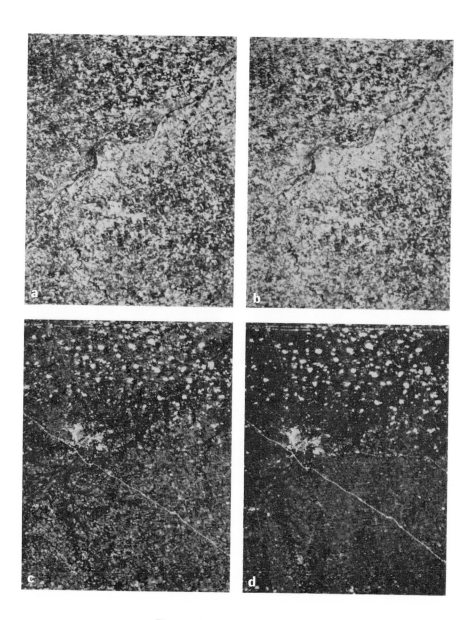

Figure 6.6. A LANDSAT frame.

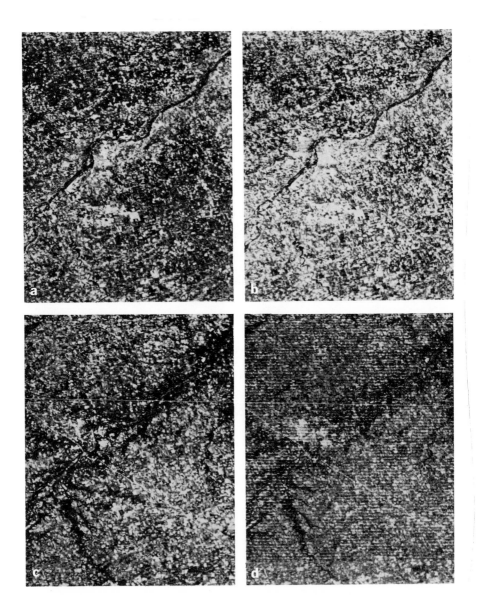

Figure 6.7. The same LANDSAT frame as in Fig. 6.6 but taken on a different day.

is required that the data be stored in a format that allows exact reconstruction of the original digital images.

The histograms of the pixel gray level values for each of the four images presented in Fig. 6.6 are shown in Fig. 6.8. It is noted that most of the 128 gray level values occur infrequently. For three of the images most gray levels fall in the range 16-48. One of the images is "lighter" with most of the levels between 48 and 64. Other LANDSAT images have other characteristics. For example, the polar ice caps are lighter and the ocean is darker.

Let us look at the differences in the gray levels of adjacent picture elements down each scan line. That is, if the picture elements from left to right along the jth row (scan line) are given by $x_1, x_2, ..., x_{3234}$, then we can map this set of 3234 integers into the new set of 3234 integers x_1, $x_2 - x_1, x_3 - x_2,..., x_{3234} - x_{3233}$. This is the difference mapping defined by Eq. (6.2-3).

Each of the original image elements x_i is one of the integers 0, 1, ..., 127. Therefore, each difference $x_i - x_{i-1}$ is an integer with value between -127 and 127. To assign a unique equal length binary code word to each picture element x_i requires 7-bit code words. To assign a unique equal length binary code word to each picture element difference $x_i - x_{i-1}$ requires 8-bit code words, assuming that the quantizer outputs a separate value for each of the above 255 differences.

The histograms of the differences $x_i - x_{i-1}$ for each of the four images of Fig. 6.6 are shown in Fig. 6.9. It is noted that all four histograms are peaked about zero (including the channel 3 image whose gray level histogram was different from those of channels 1, 2, and 4). Most of the differences fall in the range -8 to $+8$. The difference operation has mapped the input set of pixel values into a different set of numbers having much more structure (less randomness).

Suppose that we construct a code consisting of the 16 code words c_1, $c_2, ..., c_{16}$. We could, for example, use the natural code so that $c_1 = 0000$, $c_2 = 0001, ..., c_{16} = 1111$. Let us assign 14 of these 16 code words to the 14 differences $-7, -6, ..., -1, 0, +1, ..., +6$, as illustrated in Fig. 6.10. Of the two remaining code words, c_1 and c_{16}, we use c_{16} to indicate that the pixel difference was greater than $+6$ and c_1 to indicate that it was less than -7. This is a double-sided version of the shift code discussed in Section 6.2.3.4.

Any pixel difference from -127 to $+127$ can then be coded by one or more of the 16 code words if we use the following rule: If the difference $\Delta_i = x_i - x_{i-1}$ falls in the range -7 to $+6$ use the code word corresponding to the difference value. For example, -7 would be coded as c_2, -6 as c_3, etc. If the difference Δ_i is more than $+6$ we first use the shift up code word c_{16} and shift all code words up by 14 units so that they are now assigned to the differences $+7, +8, ..., +20$ as illustrated in Fig. 6.11. Any difference Δ_i falling in the range $+7, ..., +20$ is coded by using the shift up code word

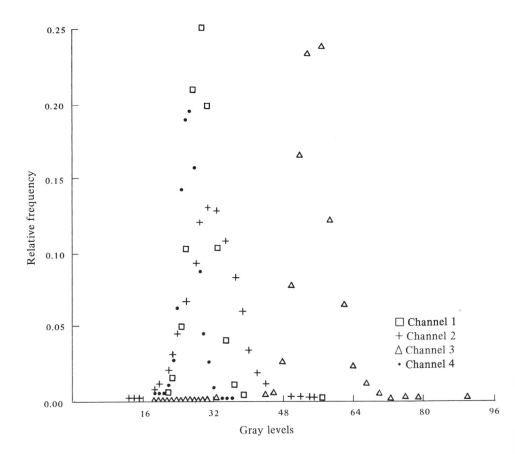

Figure 6.8. Gray level histograms for the images in Fig. 6.6.

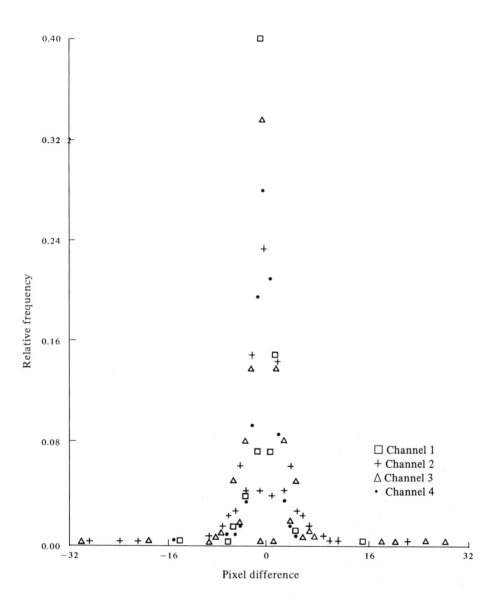

Figure 6.9. Histograms of pixel differences for the images in Fig. 6.6.

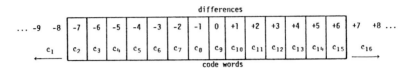

Figure 6.10. Illustration of code word assignments.

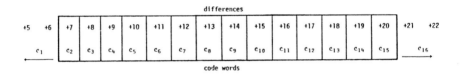

Figure 6.11. Illustration of code word assignments after shifting.

c_{16} followed by the code word corresponding to the difference. For example, $+8$ could be coded by using the sequence of two code words c_{16}, c_3. If the difference is greater than $+20$ we use the shift code word again and shift the code words up by another 14 units so that they now span the range $+21$, $+22$, ..., $+34$. The difference $+22$ would be coded by using the sequence of three code words c_{16}, c_{16}, c_3. For differences greater than $+34$ the procedure is repeated. The shift up code word is used again and again until the difference Δ_i falls in the range spanned by the 14 code words c_2, ..., c_{15}.

For a difference less than -7 we use the same procedure, except that instead of using the shift up code word c_{16} we use the shift down code word c_1 to shift the code words down by 14 units. Each difference less than -7 is coded by using the shift down code word c_1 one or more times followed by one of the code words c_2, ..., c_{15} indicating the range in which the difference value is located. For example, -23 would be coded with the sequence c_1, c_1, c_{14}.

Any difference value from -127 to $+127$ can be coded by using a sequence of one or more of the 16 code words c_1, ..., c_{15}. If we use the

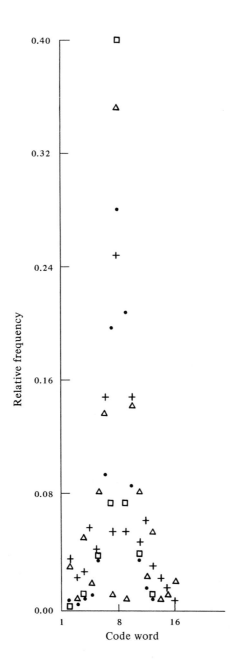

Figure 6.12. Relative frequencies of using each code word.

natural code $c_1 = 000$, $c_2 = 0001$, ..., $c_{16} = 1111$ then most differences can be coded with 4 bits because most fall in the range -7 to $+6$. Occasionally, two 4-bit code words requiring a total of 8 bits are required. On rarer occasions 12 or more bits are necessary. On the average, about 4.3 bits per picture element are required for the images in Figs. 6.6 and 6.7.

The relative frequency of using each of the 16 code words for the four images of Fig. 6.6 is presented in Fig. 6.12. The code words corresponding to differences near zero are used more often than those corresponding to differences further from zero. Consequently, an unequal length code would be more efficient than an equal length code such as the natural code used above. The code shown in Fig. 6.13 has been found to be efficient for a

c_1	1010
c_2	111 1111
c_3	11 1110
c_4	1 1110
c_5	1101
c_6	100
c_7	1100
c_8	00
c_9	0110
c_{10}	0111
c_{11}	0100
c_{12}	0101
c_{13}	1 1100
c_{14}	1 1101
c_{15}	111 1110
c_{16}	1101

Figure 6.13. A code that is nearly as efficient as the Huffman code.

large number of LANDSAT frames. The average number of bits per pixel averaged over all four channels for several different LANDSAT frames was about 3.5 bits per picture element, as illustrated in Table 6.3. This corresponds to a compression ratio of 2 relative to the 7-bit input picture elements. Table 6.3 also shows the average number of bits required by the Huffman code for the data of Fig. 6.6. It requires a fraction of a bit per picture element less than the code of Fig. 6.13.

The number of code words required to encode a LANDSAT frame using either of the codes discussed in the preceding paragraphs (equal length or unequal length) depends on the characteristics of the data since the number of times the shift words are used is strongly dependent upon variations in a scene. The equal length code has the advantage that all code words

Table 6.3. Average number of bits required to code some LANDSAT frames.

	Channel 1	Channel 2	Channel 3	Channel 4
Shift code	3.0865	3.8466	3.5894	3.3685
Huffman code	2.75175	3.5893	3.2900	2.8101

are 4-bit words and can be written on magnetic tape two per 8-bit byte, although the record length is different for different frames of data. The unequal length code has the disadvantage that it must be written as a sequence of words ranging from 2 to 7 bits in length.

6.3.2 Example 2. Contour Encoding

A digital image may be viewed as a function of two variables. The variables are the spatial coordinates and the value of the function at each coordinate is the gray level of the image at that point. Since there are a finite number of discrete levels of gray, we can visualize the function as a number of plateaus (or steps) with the plateau height equal to the gray level. Dark values correspond to low plateaus and light values correspond to high plateaus. A large area of picture elements with the same gray level would produce a large plateau. A single pixel surrounded by pixels having different gray levels would make a small plateau. Knowledge of the height, location, and shape of all the plateaus is equivalent to knowledge of the image.

The contour-encoding algorithm presented in this section reduces an image to a list of contours or plateaus. Each contour is uniquely determined by specifying (1) its gray level; (2) the location (row and column) of one pixel on its boundary, called the *initial point* (IP); and (3) a sequence of directionals that give the direction of travel as we trace around the outer extremity of the contour. The algorithm consists of two subalgorithms—an IP algorithm for locating new initial points (new contours) and a T algorithm for tracing contours after they are located. The subalgorithms are used sequentially. The IP algorithm is used to locate the first initial point on the first contour and then the T algorithm is used to trace it; the IP algorithm is then used to locate the second initial point on the second contour, and then the T algorithm is used to trace it; and so on. The IP algorithm locates all contours; none are located twice. The T algorithm traces the outer boundary of the largest connected set of elements having the same value as the initial point; it always terminates back at the initial point. For each contour, the algorithm outputs the value (gray level) and location of the initial point and the direction of travel around the boundary. All elements enclosed by the contour and having the same value as the contour are neglected.

6.3.2.1 The T Algorithm

The T algorithm employs the classical rule for finding the way out of a maze—always turn left. As illustrated in Fig. 6.14, tracing a contour implies

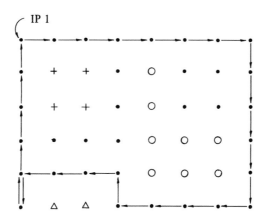

Figure 6.14. Contour No. 1 for an array with four gray levels.

defining the direction of travel between adjacent elements such that no element outside the contour and adjacent to it has the same gray level as the elements on the contour. The rule for deciding the direction of travel out of each element is given relative to the direction of entry into that element by the following *left-most-looking* (LML) rule:

LML rule: Look at the element to the left (relative to the direction of entry); if this element has the same value move to it; if not, look at the element straight ahead; if this element has the same value move to it; if not, look to the right; if this element has the same value move to it; if not, look back; if this element has the same value move to it; if not, none of the adjacent elements have the same value so that the contour consists of only one point.

The contour shown in Fig. 6.14 was traced by repeated applications of this rule, starting at the indicated IP. The direction of travel into the *first* IP is always assumed to be from the left side of the page, so that the "left relative to the direction of the entry into the IP" is "up". However, we cannot move up above the IP so we look straight ahead. This element does have the same value as the IP and so we move to it and again apply the LML rule. It is noted that the procedure terminates back at the IP. Figure 6.15

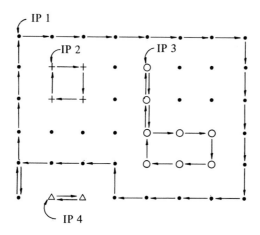

Figure 6.15. Four IPs and corresponding contours.

illustrates the four contours generated by four applications of the T algorithm, each starting at the IP shown in the figure.

The T algorithm also assigns one of four indicators to each element of the 2-dimensional array. These indicators are required by the IP algorithm described in the next section. Hence, when the 2-dimensional array is stored in memory, two extra bits must be allocated to each element in order to store the appropriate indicator. We denote the four indicators by D, A, R, and I. When the 2-dimensional data array is read into memory, each element is assigned the indicator I. As the T algorithm moves from element to element tracing out a contour it also replaces the indicator for each element on the contour with either a D, A, or R indicator according to the following *indicator assignment* (IA) rule:

IA rule: The indicator assigned to each contour element depends on the direction of travel into that element and the direction of travel out of that element, as indicated in Fig. 6.16. Some elements are passed through twice. When we pass through an element for the second time (this can be determined by checking its indicator: if it is not I we are passing it for the second time) we first determine an indicator for this pass from Fig. 6.16, but then use Fig. 6.17 to determine the indicator finally assigned to the element. (No elements are passed through more than twice.) The only exception to this rule is an IP, which always retains the indicator I.

direction of travel out of element

	↑ or →	↓ or ←
↑ or ←	A	R
↓ or →	R	D

direction of travel into element

Figure 6.16. Indicators for all possible combinations of directions of travel into and out of a pixel.

	(D,A)	(D,R)	(A,R)
indicator assigned on (first pass, second pass)	(A,D)	(R,D)	(R,A)
	(R,R)	(D,D)	(A,A)
final indicator assignment	R	D	A

Figure 6.17. Final assignments for each pair of indicators determined on the first pass and the second pass.

As an example we again consider the contour illustrated in Fig. 6.14. At each element the T algorithm applies the LML rule to determine the next direction of travel and then applies the IA rule to update the indicator for that element. The indicator stored in memory for each element on the contour shown in Fig. 6.15 is labeled next to that element in Fig. 6.18. At

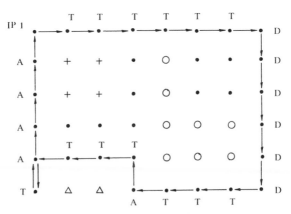

Figure 6.18. Example of indicator assignment.

this point all elements not on the contour still have indicator I. However, as contours 2, 3, and 4 are traced out, all elements on these contours (except the IPs) will be assigned an A, R, or D indicator. The IPs and all elements not lying on the contour retain the indicator I.

6.3.2.2 The IP Algorithm

The IP algorithm employs a systematic search procedure for determining initial points. The search for IPs starts with the element in the upper left hand corner (the first row, first column element) and proceeds to the right across the first row until the end of the row is reached; then the second row is searched from left to right; then the third row, and soon, we reach the element in the lower right hand corner of the data array. At this point all IPs have been located and the algorithm terminates. Hence, each element in the 2-dimensional data array is tested to determine whether or not it is in an IP.

As we move from element to element across each row in the data searching for IPs we must simultaneously compile a *comparison point list* (CPL).

Rule for construction of CPL: As we start across each row the list is empty. As we move across the row we check the indicator on each element we encounter. If the indicator is A (for add) we add the value of that element (its gray level) to the bottom of the list; if the indicator is D (for drop) we delete the *last* entry on the list; if the indicator is I or R the list is left unchanged.

At the end of a row the comparison point list will be empty because, for each row, the number of deletions is equal to the number of additions.

We are now ready to specify the test to be made on each element to determine whether or not it is an IP.

IP rule: The element under test is an IP if it meets both the following requirements: 1) its indicator is I, and 2) its value (gray level) is not equal to the value of the last entry in the CPL.

The 3-level, 14×13 array shown in Fig. 6.19 summarizes the concepts just discussed. In this case, the contour tracing algorithm reduced the image to a representation requiring eight contours.

6.3.2.3 Encoding Approach

Sequential applications of the IP and T algorithms constitute the mapping section of the encoding process illustrated in Fig. 6.1. The algorithm is initialized at the first row, first column element. This element is always an IP, say IP #1. Then we employ the T algorithm to trace out the first contour and simultaneously set the element indicators. After arriving

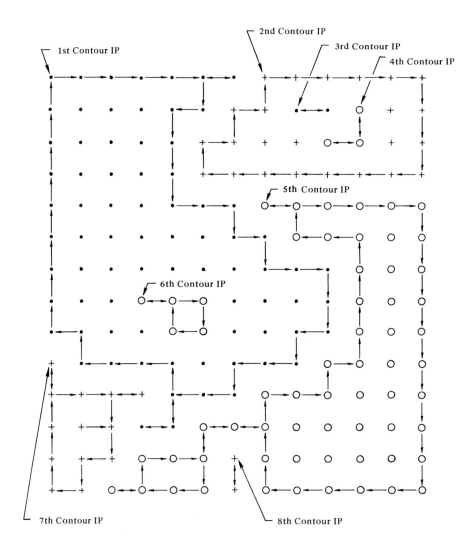

Figure 6.19. Example of contour tracer operation.

back at IP #1 we recall the IP algorithm to search for a new IP. When a new IP, say #2, is encountered we again call on the T algorithm to trace this contour and simultaneously set the appropriate element indicators. Then the IP algorithm is used to search for IP #3, and so on. Each resulting contour consists of: 1) its gray level; 2) the row number of its IP; 3) the column number of its IP; and 4) the sequence of directionals (directions of travel) around its perimeter.

One possibility for coding the contours is to use the natural code for the gray levels and the rows and columns of the initial points, and *Freeman's chain code* (Freeman [1961]) for the directionals. This chain code, illustrated in Fig. 6.20, uses the code word 00 to indicate up, 01, to the right, 10 down, and 11 to the left. For example, the sequence of directions for the

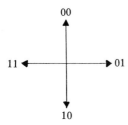

Figure 6.20. Binary coding of directionals.

third contour of Fig. 6.15 is 10, 10, 01, 01, 10, 11, 11, 00, 00, and 00. Figure 6.21(a) lists the codes used to code the contour number, the contour gray level, the row and column of the initial point, and the directionals, respectively. Figure 21.(b) shows the output of the coder using these codes to code the four contours of Fig. 6.15.

It is important for decoding purposes to know when the end of each set of directionals is reached and a new contour is to be started. This can be accomplished with no additional bits because all contours (sequences of directions) must terminate back to the starting point. We need only keep track of the cumulative number of right-left directionals and the cumulative number of up-down directionals. When they are both zero we are back at the IP and the next datum must correspond to a new contour. This rule works for all contours of length two or more. For contours of length one (single points) we use the binary pair 11; this is an impossible first directional because the pixel could not be an IP if it has the same gray level as the pixel to its left.

More sophisticated codes can be used to code the contour with fewer bits. For example, since successive gray levels are usually highly correlated the entropy of the gray level differences is usually significantly less than the entropy of the gray levels, and therefore it is possible to code the differences with fewer bits than the number required to code the gray levels directly. The same is true for the row and column numbers for the initial points. Coding the differences between the row numbers of successive contours will require fewer bits than coding the sequence of row numbers when, on the average, these are naming new IPs on each row, as is the case for many image types. A similar statement holds for the column numbers.

As indicated above, a straightforward way to code the directionals is

Contour #	Code word	Gray level	Code word	Row or column	Code word	Direction of travel	Code word
1	00	•	00	1	000	↑	00
2	01	X	01	2	001	→	01
3	10	O	10	3	010	↓	10
4	11	Δ	11	4	011	←	11
				5	100		
				6	101		
				7	110		
				8	111		

(a)

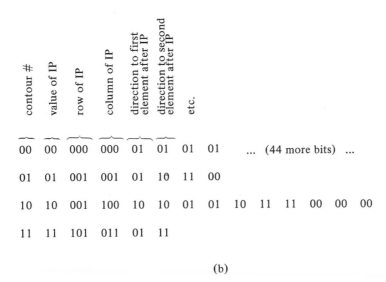

(b)

Figure 6.21. (a) A set of codes for the contour number, gray level, row, column, and direction of travel. (b) Coder output for the example in Fig. 6.15.

to use Freeman's chain code. Clearly, all possible sequences of directionals are not equally likely. Indeed, most sequences are impossible due to the constraint that the contour must terminate back at the initial point. However, it is difficult to take this structure into account. For example, the Huffman coding technique would require a different Huffman code for each different sequence length (e.g., one code for sequences of length 0, another code for sequences of length 2, etc.). Furthermore, just enumerating the possible sequences, much less determining their probabilities, gets out of hand as the sequence length increases.

The number of bits required to code a particular image depends not only on the set of codes used, but also on the number of contours, which in turn depends on the amount of detail in the image and the number of gray levels. In Fig. 6.22(d) we present some results for the three images presented in Figs. 6.22(a) through (c). The set of codes used was:

Gray levels: the gray level differences between successive gray levels were coded using a Huffman code.

Row numbers of IPs: the differences between successive row numbers (run lengths) were coded using the natural code.

Column numbers of IPs: the differences between successive column numbers were coded using a Huffman code.

Directionals: the directionals were coded using Freeman's chain code.

All Huffman codes were based on the statistics of the gray level differences averaged over the three images in Fig. 6.22. It is noted that the efficiency of the contour depends on the amount of detail in the images and also on the number of quantization levels for the pixels. In other words, for images having a large number of small contours the method is not very efficient. For images that have a few number of larger contours the efficiency is improved.

These facts are further demonstrated by experiments conducted with the binary images shown in Fig. 6.23. Gattis and Wintz [1971] used a procedure similar to the one just described to encode the contours of these images. The average number of bits needed to code the contours of the drawing, printed text, and fingerprint were 0.15, 0.15, and 0.50, respectively. As expected, the average number of bits was proportional to the complexity of the image.

6.3.2.4 Reconstruction Mapping

The reconstruction of an image from its contour code is straightforward, given an understanding of the mapping strategy. It employs the basic rules employed by the T algorithm and the IP algorithm. The reconstruction algorithm reconstructs each element in the data array starting with the first column, first row element. Since this element is IP #1 its value is known. Furthermore, the elements in the data array corresponding

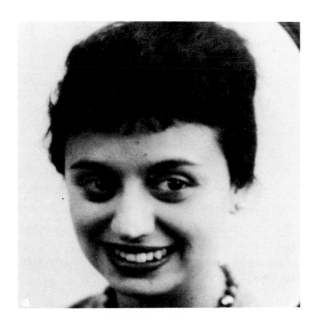

Figure 6.22. (a) Girl. (b) Cameraman. (c) Crowd. (d) Contour coding results.

number of gray levels of original	bits per pixel	average number of bits per pixel required for contour code		
		girl	cameraman	crowd
16	4	1.4	2.1	3.8
32	5	2.4	3.5	4.9
64	6	3.5	4.9	5.8
128	7	4.6	5.8	6.7

(d)

Figure 6.22. (Continued.)

Figure 6.23. (a) Section of aircraft drawing. (b) Sample of printed text. (c) Section of fingerprint.

to contour #1 can be reconstructed along with their indicators by using the system output and the IA rule. After reconstructing this contour and arriving back at IP #1 we move, element by element, across the first row, then the second row, etc. As we move across each row we also compile a CPL according to the CPL rule. Each element encountered is either an IP, an element on a contour already reconstructed, or an element not belonging to a contour. The reconstruction rule is:

1) If the element is an IP, reconstruct the contour and set the indicators.

2) If the element indicator is not I, it has already been reconstructed. Hence, move to the next element.

3) If the element is not an IP, and its indicator is I, its value is the last entry on the CPL.

6.3.3 Example 3. Run-Length Encoding for Flood Maps

River flood control in the United States is the responsibility of the United States Army Corps of Engineers (USACE). This agency monitors stream and reservoir water levels, precipitation, etc., in order to predict possible flooding conditions and to minimize flood damage by opening or closing flood gates, and raising or lowering reservoir levels. Collecting this input data over a total watershed such as the Mississippi River Valley is a severe problem.

An infrared image from a LANDSAT frame that includes a part of the Ohio River is presented in Fig. 6.24(a). Water appears dark because it absorbs electromagnetic radiation in the infrared band. For this reason it is relatively easy to classify each resolution element[†] as either "water" or "not water" with high accuracy using multispectral signature analysis techniques. The output of such a classification algorithm is presented in image form in Fig. 6.24(b) where all resolution elements classified as water are assigned a dark gray level and all nonwater resolution elements assigned a light gray level. This image was collected at a time when the river level was near normal. A LANDSAT image of the same region, but taken at a different time when the river was flooded, is shown in Fig. 6.24(c). The classification results are presented in Fig. 6.24(d). Similarly, Fig. 6.25 shows a reservoir at its normal and flooded stages along with the classification results. Processed images such as those presented in Figs. 6.24(b) and (d) and Figs. 6.25(b) and (d) are of value to the USACE, provided they can be

---

[†] In earth resources applications, pixels are sometimes called "resolution elements".

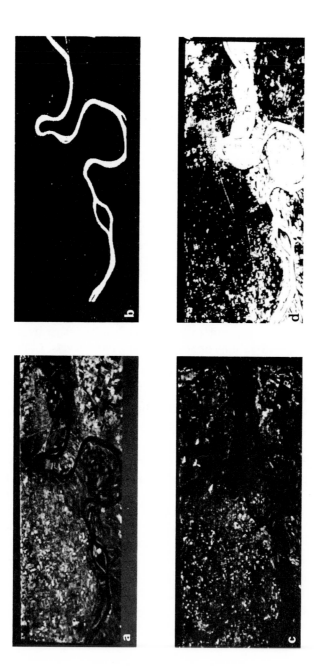

Figure 6.24. Two parts of a LANDSAT frame and corresponding classification maps.

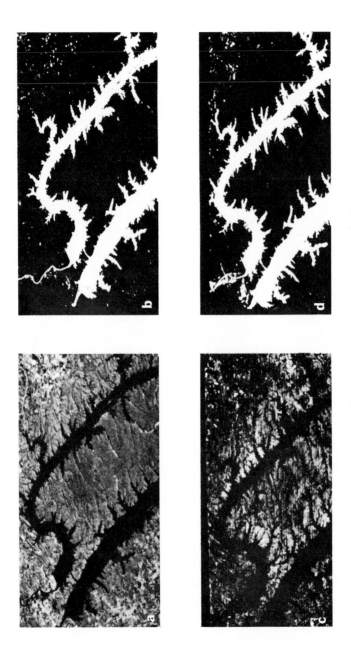

Figure 6.25. Same areas as in Fig. 6.24, but under flooded conditions.

delivered within hours. The problem is to transmit them at a reasonable cost to the USACE computer from the receiving station which may be hundreds of miles away.

6.3.3.1 One-Dimensional Run-Length Encoding (RLC)

A scan line of pixels consists of a sequence of integers (gray levels) x_1, x_2, ..., x_N. In run-length encoding we map this sequence of integers into a sequence of integer pairs (g_k, l_k) where g_k denotes gray levels and l_k denotes run lengths. The run length is the number of juxtaposed picture elements having the same gray level.

As illustrated in Fig. 6.3, we start at the leftmost pixel and set g_1 equal to x_1 and l_1 equal to the length of the run of pixels with gray level g_1. At the first gray level transition we set g_2 equal to the gray level value of second run and l_2 equal to the length of this run. This procedure is repeated until we come to the end of the scan line, at which point we repeat the procedure for the next scan line. The number of runs can vary from one (every pixel on the scan line has the same gray level) to N (no two adjacent pixels have the same value). The corresponding run lengths l_k can vary from N to 1. To accommodate the longest possible run the natural code would require $\log_2 N$ bits. If most run lengths are significantly shorter than N, this code is not very efficient because many of the code words are not used or are used infrequently. Assuming that the quantizer outputs a unique value for each mapped pair (g_k, l_k), the next problem is to choose a code for the run lengths and gray levels.

The run-length statistics for the classification results of Figs. 6.24 and 6.25 are shown in Fig. 6.26. It is noted that the run lengths are reasonably close to a power law and, as discussed in Section 6.2.3.3, the B_1 code is a good approximation to the optimal Huffman code for 2-level data satisfying this condition. Since the continuation bit can be used to code the gray levels in the case of binary data, we will assume that $C = 0$ is used for dark runs and $C = 1$ for light runs.

The B_1 code for various run lengths is shown in the last column of Table 6.4 (the center column is discussed in the next section). The data compression ratios relative to the 1 bit per pixel that would be required for straightforward coding of each element are presented in the first column of Table 6.5 for each of the classified pictures of Figs. 6.24 and 6.25. The second and third columns are discussed in the next section.

The run-length coding scheme described in the preceeding paragraphs takes into account the data structure down each scan line, but not the structure between scan lines. Stated another way, it takes into account correlations between horizontal resolution elements but not correlations between vertical resolution elements.

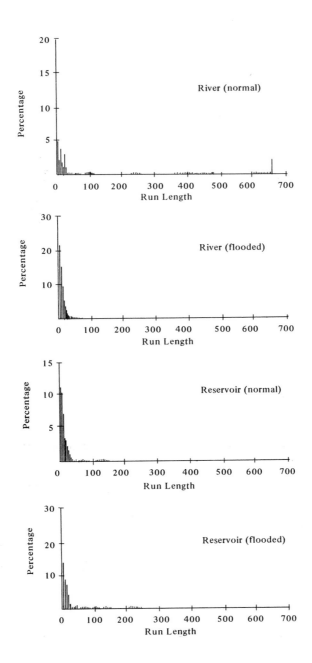

Figure 6.26. Run-length statistics for the classification results in Figs. 6.24 and 6.25.

Table 6.4. B_1 Code. $C = 0$ for dark runs; $C = 1$ for light runs in run-length coding. $C = 0$ for Δ'; $C = 1$ for Δ'' or Δ''' in PDQ and DDC coding.

Run Length	$\Delta', \Delta'', \Delta'''$	Code Word
1	0	C0
2	+1	C1
3	−1	C0C0
4	+2	C0C1
5	−2	C1C0
6	+3	C1C1
7	−3	C0C0C0
8	+4	C0C0C1
9	−4	C0C1C0
10	+5	C0C1C1
11	−5	C1C0C0
12	+6	C1C0C1
13	−6	C1C1C0
14	+7	C1C1C1
15	−7	C0C0C0C0
16	+8	C0C0C0C1
.	.	.
.	.	.
.	.	.

Table 6.5. Data compression ratios relative to 1 bit/pixel required by the original digital picture.

	RLC	PDQ	DDC
River, Normal	9.6	13.0	14.7
River, Flooded	2.9	1.2	1.4
Reservoir, Normal	6.7	5.3	6.2
Reservoir, Flooded	6.3	4.5	5.6

6.3.3.2 Two-Dimensional Run Length Encoding (PDQ and DDC)

A two-dimensional run-length encoding technique called a *predictive differential quantizer* (PDQ) is illustrated in Fig. 6.27. The PDQ mapping approach is to map the array of resolution elements into the sequence of integer pairs Δ' and Δ'', where Δ' is the difference between the starting

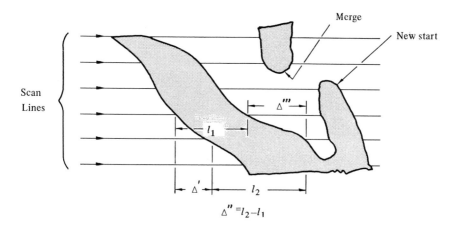

Figure 6.27. Illustration of the PDQ algorithm.

points of runs on successive lines, and Δ'' is the difference between the run lengths on successive lines together with "new start" and "merge" indicators used to denote the start and the end of each dark area. The PDQ mapping is similar to the contour algorithm described in Section 6.3.2, except that it simultaneously traces out the front and back contour edges rather than first going down the back side and then up the front side. It has the advantage that only two lines of pixels need to be stored, whereas the contour algorithm requires that the complete picture be stored. On the other hand, the PDQ approach results in more "new starts" than the contour algorithm, as illustrated in Fig. 6.27, where the black area labeled "new start" would have been included in the same contour as the black area to the left by the contour algorithm.

Assuming that the quantizer outputs a different value for each output from the mapper, the remaining problem is to code the mapper outputs Δ', Δ'', *start,* and *merge.* The histograms for the quantities Δ' and Δ'' for the classified images of Figs. 6.24 and 6.25 are shown in Figs. 6.28 through 6.31. Using the B_1 code for Δ' and Δ'' (see Table 6.4) and also for the column numbers of the new starts and merges resulted in the data compression ratios shown in the second column of Table 6.5. The B_1 code is not efficient for coding the "new start" and "merge" column numbers which are approximately uniformly distributed, but these occur only infrequently relative to the Δ' and Δ'' (many Δ' and Δ'' are coded, on the average, for each "new start" and "merge") and it is convenient to use only one code

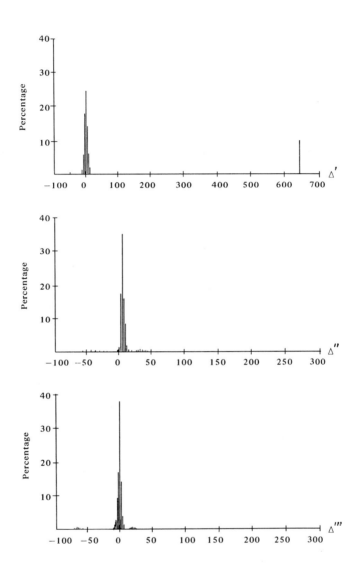

Figure 6.28. Difference histograms for Fig. 6.24(b).

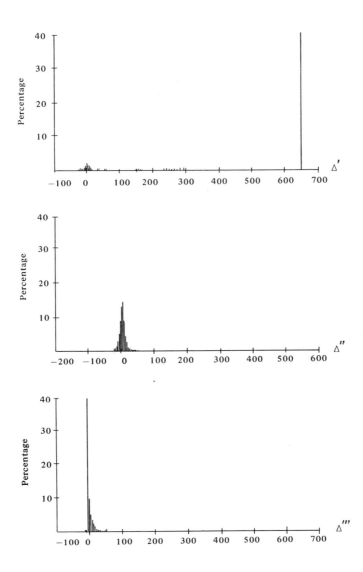

Figure 6.29. Difference histograms for Fig. 6.24(d).

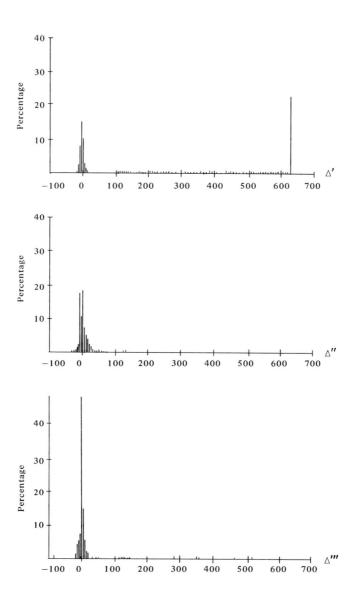

Figure 6.30. Difference histograms for Fig. 6.25(b).

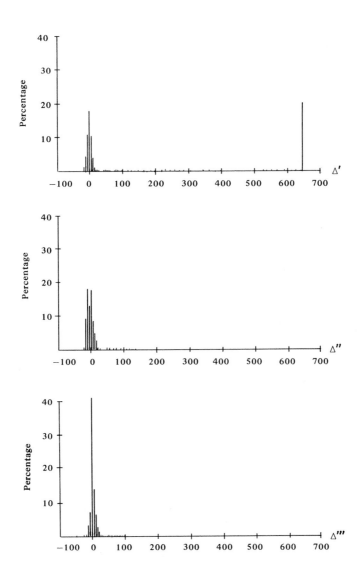

Figure 6.31. Difference histograms for Fig. 6.25(d).

for all four mapper outputs.

Another possibility is to code Δ' and Δ''' (the changes in the front and back edges of the dark areas) rather than Δ' and Δ''. This is called double delta coding (DDC). The histograms for Δ''' are also presented in Figs. 6.28 through 6.31. Using the B_1 code again we obtain the compression ratios listed in Table 6.5 under DDC. From these results we conclude that when there are a few large black areas, two-dimensional run length coding is more efficient. For many small black areas one-dimensional run length coding is more efficient.

6.4 IMAGE ENCODING RELATIVE TO A FIDELITY CRITERION

In many applications some error in the reconstructed image can be tolerated. In this section we present some examples of this case.

6.4.1 Example 1. Differential Pulse Code Modulation (DPCM)

Values of adjacent pixels are highly correlated for most images. The autocorrelation function for the cameraman picture of Fig. 6.32(a) is shown in Fig. 6.32(b). If pixel x_{i-1} is a certain gray level, then the adjacent pixel x_i along the scan line is likely to have a similar value. This is further illustrated by the histogram of pixel differences $x_i - x_{i-1}$ presented in Fig. 6.32 (c). Whereas the pixel values range over 256 different gray levels in the image, most adjacent pixel differences are in a range of about 20 gray level values. Differential pulse code modulation (DPCM) makes use of this property in the following manner. We observe a pixel x_{i-1} and based on this observed value we predict the value of the next pixel x_i. Let $\hat{x}_i$ be the predicted value of x_i, and let us subtract this value from the actual value $\hat{x}_i$ to obtain the difference $d_i = x_i - \hat{x}_i$. Assuming that the estimates are reasonably accurate, the difference $x_i - \hat{x}_i$ will, on the average, be significantly smaller in magnitude than the magnitude of the pixel x_i. Consequently, fewer quantization bins and fewer bits are required to code the sequence of differences than would be required to code the sequence of picture elements.

The problem is to estimate x_i given that we know x_{i-1}. The linear estimator that results in the least mean square estimation error, $E\{(x_i - \hat{x}_i)^2\}$, is given by

$$\hat{x}_i = \rho x_{i-1} + (1 - \rho)m \tag{6.4-1}$$

where m is the mean (average) gray level and ρ is the normalized correlation between the adjacent picture elements, that is,

(a)

Mean = 9.69
Variance = 2314
Alpha = .02981

(b)

(c)

Figure 6.32. (a) Original image. (b) Autocorrelation function. (c) Histogram of pixel differences.

$$\rho = \frac{E\{x_i x_{i-1}\}}{E\{x_i^2\}} \tag{6.4-2}$$

The estimate $\hat{x}_i$ in Eq. (6.4-1) can be interpreted as a weighted average of the preceding pixel x_{i-1} and the mean of x_i. The weights depend on the correlation coefficient ρ. When the pixel values are highly correlated ρ approaches 1 and $(1-\rho)$ approaches 0, in which case the estimate is based primarily on the value of x_{i-1}. When the pixel values are not very correlated the opposite condition holds and the estimate is based primarily on the mean value. For properly sampled images, ρ lies typically between 0.85 and 0.95.

It is easy to show that the variance of the difference

$$d_i = x_i - \hat{x}_i \tag{6.4-3}$$

is given by

$$\sigma_{d_i}^2 = (1 - \rho^2)\sigma_{x_i}^2 \tag{6.4-4}$$

where $\sigma_{x_i}^2$ is the variance of x_i. It can also be shown that the d_i are uncorrelated; that is, the mapping from the x_i to the d_i produces uncorrelated coefficients. It is noted that, if $\rho = 1$, this mapping is identical to the difference mapping defined by the transformation given in Eq. (6.2-3). Stated another way, the difference mapping defined by Eq. (6.2-3) corresponds to saying that $\hat{x}_i$ is the same as x_{i-1}.

The remaining problems are to quantize and code the differences d_i. A system block diagram for a typical DPCM encoder-decoder pair is shown in Fig. 6.33. The predictors in the encoder and decoders are identical. Both simply delay the input by an amount of time equal to the time between samples (the inverse of the sampling rate) and scale this delayed input by a constant α. This factor is called the *prediction coefficient* and is chosen in accordance with Eq. (6.4.2); that is, the optimum value for the prediction coefficient α is the correlation coefficient ρ.

As indicated above, the mapping in this case consists of forming the difference given in Eq. (6.4-3). Without the quantizer in the loop, the estimates given by the encoder predictor would be based on the exact differences d_i. In addition, the decoding process would be simply the inverse mapping. With the quantizer in the loop, the estimates are based on the *quantized* differences, and the decoding process is no longer exact. Suppose, for example, that the quantizer can have only eight output levels spaced uniformly. The decoder would incur relatively little error on slowly varying (small differences) scan lines. On rapidly changing scan lines, however, the difference between adjacent pixels would become large and the decoder would not be able to follow the input because the largest difference out the

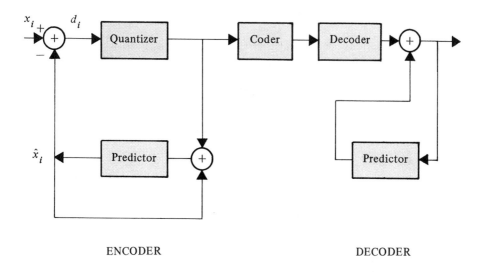

ENCODER DECODER

Figure 6.33. Block diagram of DPCM system.

the quantizer corresponds to a step of plus or minus four quantizer bin widths. This inability to follow rapid gray level variations is called *slope overload*, and it results in a smearing (blurring) of sharp edges in the reconstructed image. The encoder can be made to respond more rapidly to to a rapidly changing input by increasing the bin width, but this occurs at the expense of increased error when the signal is slowly varying. This error is called *granular noise* because smooth sections of the reconstructed image appear "grainy". Consequently, the quantizer bin width is chosen to provide a best compromise between these two sources of error. Both slope overload and granular noise can be reduced by increasing the number of quantization levels at the expense of more bits to code the increased number of levels.

Some examples of DPCM coded-decoded pictures are presented in Figs. 6.34(a) through (c). The prediction coefficient was set equal to the correlation coefficient in all three cases. The differences were quantized with an exponential quantizer matched to the difference statistics of the image, and the natural code was used to code the quantizer output. At 3 bits per pixel very little distortion is noticeable. At 2 bits per pixel we observe both smeared gray level edges and granular noise in the background (sky). At

Figure 6.34. Example of DPCM coding of Fig. 6.32(a) and then decoding the result. (a) 3 bits/pixel. (b) 2 bits/pixel. (c) 1 bit/pixel. (d) Effect of channel error with 3 bits/pixel and a signal-to-noise ratio of 13.7.

1 bit per pixel severe distortions are evident. DPCM with a 1-bit quantizer is called *delta modulation*.

In many applications the bits out of the encoder are transmitted over a digital data transmission system (called a channel) to the decoder. Such channels are not usually perfect in the sense that errors occur; that is, a bit may be put into the channel as a 1, but come out as a 0. The effect of channel bit errors on the 3-bit image of Figure 6.34(a) is illustrated in Figure 6.34(d). It can be shown that with a prediction coefficient of α the duration of the error streak is proportional to $1/(1-\alpha^2)$. For $\alpha = 1$, the errors persist until some corrective action is taken, as illustrated in Fig. 6.35 (a) where we reinitialized at the start of each scan line; that is, the output of the predictor was set equal to the mean gray level for the first pixel of each scan line so that an error in the previous line would not persist past the end of that line. One way to counter this effect is to replace each scan line containing an error with the previous scan line. The image in Fig. 6.35 (b) was obtained by applying this line replacement method to the image of Fig. 6.35(a).

Figure 6.35. Illustration of (a) line initialization, and (b) line replacement. In both cases α was 1, the signal-to-noise ratio was 19.5, and 3 bits/pixel were used.

6.4.2 Example 2. Transform Encoding

Image transforms such as the ones discussed in Chapter 3 are often used as the mapping function in the encoding procedure described in Fig. 6.1. An encoding method which employs this type of mapping is appropriately referred to as a *transform encoding* technique.

In transform encoding we first subdivide a given $N \times N$ image into a number of subarrays. For one-dimensional encoding the subarrays are of size $1 \times n$ with $n < N$, and each "subimage" may be interpreted as an n-dimensional vector. In two-dimensional encoding the subimages are usually $n \times n$ square arrays of pixels, with $n < N$. After dividing the image into subimages we code each subimage as a unit, independently of all other subimages. In *nonadaptive transform encoding* we use the same encoder for all subimages. In *adaptive transform encoding* we can choose the one that works best in some sense for the particular subimage content.

The purpose of the transform mapping is to reduce the correlation between pixels. The motivation behind this approach is to improve encoding efficiency by processing the transformed coefficients independently of each other. To illustrate this point we present a simplified example using a hypothetical 3-bit (8 gray-level) image. Consider a one-dimensional transform encoder with $n = 2$ so that the image is first subdivided into 1×2 arrays, where the two pixels in each 1×2 array are adjacent. In other words, the first array consists of the first and second pixels on the first scan line; the second array consists of the third and fourth pixels on the first scan line, and so on. Let the vector $x = (x_1, x_2)'$ represent two adjacent picture elements, and let us make a scatter plot of the gray level value of x_1 against the value gray level of x_2, as illustrated in Fig. 6.35(a). Since each pixel has any of 8 gray levels there are 64 possible combinations of x_1 and x_2. However, all combinations are not equally likely. It is unlikely that x_1 will have a high value and x_2 a low value, and vice versa. Since adjacent pixels are more likely to have nearly the same gray level the most likely combinations are the ones in the vicinity of $x_1 = x_2$, that is, those in the shaded area of Fig. 6.35(a).

Now suppose we rotate the coordinate system, as illustrated in Fig. 6.35(b). In the new coordinate system the more likely values are not in the vicinity of $y_1 = y_2$ but are lined up with the y_1 axis. Hence, the variables y_1 and y_2 are "more independent" than were x_1 and x_2; that is, y_2 is likely to be small independently of the value of y_1. Rotating the coordinate system also rearranged the variances. The total is the same, $\sigma_{y_1}^2 + \sigma_{y_2}^2 = \sigma_{x_1}^2 + \sigma_{x_2}^2$, but whereas both initial elements had the same variance ($\sigma_{x_1}^2 = \sigma_{x_2}^2$) more of the variance is now in the first component of the transformed space ($\sigma_{y_1}^2 > \sigma_{y_2}^2$). Finally, we note that given the coefficients y_1 and y_2 we can perform the inverse rotation to obtain the pixels x_1 and x_2.

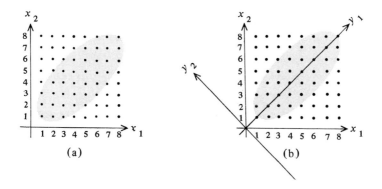

Figure 6.36. (a) Correlated pixels. (b) New coordinate system (y_1, y_2) for eliminating correlation.

The same procedure can be used for an $n \times n$ array of pixels, each of which is quantized to one of 2^K gray levels. In this case an n^2-dimensional coordinate system is required with each coordinate labeled with the values 1, 2, ..., 2^K. Each of the $2^{n^2 K}$ points corresponds to one of the $2^{n^2 K}$ possible $n \times n$ subimages.

6.4.2.1 One-Dimensional Transformations

Rotating the n-dimensional coordinate system of the n pixels corresponds to arranging the pixels into the n-vector, $\mathbf{x} = (x_1, x_2, ..., x_n)'$, and performing the linear transformation

$$\mathbf{y} = \mathbf{A}\mathbf{x} \qquad (6.4\text{-}5)$$

where $\mathbf{A}$ is an $n \times n$ unitary matrix, with elements a_{ki}, $k,i = 1, 2, ..., n$, that determines the rotation, and $\mathbf{y} = (y_1, y_2, ..., y_n)'$ is an n-vector. Since for unitary matrices $\mathbf{A}^{-1} = \mathbf{A}'$, the inverse rotation is accomplished by the inverse transformation

$$\mathbf{x} = \mathbf{A}'\mathbf{y} \qquad (6.4\text{-}6)$$

where $\mathbf{A}'$ is the transpose of $\mathbf{A}$.

According to Eq. (6.4-5), each coefficient y_k is a linear combination of all pixels, that is,

$$y_k = \sum_{i=1}^{n} a_{ki} x_i \qquad (6.4\text{-}7)$$

for $k = 1, 2, ..., n$.

Similarly, Eq. (6.4-6) gives each pixel as a linear combination of all the coefficients:

$$x_i = \sum_{k=1}^{n} b_{ik} y_k \qquad (6.4\text{-}8)$$

for $i = 1, 2, ..., n$.

With a slight change in notation, we see that Eqs. (6.4-7) and (6.4-8) are identical in form to the general transform relations given in Eqs. (3.5-1) and (3.5-2), with a_{ki} and b_{ki} being the forward and inverse transformation kernels, respectively.

For our purpose the best transformation would be one that results in statistically independent variables y. This transformation cannot be determined for two reasons. First, it evidently depends on very detailed statistics (the joint probability density function of the n pixels) which have not been deduced from basic physical laws and which cannot be measured. Second, even if the joint density function of the n pixels were known, the problem of determining a reversible transformation that results in independent coefficients is unsolved. The closest we can get with linear transformations to a transformation that produces independent coefficients is the one that produces uncorrelated coefficients. The resulting coefficients are uncorrelated but *not* necessarily statistically independent.

As indicated in Section 3.5.5, the transformation matrix **A** that produces uncorrelated coefficients is one whose rows are formed by the eigenvectors of the covariance matrix of the original pixel vectors. In this case, the samples used to obtain the covariance matrix would be all the n-dimensional vectors into which the given $N \times N$ image was decomposed. Equation (3.5-56) provides a convenient method for estimating the covariance matrix of a finite number of vector samples.

6.4.2.2 Two-Dimensional Transformations

For two-dimensional transformations the $n \times n$ array of pixels are arranged in the form of an $n \times n$ matrix **X** with elements x_{ij}, $i, j = 1, 2, ...,n$, and then transformed into an $n \times n$ matrix **Y** with elements y_{kl}, $k,l = 1, 2, ..., n$.

The general form of the transformation that maps the elements of **X** into the elements of **Y** is given by the relation

$$y_{kl} = \sum_{i=1}^{n} \sum_{j=1}^{n} x_{ij} a_{ijkl} \qquad (6.4\text{-}9)$$

for $k,l = 1, 2, ..., n$. With a change in notation, we see that Eq. (6.4-9) is identical in form to Eq. (3.5-3), with a_{ijkl} being the forward transformation kernel.

The inverse transformation gives each original pixel as a linear combination of the coefficients; that is,

$$x_{ij} = \sum_{k=1}^{n} \sum_{l=1}^{n} y_{kl} b_{ijkl} \qquad (6.4\text{-}10)$$

for $i, j = 1, 2, ..., n$, where b_{ijkl} is the inverse transformation kernel.

As indicated in Section 3.5, numerous types of transformations can be expressed in the form of Eqs. (6.4-9) and (6.4-10), depending on the choice of the kernel. The Fourier, Hadamard, and Hotelling transforms are the most popular for transform coding applications. The Fourier and Hadamard transforms fit directly in the format of Eqs. (6.4-9) and (6.4-10). When using the Hotelling transform, however, care must be exercised in interpreting the notation in these equations. The reason for this is best illustrated by a simple example.

Consider the problem of transforming a 2 × 2 subimage array

$$\mathbf{X} = \begin{bmatrix} x_{11} & x_{12} \\ x_{21} & x_{22} \end{bmatrix}$$

Expansion of Eq. (6.4-9) yields

$$y_{11} = x_{11}a_{1111} + x_{12}a_{1211} + x_{21}a_{2111} + x_{22}a_{2211}$$

$$y_{12} = x_{11}a_{1112} + x_{12}a_{1212} + x_{21}a_{2112} + x_{22}a_{2212}$$

$$y_{21} = x_{11}a_{1121} + x_{12}a_{1221} + x_{21}a_{2121} + x_{22}a_{2221}$$

$$y_{22} = x_{11}a_{1122} + x_{12}a_{1222} + x_{21}a_{2122} + x_{22}a_{2222}$$

Computation of the kernel values for, say, the Fourier or Hadamard transform is straightforward. From Section 3.5.1 we have, for example, that the Fourier kernel is given by

$$a_{ijkl} = \frac{1}{N} \exp\left[-j2\pi(ik + jl)\right]$$

with $N = 2$ in this case.

Interpretation of the kernel values for the Hotelling transform is different because, as indicated in Section 3.5.5 and again in Section 6.4.2.1, this transform is expressed in the *vector* form $\mathbf{y} = \mathbf{Ax}$. For a two-dimensional problem, the 2 × 2 subimage $\mathbf{X}$ can be expressed in vector form, as follows:

$$\mathbf{x} = \begin{bmatrix} x_{11} \\ x_{12} \\ x_{21} \\ x_{22} \end{bmatrix}$$

The transformation matrix $\mathbf{A}$ is formed from the eigenvectors of the covariance matrix of all the $\mathbf{x}$'s extracted from a given image. Suppose that $\mathbf{A}$ has the form

$$\mathbf{A} = \begin{bmatrix} e_{11} & e_{12} & e_{13} & e_{14} \\ e_{21} & e_{22} & e_{23} & e_{24} \\ e_{31} & e_{32} & e_{33} & e_{34} \\ e_{41} & e_{42} & e_{43} & e_{44} \end{bmatrix}$$

where e_{ij} is the jth component of the ith eigenvector. The transformed subimage $\mathbf{Y}$, if expressed in vector form, is

$$\mathbf{y} = \begin{bmatrix} y_{11} \\ y_{12} \\ y_{21} \\ y_{22} \end{bmatrix}$$

Since $\mathbf{y} = \mathbf{Ax}$, we have

$$y_{11} = x_{11}e_{11} + x_{12}e_{12} + x_{21}e_{13} + x_{22}e_{14}$$

$$y_{12} = x_{11}e_{21} + x_{12}e_{22} + x_{21}e_{23} + x_{22}e_{24}$$

$$y_{21} = x_{11}e_{31} + x_{12}e_{32} + x_{21}e_{33} + x_{22}e_{34}$$

$$y_{22} = x_{11}e_{41} + x_{12}e_{42} + x_{21}e_{43} + x_{22}e_{44}$$

Thus, we see that the *form* of this expansion is identical to the one obtained from Eq. (6.4-9), since each y_{kl} is given as a linear combination of all the x_{ij}, for $i, j = 1, 2, ..., n$. It is noted, however, that the elements of $\mathbf{A}$ do not follow the same notation used for the kernels in Eq. (6.4-9). In other words, the Hotelling transform can be used in a two-dimensional formulation, as long as the elements in the transformation are interpreted properly. Similar comments hold for the inverse transformation.

6.4.2.3 Basis Images

Another interpretation of Eq. (6.4-10) is possible. Let us write that equation in the form

$$\mathbf{X} = \sum_{k=1}^{n} \sum_{l=1}^{n} y_{kl} \mathbf{B}_{kl} \tag{6.4-11}$$

and interpret this as a series expansion of the $n \times n$ subimage $\mathbf{X}$ into n^2 $n \times n$ basis images:

$$\mathbf{B}_{kl} = \begin{bmatrix} b_{kl11} & b_{kl12} & \cdots & b_{kl1n} \\ b_{kl21} & b_{kl22} & \cdots & b_{kl2n} \\ \vdots & \vdots & \ddots & \vdots \\ b_{kln1} & b_{kln2} & \cdots & b_{klnn} \end{bmatrix} \tag{6.4-12}$$

with the y_{kl} for $k,l = 1, 2, ..., n$, being the coefficients (weights) of the expansion. Hence, Eq. (6.4-11) gives the image $\mathbf{X}$ as a weighted sum of the basis images $\mathbf{B}_{kl}$. The coefficients of the expansion are given by Eq. (6.4-9), which may be written in the form

$$\mathbf{y} = A_{kl}\mathbf{X} \tag{6.4-13}$$

where $\mathbf{A}_{kl}$ is formed in the same manner as $\mathbf{B}_{kl}$, except that the forward kernel is used.

Figure 6.37(a) shows the Hadamard basis images for $N = 256$ and $n = 16$. Since the inverse Hadamard kernel b_{ijkl} depends only on the values of i, j, k, and l, all $N \times N$ images that are subdivided into $n \times n$ subimages have the same set of Hadamard basis images $\mathbf{B}_{kl}$. Computation of $\mathbf{B}_{kl}$ for the Hotelling transformation, on the other hand, depends on the inverse of the transformation matrix $\mathbf{A}$. In other words, the b_{ijkl} in Eq. (6.4-10) depend in this case on the values of $\mathbf{A}^{-1}$ in the same manner shown above for the forward transform. Since the entries in $\mathbf{A}^{-1}$ in turn depend on the covariance matrix of the subimages, the $\mathbf{B}_{kl}$ for the Hotelling transform are different for different images. The example shown in Fig. 6.37(b), for example, corresponds to the cameraman picture with $n = 16$.

We have already noted that if a set of subimages is transformed such that the coefficients y_{kl} are "more independent" than the original pixels x_{ij}, then the variances of the coefficients are in general not equal. Therefore, we can index the basis images such that the terms in Eq. (6.4-11) are ordered according to the variances of the coefficients. In this manner, successive terms contribute proportionally less and less, on the average, to the total. Indeed, for some choices of basis images the coefficients become

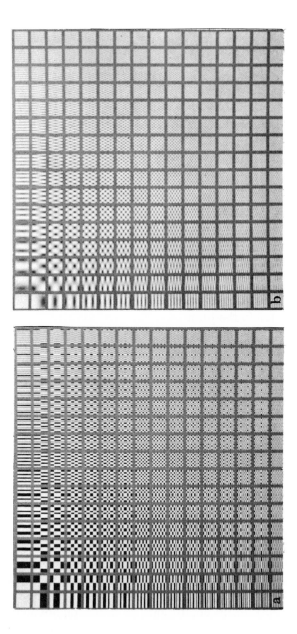

Figure 6.37. (a) Hadamard basis images. (b) Hotelling basis images.

insignificant after the first, say, η terms so that an approximate representation for the original image can be obtained by truncating the series after the first η terms, that is,

$$\mathbf{X} = \sum_{k=1}^{n} \sum_{l=1}^{n} y_{kl} \mathbf{B}_{kl} \approx \sum_{k=1}^{\eta} \sum_{l=1}^{\eta} y_{kl} \mathbf{B}_{kl} = \hat{\mathbf{X}} \qquad (6.4\text{-}14)$$

The mean-square approximation error between the original $\mathbf{X}$ and the approximate image $\hat{\mathbf{X}}$ is given by

$$e_{\text{ms}} = E\left\{ \|\mathbf{X} - \hat{\mathbf{X}}\|^2 \right\}$$

$$= E\left\{ \left\| \sum_{k=1}^{n} \sum_{l=1}^{n} y_{kl} \mathbf{B}_{kl} - \sum_{k=1}^{\eta} \sum_{l=1}^{\eta} y_{kl} \mathbf{B}_{kl} \right\|^2 \right\}$$

$$= E\left\{ \left\| \sum_{k=\eta+1}^{n} \sum_{l=\eta+1}^{n} y_{kl} \mathbf{B}_{kl} \right\|^2 \right\}$$

$$= \sum_{k=\eta+1}^{n} \sum_{l=\eta+1}^{n} \sigma_{y_{kl}}^2 \qquad (6.4\text{-}15)$$

where $\|\mathbf{X} - \hat{\mathbf{X}}\|$ is the norm of the matrix difference $(\mathbf{X} - \hat{\mathbf{X}})$, and the last step follows because the basis images are orthonormal. Equation (6.4-15) states that the mean square approximation error is given by the sum of variances of the discarded coefficients.

We now pose the following problem: What set of basis images minimizes the mean square error by packing the most variance into the first η coefficients? The solution to this problem is the same as the solution to the seemingly unrelated problem of determining the set of basis images that produce uncorrelated coefficients. The Hotelling transformation: (a) produces uncorrelated coefficients; (b) minimizes the mean square approximation error; and (c) packs the maximum amount of variance into the first η coordinates (for any η).

The cameraman picture was divided into 16 × 16 subimages and each subimage was expanded in a Hotelling, Fourier, and Hadamard series expansion. The sample variances of the coefficients are presented in Fig. 6.38(a). It is noted from Fig. 6.38(b) that all three transformations are approximately equally efficient in packing the variances into lower-order coefficients.

Some approximations to the cameraman picture using the Hotelling, Hadamard, and Fourier transforms are presented in Figs. 6.39 and 6.40. These pictures were obtained by dividing the image into 16 × 16 subimages, representing each subimage with its expansion in terms of the 16 × 16 basis images, and truncating the expansion after η terms. As shown in Fig. 6.38, half of the terms ($\eta = 128$) can be discarded with no visible degradation in image quality, although some mean-square error is incurred. Truncating the

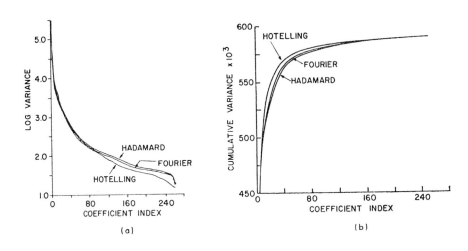

Figure 6.38. (a) Variances of the coefficients. (b) Cumulative sums of the variances.

expansion after 64 terms (Fig. 6.40) results in a visible blurring or lowpass filtering effect. This phenomenon can be explained by considering the basis images shown in Fig. 6.37. It is noted in those images that increasing index corresponds to increasing frequency content (i. e., the complexity of the basis images increases as a function of increasing index). Hence, truncating the expansion at η terms corresponds to discarding all image energy at frequencies higher than those corresponding to the first η basis images. It is also noted that for small η the subimage edges are visible. Figure 6.41 shows the error between the reconstructed image of Fig. 6.39(b) and the original cameraman image. It is evident that the largest errors occur in the high detail (high frequency) parts of the image and at the edges of the basis images.

6.4.2.4 Quantizing the Coefficients

In our discussion of image encoding thus far, we have assumed that the quantizer is capable of assigning a unique value to each output from the mapper. This is not an unreasonable assumption for the methods discussed earlier in this chapter because we were dealing with a manageable number of distinct mapper outputs. In difference mapping, for example, the number of distinct differences out of the mapper is of the same order as the number of levels in the image. In transform mapping, on the other hand, the number of distinct output values may be infinite. For instance, each component y_{kl} out

Figure 6.39. Reconstructed images obtained by retaining the first 128 of the $n^2 = 256$ terms in Eq. (6.4-15) for each of the 256 16 $\times$ 16 subimages of the original 256 $\times$ 256 cameraman picture. (a) Hotelling transform: $e_{ms} = 0.34\%$. (b) Fourier transform: $e_{ms} = 0.45\%$. (c) Hadamard transform: $e_{ms} = 0.49\%$.

Figure 6.40. Reconstructed images obtained by retaining the first 64 of the $n^2 = 256$ terms in Eq. (6.4-15) for each of the 256 16 X 16 subimages of the original cameraman picture. (a) Hotelling transform: $e_{ms} = 0.34\%$. (b) Fourier transform: $e_{ms} = 0.45\%$. (c) Hadamard transform: $e_{ms} = 0.49\%$.

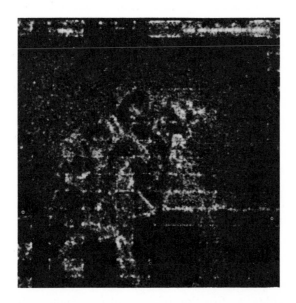

Figure 6.41. Error between Fig. 6.39(b) and the original cameraman picture.

of the Hotelling transform is a linear combination of all the pixels in the input subarray. Although the input pixels may only range over L gray levels, the linear combination which produces y_{kl} is formed by weighting each pixel with an eigenvector component. Since these components depend on the covariance matrix of the input subarray, it is not difficult to see that each y_{kl} can, in theory, assume an infinite number of different values. In practice, the number is finite because of computer limitations, but even so, the range of values obtainable in a typical computer is so large it becomes necessary to use a quantization strategy for the mapper outputs.

As illustrated in Fig. 6.38, the variance of the output coefficients vary widely. It would be inefficient, therefore, to use the same quantizer for all coefficients. In other words, if the quantizer output levels are adjusted to span the range of the coefficients with the largest variance, then the coefficients with much smaller variances would fall in a much smaller range, with the result that most of the quantizer levels would not be used. This effect can be reduced by first scaling each coefficient by the inverse of its standard deviation to form the normalized coefficients $y_{kl} = y_{kl}/\sigma_{kl}$, all of which have unit variance and can be efficiently quantized with the same quantizer.

Using the same quantizer for each normalized coefficient y_{kl} and the natural or gray code to assign equal length code words to all quantizer output levels results in each coefficient requiring the same number of bits. Since the coefficients with the larger variances generally contribute significantly more to the reconstructed image than the coefficients with the smaller variances, the total distortion due to quantizing the coefficients may be lessened by alloting more quantization levels and/or bits to the coefficients with the larger variances and proportionally fewer to the coefficients with the smaller variances. Also, since each coefficient corresponds to a particular frequency band, and the sensitivity of the human visual system to distortion is dependent on the frequency of the distortions, better subjective quality may be obtained in some cases by alloting more quantization levels and/or bits to those coefficients corresponding to the frequencies to which the eye is most sensitive in a given image.

Another quantization strategy is to choose the quantized levels so that they minimize the total quantizer mean-square error. For an $n \times n$ array this error is defined as

$$e_q = E \left\{ \sum_{k=1}^{n} \sum_{l=1}^{n} (y_{kl} - \hat{y}_{kl})^2 \right\} \qquad (6.4\text{-}16)$$

where $\hat{y}_{kl}$ is the quantized value of y_{kl}. Equation 6.4-16) depends on the joint probability density function of y_{kl} and $\hat{y}_{kl}$. Since each coefficient is a linear combination of n^2 pixels, the central limit theorem indicates that the distributions of the y_{kl}'s tend toward a Gaussian density since some of the pixels are more or less independent. Indeed, histograms for the coefficients for various transformations have been constructed and found to be roughly "bell-shaped". This effect becomes more pronounced with array size.

Panter and Dite [1951] and Max [1960] investigated quantization strategies that minimize the mean-square error of a *single* coefficient. They found that if the probability density function of y_{kl} is uniform, then a quantizer with uniformly-spaced output levels is optimum. For other distributions the mean square error can be decreased by using a nonuniform quantizer with the spacing between output levels decreased in regions of high probability and increased in regions of low probability. For a Gaussian distribution the nonuniform quantizer can be 20–30% more efficient than the uniform quantizer.

Quantization strategies for minimizing the *total* mean-square error were investigated by Huang and Schultheiss [1963] who determined the optimum allocation of a total of M bits to the n^2 coefficients when using a natural code for each coefficient. That is, all of the code words for each coefficient have the same length, but different coefficients have a different number of quantization levels and, therefore, require a natural code of a different length. They found that the number of bits m_{kl} used to code coefficient y_{kl} should be proportional to $\log \sigma_{kl}^2$. They give an algorithm for computing the m_{kl}, $k,l = 1, 2, ..., n$, such that Eq. (6.4-16) is minimized for given M and the set of variances σ^2_{kl} $k,l = 1, 2, ..., n$. This technique, called *block quantization*, is significantly more efficient than using the same number of bits $m_{kl} = M/n^2$ for all coefficients. Its disadvantage lies in the implementation problems inherent in handling binary words of unequal lengths.

Block quantization simply means using a different quantizer for each coefficient. Each quantizer can have a different number of quantization bins and a different spacing between the bins. The two images shown in Fig. 6.42 were obtained by subdividing the cameraman picture into 16 × 16 subimages. Each subimage was expanded into a set of Fourier basis images and the expansion truncated at $\eta = 128$ terms. The 128 retained coefficients y_{kl} were quantized using the two basic approaches discussed above. The results shown in Fig. 6.42(a) were obtained by normalizing the coefficients (i.e., $y_{kl} = y_{kl}/\sigma_{kl}$) and using a 16-level, uniform quantizer. The Fourier expansion was then formed with the resulting coefficients to yield the reconstructed image. The same procedure was used to obtain the image in Fig. 6.42(b), except that the number of levels used to quantize the 128 coefficients was made proportional to the quantity $\log\sigma^2_{kl}$. The two results are of approximately equal subjective quality, although the error (computed by averaging $(y_{kl} - \hat{y}_{kl})^2$ over all subimages) is much larger for Fig. 6.42(a). The *spatial* error between the original and Fig. 6.42(b) is shown in Fig. 6.42 (c).

Figure 6.43(a) shows the results of an identical experiment except that four quantization levels were used for each of the normalized coefficients. Figure 6.43(b) shows the results of using the same total number of bits but with the block quantization algorithm. The latter technique makes much more efficient use of the bits and this results in a picture of significantly better subjective quality and less mean-square error.

6.4.2.5 Coding Considerations

Coding performance for a transform encoder depends primarily on: (1) the transformation; (2) the quantization stragegy; (3) the subpicture size; (4) the subpicture shape.

Transformation. The best transformation from both a mean-square error and subjective quality viewpoint is the Hotelling transformation, but it is closely followed by the Fourier and Hadamard transformations. Each is separated by .1 or .2 bits/pixel for $n = 8$ or 16. For $n = 4$ the performances are essentially the same.

Quantization strategy. Both mean-square error and subjective quality are quite sensitive to the efficiency with which bits are used to code the coefficients. The simplest strategy is to form the normalized coefficients $y_{kl} = y_{kl}/\sigma_{kl}$ and use the same quantizer (number of bits) for each

Figure 6.42. Reconstructed images obtained by quantizing the 128 coefficients of Fig. 6.39(b) using an average of 2 bits/pixel (4 bits per retained coefficient). (a) All 128 normalized coefficients quantized to 16 levels: error = 2.09%. (b) The 128 coefficients block quantized: error = 0.78%. (c) Magnitude of spatial error between Figs. 6.39(b) and 6.42(c).

Figure 6.43. Reconstructed images obtained by quantizing the 128 coefficients of Fig. 6.39(b) using an average of 1 bit/pixel (2 bits per retained coefficient). (a) All 128 coefficients quantized to 4 levels: error = 8.68%. (b) The coefficients block quantized: error = 2.21%.

coefficient. If η coefficients are retained and m bits used to code each coefficient then a total of $m\eta/n^2$ bits/pixel are required. For good quality reproductions, approximately half the coefficents should be retained, in which case 7 bits/coefficient must be used. Hence, $m\eta/n^2 \approx m(n^2/2)n^2 = m/2 = 3.5$ bits/pixel are required. If the normalized coefficients are quantized with the same 7 or 8-bit quantizer, but only the $m_{kl} \sim \log\sigma^2_{y_{kl}}$ most significant bits retained (block quantization) the same quality pictures can be obtained with a savings of about 1 bit/pixel. Sometimes a further .1 bit/pixel can be saved by choosing the bit assignments to give the best subjective quality. A further saving of .2 to .3 bit/pixel can be achieved by quantizing the coefficients with the same quantizer and using a Huffman code to assign code words of unequal lengths to the quantizer output levels.

Subpicture size. Mean-square error performance should improve with increasing n since the number of correlations taken into account increases with n. However, most images contain significant correlations between pixels for only about 20 adjacent pixels, although this number is strongly dependent on the amount of detail in the picture. Hence, a point of diminishing returns is reached and $n > 16$ is not warranted. Even smaller n, say $n = 8$, does not significantly increase the error.

This argument does not appear to apply when subjective quality is the criterion of goodness. The subjective quality appears to be essentially independent of n for $n \geqslant 4$. Since the number of computations per pixels is proportional to n, 4×4 is a reasonable choice of subpicture size.

Subpicture shape. Transforming two-dimensional $n \times n$ arrays of pixels yields better performance than one-dimensional arrays of pixels but the gain is surprisingly small—about .2 bits/pixel.. However, a larger n is required for one-dimensional arrays. For example, if two-dimensional arrays of size 4×4 are reasonable, one-dimensional arrays of 1×16 arrays are required to obtain comparable results.

6.4.2.6 Effect of Bit Errors in the Coefficients

Since the decoder reconstructs pixels from linear combinations of the coefficients, an error in a coefficient leads to errors in all the pixels reconstructed from it. If the complete image is transformed as a unit, (i.e., $n = N$), then one or more errors in the coefficients results in some error in all the reconstructed pixels. If the $N \times N$ image is first divided into $n \times n$ subimages and each subimage coded independently, then only the subimages with errors are effected.

The images presented in Fig. 6.44 we obtained by starting with the image of Fig. 6.42(b) and making random bit errors in the coefficients at error rates of one error per 100 and 1000 bits, respectively. The structure of the blocks containing errors can be explained by recalling that each resulting subimage is a weighted sum of basis images. Hence, changing the

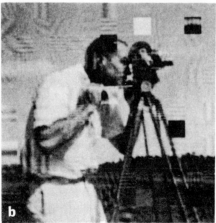

Figure 6.44. Reconstructed pictures obtained after dividing the picture into 16 × 16 subpictures, expanding each subpicture into a Fourier series expansion and retaining the first 128 terms, block quantizing the coefficients using 2 bit/pixel, and making (a) Bit error rate = 10^{-3}. (b) Bit error rate = 10^{-2}.

value of a coefficient results in the corresponding basis image receiving the wrong weight in the reconstruction process.

6.4.2.7 Adaptive Transform Encoding

Transform encoders can be made to adapt to local image structure by allowing a number of modes of operation and, for each subimage, choosing the mode that is most efficient for that subimage. Bookkeeping information that indicates which mode was used must be coded along with the subimage pixels. In general, increasing the number of modes decreases the number of bits required to code the subimage pixels, but increases the number of bits required to code the bookkeeping information.

Since the Hotelling transformation is matched to the subimage statistics, one might be tempted to use different transformations for subimages with different statistics. However, if we segregate the cameraman subimages with different statistics into different groups and compute the Hotelling transformation matrix for each group, we would find them strikingly similar. Hence, using different transformations for different subimages is not generally warranted. On the other hand, once the transformation is chosen (whether it be Hotelling, Fourier, Hadamard, or other) significant encoding efficiency can be achieved by adapting to the coefficients generated for each subpicture. There are a number of schemes for accomplishing this, but most are variations of the following three methods.

Method 1. Compute all n^2 coefficients (ordered according to their variances) and determine the smallest η for which the quantity

$$\sum_{k=1}^{\eta} \sum_{l=1}^{\eta} |y_{kl}|^2 \bigg/ \sum_{k=1}^{n} \sum_{l=1}^{n} |y_{kl}|^2$$

exceeds a predetermined threshold, (e.g., 0.99). Code the first η coefficients and code the number η. Making the threshold dependent on the average subimage brightness improves the subjective quality of the image due to the properties of the human visual system discussed in Section 2.1.3.

Method 2. Compute all n^2 coefficients and retain all that exceed a predetermined threshold. Code the retained coefficients and code which coefficients are retained using, for example, a run length code. As in Method 1 the threshold can depend on the average brightness.

Method 3. A pattern recognition algorithm (see Tou and Gonzalez [1974]) is first used to classify each subimage into one of three types, according to the gray level and the amount of detail in the subpicture. The three types of subimages are: (1) high detail; (2) low detail and darker than average; (3) low detail and brighter than average. The Hotelling transformation is used to map the subimage array into a set of coefficients. A different block quantizer and the Huffman code are used to complete the encoding of each of the three types of images. For the high-detail subimages, twice as many coefficients are retained as for the two low-detail types. Of the two low-detail types, a finer quantizer is used for the dark images, since the human visual system is more sensitive to errors in dark regions.

The results shown in Fig. 6.45 were obtained by adaptive transform encoding using Method 3. Figures 6.45(a) and (b) were encoded using an average of 0.59 and 1.13 bits per pixel, respectively. For the original 64-level image of Fig. 6.45(c), the encoded image shown in Fig. 6.45(d) was obtained using 0.86 bits per pixel. These results indicate that adaptive transform coding can yield reasonably good quality pictures at rates below 1 bit per pixel.

6.4.3 Example 3. Hybrid Encoding for RPV TV

Remotely piloted vehicles (RPVs) are small aircraft that operate without a pilot. In certain military applications RPVs are more effective than piloted aircraft because they are smaller and, therefore, provide less radar cross section and present less of a target. They are also less expensive because pilot support equipment is not required. A television camera inside the RPV looks ahead and transmits a continuous image to the "pilot" who observes a TV monitor some distance away. The pilot "flies" the RPV by remote control signals. A squadron of 20 RPVs would require 20 simultaneous TV transmissions. Significant data compression is required to reliably transmit this number of TV channels over long distances with low power, inexpensive transmitters, and antennas that operate in a hostile (jamming) environment. A typical specification calls for a transmission bit rate of less than 0.5 M bit/sec. This corresponds to a data compression ration of approximately 60 to 1.

One possibility is to transmit only every 10th frame; that is, transmit 3 frames per second rather than the 30 frames per second generated by the TV camera. This is about the minimum number of frames per second required by the pilot to maintain adequate control of the RPV due to the picture "jumping" because of RPV attitude fluctuations. (If RPV attitude information is also transmitted and used to keep the picture "centered" on the monitor, then frame rates of less than 1 frame per second can be tolerated). A frame rate reduction from 30 frames per second to 3 frames per second yields a compression ratio of 10 to 1.

The remaining problem is to achieve a 6 to 1 compression by encoding each transmitted frame, which allows us about 1 bit per pixel, assuming that the original TV picture is digitized using 6 bits. The encoder must, of course, satisfy rather severe weight, volume, and power requirements along with reasonable cost, say, 2 lbs, 25 in^3, and 10 watts. These specifications can be met with hybrid encoding techniques.

Hybrid encoding, due to Habibi [1974], combines the concepts of transform encoding and DPCM into a system that achieves essentially the

Figure 6.45. Illustration of adaptive transform encoding performance. (a) Original image. (b) Decoded image at 0.89 bits/pixel. (c) Original image. (d) Decoded image at 0.86 bits/pixel.

Figure 6.45. (Continued.)

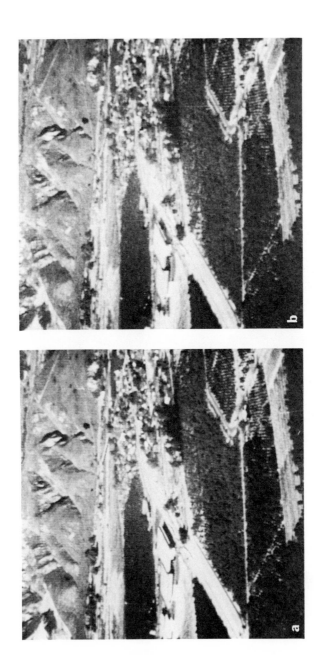

Figure 6.46. (a) Original image. (b) through (d) Decoded images using 2, 1, and 1/2 bits per pixel, respectively.

Figure 6.46. (Continued.)

same performance as two-dimensional transform encoding, but it is easier to implement. Hybrid encoding uses one-dimensional transform encoding in the horizontal direction and DPCM in the vertical direction.

Each scan line is first subdivided into sub-blocks containing n pixels each. Each of the resulting $1 \times n$ arrays is processed using a transformation such as, for example, the Hotelling, Hadamard, or discrete cosine transform. The purpose of the transformation is to reduce the correlation (i.e., redundancy) in the horizontal direction. Successive scan lines are also high correlated in most images. Thus, since the pixels in a $1 \times n$ subarray are correlated with the pixels in the corresponding subarray in the next line, we also expect the coefficients produced by the transformation to be correlated. This correlation between coefficients in the vertical direction, can be diminished by differencing the coefficients using DPCM. Since the coefficients generally have different variances, it follows that the differences between coefficients will also have different variances. In order to efficiently code the coefficient differences we can use the block quantization method mentioned in the previous section.

As an illustration of hybrid encoding, consider the image shown in Fig. 6.46(a). This image was encoded using the discrete cosine transformation (with $n = 32$) in the horizontal direction and DCPM in the vertical direction. The differences were processed using the block quantization bit assignments shown in Table 6.6. The decoded images for 2, 1 and 1/2 bits per pixel are presented in Figs. 6.46(b) through (d). The 2-bit image has only a very small degradation relative to the original. The one-bit image has more degradation. The 1/2-bit image not only has very noticeable degradation, but the edges of the 1×32 blocks are also apparent.

The same experiment was performed on the original image presented in Fig. 6.47(a). The decoded images are presented in Figs. 6.47(b) through (d) for 2 bits, 1 bit, and 1/2 bit, respectively. Careful observation of these images leads to the same conclusion as for the images shown in Fig. 6.46.

6.5 USE OF THE MAPPER OUTPUTS AS FEATURES

The encoding problem discussed in the previous sections deals with data compaction, subject to the constraint that a reconstructed image be either the same as, or a reasonable facsimile of, the original. In some applications, such as pattern recognition by machine, interest also lies on data compaction, but the constraint is one of preserving only enough information to allow an image, or parts of an image, to be classified into one of several categories or pattern classes. Without a human operator in the loop, reconstruction of the original image is no longer an essential requirement. The problem of reducing the representation of an image to a small number of

Table 6.6. Bit allocation for coefficient differences.

Coefficient Differences	2 bits	1 bit	1/2 bit
1	4	3	3
2	4	3	2
3	3	2	2
4	3	2	1
5	3	2	1
6	3	2	1
7	3	2	1
8	3	2	1
9	3	2	1
10	3	2	1
11	3	2	1
12	2	1	1
13	2	1	0
14	2	1	0
15	2	1	0
16	2	1	0
17	2	1	0
18	2	1	0
19	2	1	0
20	1	0	0
21	1	0	0
22	1	0	0
23	1	0	0
24	1	0	0
25	1	0	0
26	1	0	0
27	1	0	0
28	1	0	0
29	1	0	0
30	1	0	0
31	1	0	0
32	1	0	0

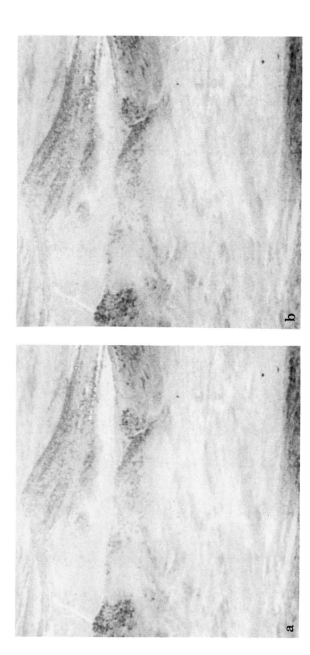

Figure 6.47. (a) Original image. (b) through (d) Decoded images using 2, 1, and 1/2 bits per pixel, respectively.

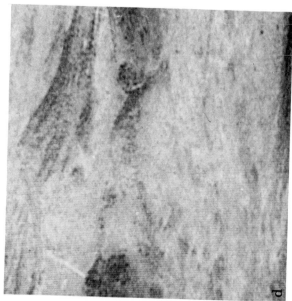

Figure 6.47. (Continued.)

components carrying enough discriminating information is referred to as *feature extraction*.

Many of the mapping techniques used in the encoding process described in Fig. 6.1 are also often used for feature extraction. In other words, in this type of application we are generally not interested in quantizing or coding techniques. Instead, the mapper outputs are used as features which are input directly into a pattern recognition device. Of the methods normally associated with an encoder mapping operation, the Hotelling transform is one of the most often used for feature extraction. In this section we illustrate uses of this transform for extracting feature information from multispectral imagery.

Satellite and airborne multispectral sensors provide data in the form of several images of the same area of the Earth's surface, but taken through different spectral windows or bands. The number of spectral bands varies, but typically ranges from four, as with the NASA LANDSAT, to twenty or more for many aircraft-borne sensors. The spectral bands typically exhibit high interband correlations so that a significant amount of redundancy exists between the spectral images. These correlations, coupled with the large quantities of data, lead to the consideration of efficient methods of information extraction for user analysis purposes.

An example of some aircraft multispectral imagery is presented in Fig. 6.48. These data were taken by a 6-band multispectral scanner flown over Tippecanoe County, Indiana, at a mean altitude of 3,000 feet. The six images shown in Fig. 6.48 correspond to the six bands in the scanner. These images, which are 384 by 239 pixels and represent an area of approximately 0.9 by 0.7 mi are typical of multispectral data taken over agricultural terrain. The scanner bands used are listed in Table 6.7. By comparing Fig. 6.48 with the bands in this table we see that vegetation, which is dark in the visible bands (1, 2, 3), appears bright in the infrared bands (4, 5). By contrast, the

Table 6.7. Channel number and wavelengths.

Channel	Wavelength Band (μm)
1	0.40–0.44
2	0.62–0.66
3	0.66–0.72
4	0.80–1.00
5	1.00–1.40
6	2.00–2.60

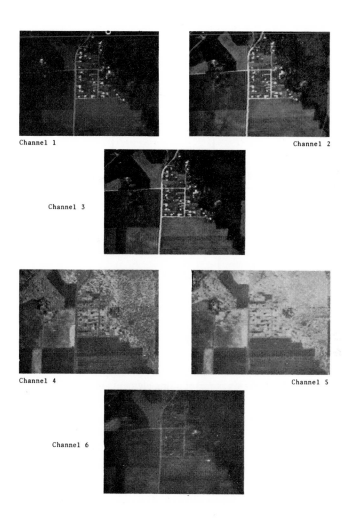

Figure 6.48. Six spectral images from an airborne scanner. (Courtesy of the Laboratory for Applications of Remote Sensing, Purdue University).

the roof tops and roads appear light in the visible bands and dark in the infrared bands.

Suppose that the six digital images presented in Fig. 6.48 are stacked one behind the other as illustrated in Fig. 6.49. We can then form one 6-element column vector $x = (x_1, x_2, ..., x_6)'$ for each ground pixel. In other

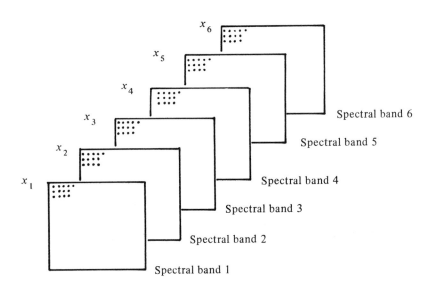

Figure 6.49. Obtaining six corresponding elements to form a pixel vector.

words, the elements of this vector correspond to the same point on the ground, but give the gray level in each of the six different bands. If we were to compute the average value for each x_i and the correlation between every pair of elements x_i and x_j over the entire image, we would in general find that the elements have different means and are highly correlated. As indicated in Section 3.5.5, the components can be normalized about their means and their correlation eliminated by use of the Hotelling transformation

$$y = A(x - m_x) \qquad (6.5\text{-}1)$$

where m_x is the mean vector population and the rows of A are formed from the normalized eigenvectors of the covariance matrix C_x. In the present application there were 384×239 vectors x available for computing the mean vector and covariance matrix. The eigenvalues of C_x are listed in Table 6.8.

Table 6.8. Eigenvalues of the covariance matrix of the images shown in Figure 6.50.

λ_1	λ_2	λ_3	λ_4	λ_5	λ_6
3210	931.4	118.5	83.88	64.00	13.40

The transformation given in Eq. (6.5-1) was applied to the images in Fig. 6.48. The results are shown in Fig. 6.50. These images are referred to as the *principal component images* because, as indicated in Section 3.5.5, the Hotelling transform picks out the orthogonal components with the largest variance. The energy packing property of this principal components transformation manifests itself as a significant increase in contrast (variance) in the first images, with monotonically decreasing contrast as their variance decrease. In fact, as indicated by the values in Table 6.8, 97% of the variance is contained in the first three images.

In some pattern recognition applications the problem is to classify each ground resolution element into one of several classes, based on its multispectral representation x. Interest in automatic pattern recognition techniques is motivated by the large quantities of data generated by the scanner as it moves along the flight path. Since the amount of computation is dependent on the dimensionality of x, the usual procedure is to choose a subset of the elements of this vector for processing by the pattern recognition machine. The problem of selecting, *a priori*, the best subset is generally difficult, and is usually based on the combination of (1) statistical interclass distance measures (resulting from an exhaustive search of all possible combinations of the spectral bands) to form a subset of given dimension; and (2) an intuitive selection based on known characteristics of the areas in question.

Computer classification results using the best subset of the original spectral channel vector x and using a subset of the principal components vector y are presented in Fig. 6.51. A Gaussian maximum likelihood decision rule (Tou and Gonzalez [1974]) was used to classify selected areas of the data set into one of six classes (various types of vegetation, roads, etc.). The sixth class was a null class into which all points having classification error probabilities greater than a specified threshold were placed. All points in this null class were considered errors and were used as such in computing classification accuracy.

The abscissa in Fig. 6.51 is the number of features (spectral channels or principal components) used in the classification. For each number, m, of features, an exhaustive search was conducted to determine the m-member set of features giving the highest percent correct classification results. The

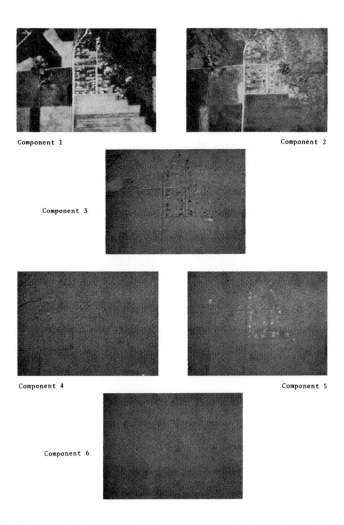

Figure 6.50. Six principal component images computed from the data in Fig. 6.48. (Courtesy of the Laboratory for Applications of Remote Sensing, Purdue University).

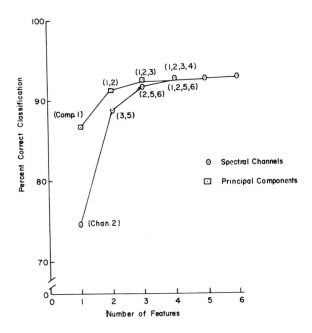

Figure 6.5l. Classification accuracy versus number of features using the spectral bands and the principal components.

features used are listed in Table 6.9. This table and Fig. 6.51 point out the feature selection advantages inherent in the principal components. For each value of m, the first m principal components are the best selection. In addition, for small values of m, the principal components contain more class separability information than any group of m original spectral channels.

Table 6.9. Principal components and corresponding spectral channels.

n	Principal Components	Spectral Channels
1	1	2
2	1,2	3,5
3	1,2,3	2,5,6
4	1,2,3,4	1,2,5,6
5	1,2,3,4,5	1,2,3,5,6
6	1,2,3,4,5,6	1,2,3,4,5,6

Results similar to the above have also been obtained with satellite multispectral data. The channels and corresponding sample images from the Apollo 9-S065 experiment over Imperial Valley, California, are shown in Table 6.10 and Fig. 6.52, respectively. For this particular data set, x was a 3-dimensional vector. The transformation into principal components was carried out as in the preceding example. The eigenvalues are listed in Table 6.11 and the principal-component images are shown in Fig. 6.53. In this case the first two eigenvalues represented approximately 93% of the total data variance.

Table 6.10. Channel numbers and corresponding wavelengths.

Channel	Wavelength Band (μm)
1	0.47–0.61
2	0.68–0.89
3	0.59–0.71

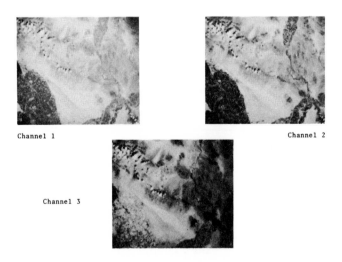

Channel 1

Channel 2

Channel 3

Figure 6.52. Three spectral images from a satellite scanner.

Table 6.ll. Eigenvalues of the images shown in
Fig. 652.

λ_1	λ_2	λ_3
1689	512.5	158.0

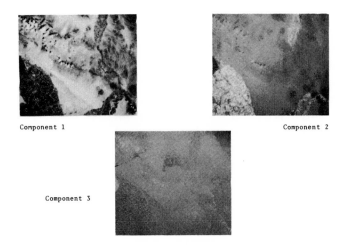

Component 1   Component 2

Component 3

Figure 6.53. Three principal component images corresponding to the data in Fig. 6.52.

6.6 CONCLUDING REMARKS

The success of any encoding technique is ultimately dependent on how well it matches the structure (i.e., departure from randomness) in a given image. The ideal approach in designing an efficient encoder is first to determine the structure of the data and then choose a method which best fits that structure. Since structural properties inherent in pictorial data are not well understood, however, the design and implementation of an image encoder often involves a certain amount of experimentation. The concepts introduced in this chapter are representative of available techniques which have been found of practical value in image processing applications.

In some applications, such as encoding images that exhibit regularity (rivers, man-made objects, etc.), the structural properties of interest often manifest themselves in the form of boundaries. Encoding strategies that take this type of structure into account include the contour-encoding approach developed in Section 6.3.2. In other applications, such as encoding satellite imagery, the structure is not so obvious and a typical approach is to let the statistical information in the image dictate the choice of encoding technique. The Hotelling transformation is an example of this approach.

When encoded images are to be ultimately used for human viewing and interpretation, one must take into account the effects produced by the encoding approach on the visual system. The sensitivity of the human visual system to errors in a reconstructed image depends on such factors as the frequency spectrum of the error, the gray-level content, and the amount of detail in the image. Hence, it is possible to increase the efficiency of an encoder by allowing distortions that minimize degradations in subjective quality. Transform encoding is an example where some aspects of both statistical and psychovisual properties are taken into account. Transform encoders perform a sequence of two operations, the first of which is based on statistical considerations and the second on psychovisual considerations. The first operation is a linear transformation whose objective is to reduce the statistical dependence of the pixels. The second operation is to individually quantize and code each of the resulting coefficients. The number of bits required to code the coefficients depends on the number of quantizer levels, which is dictated by the sensitivity of the visual system to the subjective effect of the quantizer error.

REFERENCES

The book by Abramson [1963] contains an excellent introduction to information theory and source encoding. Two special issues of *Proc. of the IEEE* [1967, 1972] are also of interest as general references.

The paper by Frendendall and Behrend [1960] deals with procedures for evaluating subjective image quality (Section 6.1). Additional reading for Section 6.2 are Abramson [1963], Wintz [1972], Huffman [1952], Meyer, Rosdolsky and Huang [1973], Max [1961], Panter [1951], Schreiber [1967], and Duan and Wintz [1974].

Section 6.3.1 is based on the report by Duan and Wintz [1974]. References for the contour-encoding algorithm developed in Section 6.3.2 are Graham [1967], and Wilkins and Wintz [1970]. For additional references on Section 6.3.3 see Chen and Wintz [1976].

Additional reading on the material in Section 6.4.1 may be found in Huang [1965] and Essman and Wintz [1973]. Section 6.4.2 is based on the work of Habibi (see Habibi and Wintz [1971]). Additional reading for

Section 6.43 may be found in the paper by Habibi [1974].

For references on the Hotelling transform (Section 6.5) see the Reference section at the end of Chapter 3. The examples given in Section 6.5 are based on the paper by Ready and Wintz [1973].

Other references related to the material in this chapter are Huang [1966], Schreiber [1967], Huang and Schulthesis [1963], Stevens [1951], Gattis and Wintz [1971], Schreiber [1956], Proctor and Wintz [1971], Tasto and Wintz [1971, 1972], Wood [1969], Gish and Pierce [1968], Sakrison and Algazi [1971], Habibi [1971], Wilkins and Wintz [1970], and Kramer and Mathews [1956].

IMAGE SEGMENTATION AND DESCRIPTION

The whole is equal to the sum of its parts.
Euclid

The whole is greater than the sum of its parts.
Max Wertheimer

The methods discussed in the previous chapters deal primarily with digital image processing at a global level. In this chapter, we focus attention on more detailed descriptions of image components and their interrelations. This type of processing approach is fundamental in such areas as automatic scene analysis and pattern recognition, where one wishes to use a machine to extract information that requires detailed knowledge of image content at the object level.

The following discussion is organized into four principal topics: (1) segmentation, (2) regional descriptions, (3) relational descriptions, and (4) descriptions of similarity. Figure 7.1 illustrates the hierarchichal order of these operations. We consider segmentation to be the most primitive (but not necessarily the easiest) process. The objective of segmentation techniques is to partition a given image into regions or components. Regional description procedures are used to characterize the individually-segmented regions. Relational description procedures deal with the organization of these regions into a meaningful structure. Finally, the last level in the hierarchy deals with the problem of establishing measures of similarity between regions of an image or between sets of different images.

7.1 SEGMENTATION

The purpose of segmentation is to partition the image space into meaningful regions. The definition of a "meaningful region" is a function of the problem being considered. For example, in an image derived from viewing a three-dimensional scene, the objective of segmentation might be to identify regions corresponding to objects in the scene. In aerial recognizance

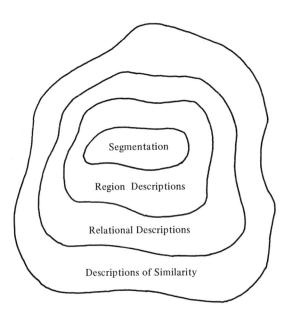

Figure 7.1. Hierarchical ordering of procedures for image description.

applications, on the other hand, a more practical objective is to extract regions corresponding to industrial, residential, agricultural, or natural terrain.

In this section we consider the segmentation problem both as a point- and a region-dependent process. The first category deals with methods which are based on examining an image on a pixel-by-pixel basis. The second category, as the name implies, deals with techniques which utilize the image information in a prescribed neighborhood. In both cases, we may view the problem as a decision-making or pattern recognition process whose objective is to establish boundaries between regions.

In its simplest form, the decision process utilizes a single variable (e.g., brightness of a pixel). The problem, however, is often simplified by having available more than one variable. If we denote the available variables by x_1, x_2, ..., x_n, we may express this information in vector form:

$$\mathbf{x} = \begin{bmatrix} x_1 \\ x_2 \\ \cdot \\ \cdot \\ \cdot \\ x_n \end{bmatrix} \qquad (7.1\text{-}1)$$

These variables can, for example, be the gray levels of points in the neighbor-hood of a given image coordinate or the spectral components of a single pixel.

Pictorial segmentation differs from the general recognition problem in several ways. One of the most important differences is that the solution regions in an image can be visualized and, therefore, verified to a high degree of accuracy. It is usually even possible to superimpose the segmentation results on an image of the original scene to determine the effectiveness of the method. This ability has lead to two general approaches to segmentation, as shown in Fig. 7.2. The first approach is to locate boundaries or edges of regions. The second is to group points into similar regions, which also deter-mines the boundary. The two approaches are similar to defining a line in terms of points or as the intersection of two surfaces.

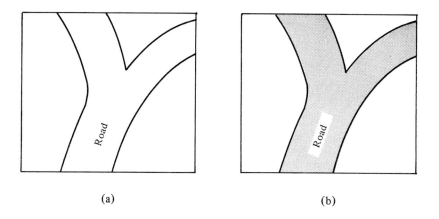

(a)   (b)

Figure 7.2. Two approaches to segmentation.

7.1.1 Point-Dependent Techniques

7.1.1.1 Gray-Level Thresholding

A simple approach that is often useful for segmenting an image consists of dividing the gray scale into bands and using thresholds to deter-mine regions or to obtain boundary points.

As an introduction to this technique, suppose that the gray levels in a given image $f(x, y)$ have the histogram shown in Fig. 7.3(a). Based on the

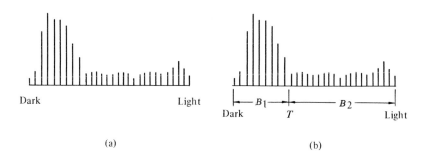

Figure 7.3. Histogram thresholding.

discussion in Section 4.2, we conclude from the histogram that a large number of pixels in $f(x, y)$ are dark, with the remaining pixels being distributed fairly even in the remaining portion of the gray scale. This histogram behavior is characteristic of images consisting of gray objects superimposed on a dark background. To outline the boundary between objects and the background, we divide the histogram into two bands separated by a threshold T, as shown in Fig. 7.3(b). The objective is to select T so that band B_1 contains, as closely as possible, levels associated with the background, while B_2 contains the levels of the objects. As the image is scanned, a change in gray level from one band to the other denotes the presence of a boundary. In order to detect boundaries in both the horizontal and vertical directions, two passes through $f(x, y)$ are required. Once B_1 and B_2 have been selected, the procedure is as follows:

Pass 1. For each row in $f(x, y)$ (i.e., $x = 0, 1, ..., N-1$), create a corresponding row in an intermediate image $g_1(x, y)$ using the following relation for $y = 1, 2, ..., N-1$:

$$g_1(x, y) = \begin{cases} L_E & \text{if the levels of } f(x, y) \text{ and } f(x, y-1) \\ & \text{are in different bands of the gray scale} \\ \\ L_B & \text{otherwise} \end{cases} \tag{7.1-2}$$

where L_E and L_B are specified edge and background levels, respectively.

Pass 2. For each column in $f(x, y)$ (i.e., $y = 0, 1, ..., N-1$), create a corresponding column in an intermediate image $g_2(x, y)$ using the following relation for $x = 1, 2, ..., N-1$:

$$g_2(x, y) = \begin{cases} L_E & \text{if the levels of } f(x, y) \text{ and } f(x - 1, y) \\ & \text{are in different bands of the gray scale} \\ \\ L_B & \text{otherwise} \end{cases} \quad (7.1\text{-}3)$$

The desired image, consisting of the points on the boundary of objects different (as defined by T) from the background, is obtained by using the following relation for $x, y = 0, 1, ..., N-1$:

$$g(x, y) = \begin{cases} L_E & \text{if either } g_1(x, y) \text{ or } g_2(x, y) \\ & \text{is equal to } L_E \\ \\ L_B & \text{otherwise} \end{cases} \quad (7.1\text{-}4)$$

Example: The gray-level thresholding technique just described is illustrated in Fig. 7.4. The original image, Fig. 7.4(a), is a 256-level picture of the Sombrero (Spanish for hat) Nebula. The histogram of this image, shown in Fig. 7.4(b), contains two prominent peaks, one in the dark, and one in the light portion of the gray scale. The first peak corresponds to the background; the second corresponds to the light tones in the image itself. Figure 7.4(c) was obtained by using Eqs. (7.1-2) through (7.1-4) with $B_1 = B_2$ (i.e., $T = 128$), $L_E = 0$, and $L_B = 255$. Figure 7.4(d) was formed by superimposing the edges on Fig. 7.4(a).   □

The above procedure is easily generalized to more gray-level bands. In fact, since relations (7.1-2) and (7.1-3) are based only on a *change* in gray-level band from one pixel to the next, the basic technique for establishing boundary points would be the same if more than two bands were considered. One possible extension would be to code the edge points with different gray levels, depending on the band in which the change took place.

By thresholding more gray-level bands, it is possible to increase the power of the edge extraction technique. The problem, of course, is where to place the thresholds. One approach is to specify the number and location of the thresholds by trial and error. This method is satisfactory if the number of different images to be processed is small. In situations where automatic setting of the thresholds (as in machine perception applications) is required, the problem becomes one of characterizing a given histogram in some invariant manner. This problem is discussed next.

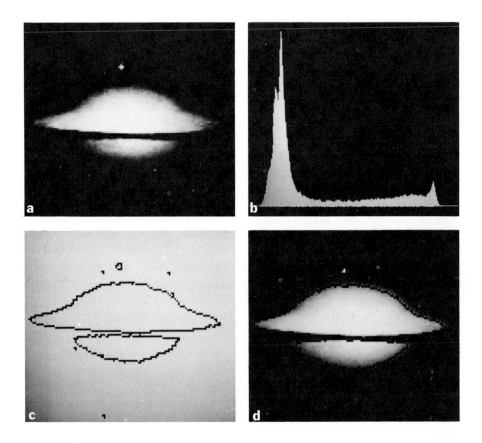

Figure 7.4. Edge extraction by gray-level thresholding. (a) Picture of Sombrero Nebula. (b) Histogram. (c) Edge image obtained with $T = 128$. (d) Edges superimposed on original.

7.1.1.2 Optimum Thresholding

Suppose it is known a priori that an image contains only two principal brightness regions. The histogram of such a picture may be considered as an estimate of the brightness probability density function. This overall density function would be the sum or mixture of two unimodal densities, one for the white and one for the dark regions in the image. Furthermore, the mixture parameters would be proportional to the areas of the picture of each brightness. If the form of the densities is known or assumed, then it is possible to determine an optimum threshold (in terms of minimum error) for segmenting the image into the two brightness regions.

Suppose that an image contains two values combined with additive Gaussian noise. The mixture probability density function is given by

$$p(x) = P_1 p_1(x) + P_2 p_2(x) \tag{7.1-5}$$

which, for the Gaussian case, is

$$p(x) = \frac{P_1}{\sqrt{2}\,\pi\sigma_1} \exp \frac{(x - \mu_1)^2}{2\sigma_1^2} + \frac{P_2}{\sqrt{2}\,\pi\sigma_2} \exp \frac{(x - \mu_2)^2}{2\sigma_2^2} \tag{7.1-6}$$

where μ_1 and μ_2 are the mean values of the two brightness levels, σ_1 and σ_2 are the standard deviations about the means, and P_1 and P_2 are the a priori probabilities of the two levels. Since the constraint

$$P_1 + P_2 = 1 \tag{7.1-7}$$

must be satisfied, the mixture density has five unknown parameters. If all the parameters are known, the optimum threshold is easily determined.

Suppose that the dark regions correspond to the background and the bright regions correspond to objects. In this case $\mu_1 < \mu_2$ and we may define a threshold, T, so that all pixels with gray level below T are considered background points and all pixels with level above T are considered object points. The probability of (erroneously) classifying an object point as a background point is

$$E_1(T) = \int_{-\infty}^{T} p_2(x)\,\mathrm{d}x \tag{7.1-8}$$

Similarly, the probability of classifying a background point as an object point is

$$E_2(T) = \int_{T}^{\infty} p_1(x)\,\mathrm{d}x \tag{7.1-9}$$

Therefore, the overall probability of error is given by

$$E(T) = P_2 E_1(T) + P_1 E_2(T) \tag{7.1-10}$$

To find the threshold value for which this error is minimum, we may differentiate $E(T)$ with respect to T (using Liebnitz's rule) and equate the result to zero. The result is

$$P_1 p_1(T) = P_2 p_2(T) \tag{7.1-11}$$

Applying this result to the Gaussian density gives, after taking logarithms and simplifying, a quadratic equation,

$$AT^2 + BT + C = 0 \qquad (7.1\text{-}12)$$

where

$$A = \sigma_1^2 - \sigma_2^2$$

$$B = 2(\mu_1 \sigma_2^2 - \mu_2 \sigma_1^2) \qquad (7.1\text{-}13)$$

$$C = \sigma_1^2 \mu_2^2 - \sigma_2^2 \mu_1^2 + \sigma_1^2 \sigma_2^2 \ln(\sigma_1 P_1 / \sigma_2 P_2)$$

The possibility of two solutions indicates that it may require two threshold values to obtain the optimum solution

If the variances are equal, $\sigma^2 = \sigma_1^{\,2} = \sigma_2^{\,2}$, a single threshold is sufficient:

$$T = \frac{\mu_1 + \mu_2}{2} + \frac{\sigma^2}{\mu_1 - \mu_2} \ln\!\left(\frac{P_2}{P_1} \right) \qquad (7.1\text{-}14)$$

If the prior probabilities are equal, $P_1 = P_2$, the optimum threshold is just the average of the means. The determination of the optimum threshold may be accomplished easily for other unimodal densities of known form, such as the Raleigh and log-normal densities.

To estimate the parameters from a histogram of an image one may use a maximum likelihood or minimum mean-square error approach. For example, the mean-square error between the mixture density, $p(x)$, and the experimental histogram $h(x_i)$, is

$$M = \frac{1}{N} \sum_{i=1}^{N} \{ p(x_i) - h(x_i) \}^2 \qquad (7.1\text{-}15)$$

where an N-point histogram is available.

In general, it is not a simple matter to determine analytically parameters which minimize this mean-square error. Even for the Gaussian case, the straightforward computation of equating the partial derivatives to zero leads to a set of simultaneous transcendental equations which usually can be solved only by numerical procedures. Since the gradient is easily computed, a conjugate gradient or Newton's method for simultaneous nonlinear equations may be used to minimize M. With either of these iterative methods, starting values must be specified. Assuming the a priori probabilities to be equal may be sufficient. Starting values for the means and variances may be determined by detecting modes in the histogram or simply dividing the histogram into two parts about its mean value, and computing means and variances of the two parts to be used as starting values.

Example: As an illustration of optimum threshold selection we consider in the following discussion an approach developed by Chow and Kaneko [1972] for outlining boundaries of the left ventricle in cardioangiograms (i.e., x-ray pictures of a heart which has been injected with a dye).

Before thresholding, the images were first preprocessed by: (1) taking the logarithm of every pixel to invert the exponential effects caused by radioactive absorption; (2) subtracting two images which were obtained before and after the dye agent was applied in order to remove the spinal column present in both images; and (3) averaging several angiograms to remove noise (see Section 4.3.3). Figure 7.5 shows a cardioangiogram before and after preprocessing (the regions marked A and B are explained below).

In order to compute the optimum thresholds, each preprocessed image was subdivided into 7 × 7 regions (the original images were of size 256 × 256) with 50% overlap. Each of the 49 resulting regions contained 64 × 64 pixels. Figures 7.6(a) and 7.6(b) are the histograms of the regions marked A and B in Fig. 7.5(b). It is noted that the histogram for region A is very clearly bimodal, indicating the presence of a boundary. The histogram for region B, on the other hand, is unimodal, indicating the absence of two markedly distinct regions.

After all 49 histograms were computed, a test of bimodality was performed to reject the unimodal histograms. The remaining histograms were then fitted by bimodal Gaussian density curves [see Eq. (7.1-6)] using a conjugate gradient hill-climbing method to minimize the error function given in Eq. (7.1-15). The ×'s and 0's in Fig. 7.6(a) are two fits to the histogram shown in black dots. The optimum thresholds were then obtained by using Eqs. (7.1-12) and (7.1-13).

At this stage of the process only the regions with bimodal histograms were assigned thresholds. The thresholds for the remaining regions were obtained by interpolating the original thresholds. After this was done, a second interpolation was carried out in a point-by-point manner using neighboring threshold values so that, at the end of the procedure, every point in the image was assigned a threshold. Finally, a binary decision was carried out for each pixel using the rule

$$f(x, y) = \begin{cases} 1 & \text{if } f(x, y) \geqslant T_{xy} \\ \\ 0 & \text{otherwise} \end{cases}$$

where T_{xy} was the threshold computed at location (x, y) in the image. Boundaries were then obtained by taking the gradient of the binary picture. The results are shown in Fig. 7.7, in which the boundary was superimposed

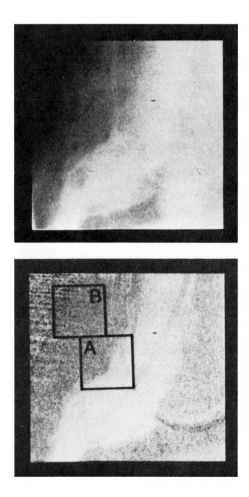

Figure 7.5. A cardioangiogram before and after processing. (From Chow and Kaneko [1972].)

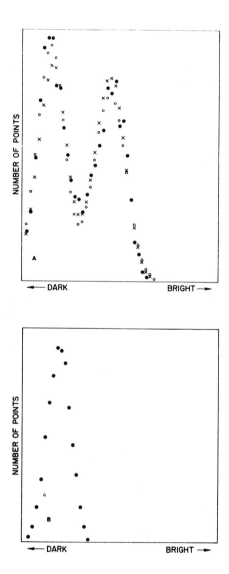

Figure 7.6. Histograms (black dots) of regions A and B in Fig. 7.5(b). (From Chow and Kaneko [1972].)

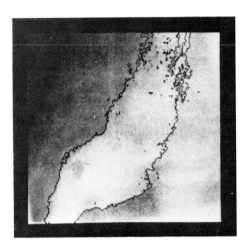

Figure 7.7. Cardioangiogram showing superimposed boundaries. (From Chow and Kaneko [1972].)

on the original image.   □

7.1.1.3 Use of Several Variables

Thus far we have been concerned with thresholding a single intensity variable. In some cases, a sensor might make available more than one variable to characterize each pixel in an image. A notable example is color photography where red, green, and blue components are used to form a composite color image (see Section 4.6). In this case, each pixel is characterized by three values and it becomes possible to construct a three-dimensional histogram. The basic procedure is the same as that used for one variable. For example, given three 16-level images corresponding to the RGB components, we form a $16 \times 16 \times 16$ grid (cube) and insert in each cell of the cube the number of pixels whose RGB components have intensities corresponding to the coordinates defining the location of that particular cell. Each entry can then be divided by the total number of pixels in the image to form a normalized histogram.

The concept of thresholding now becomes one of finding clusters of points in three-dimensional space. Suppose, for example, that we find K significant clusters of points in the histogram. The image can be segmented by assigning one intensity to pixels whose RGB components are closer to one cluster and another intensity to the other pixels in the image. This concept is easily extendable to more components and, certainly, to more clusters. The principal difficulty is that cluster seeking becomes as increasingly complex task as the number of variables is increased. The reader interested in pursuing techniques for cluster seeking can consult, for example, the book by Tou and Gonzalez [1974].

Example: As an illustration of the multivariable histogram approach, consider the images shown in Fig. 7.8. Figure 7.8(a) is a monochrome picture of a color photograph. The original color image was composed of three 16-level RGB images. For our purposes, it is sufficient to point out that the scarf was a vivid red and that the hair and facial colors were light and different in spectral characteristics from the window and other background features.

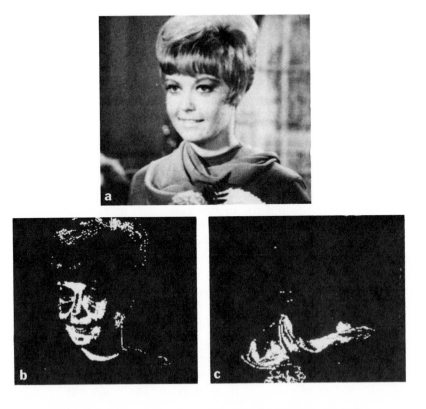

Figure 7.8. Segmentation by multivariable histogram approach.

Figure 7.8(b) was obtained by thresholding about one of the histogram clusters. It is important to note that the window, which in the monochrome picture is close in intensity to the hair, does not appear in the segmented image because of our use of multispectral characteristics to separate these two regions. Figure 7.8(c) was obtained by thresholding about a cluster close to the red axis. In this case only the scarf and part of a flower (which was also red) appeared in the segmented result. The threshold used to obtain both results was a distance of one cell. Thus, any pixel whose components were outside the cell enclosing the center of the cluster in question was classified as background (black). Pixels whose components placed them inside the cell were coded white. □

7.1.2 Region Dependent Techniques

The alternative to segmenting an image into regions of similar characteristics based on point properties, is to permit regional properties to be used for segmentation. In terms of the decision process involved in assigning each point to a region, the extension to local regional properties permits an increase in the dimensionality of the feature vector used in making the decision. It is interesting to note that boundary determination is inherently a region-dependent process. As shown in Fig. 7.9(a), for example, an *edge* may be defined as a difference in image characteristics within a local region. A *line* may be considered as a pair of edges of finite width with a common characteristic in the region between them, as shown in Fig. 7.9(b). Finally, a region may be considered as a finite area bounded by a closed edge, as illustrated in Fig. 7.9(c). In visual images, the method used most often for establishing region characteristics are differences in gray level content, but this is not the only feature which may be used for this purpose. *Texture*, for instance, may be considered as a global repetition of a basic pattern defined over a local region, and differences in texture over such a region can sometimes be used to identify edges. Other properties, such as color differences, are also often useful identifiers.

In this section we consider three region-dependent approaches to the problem of segmentation. The first method is based on detecting transitions between regions; the second is based on growing regions by grouping adjacent areas of similar characteristics; and the third is based on the concept of region clustering.

7.1.2.1 Template Matching

The concept of template matching has found wide acceptance in segmentation applications because of its simplicity. In terms of a digital image, a *template* (also called a *mask* or *window*) is an array designed to detect some invariant regional property.

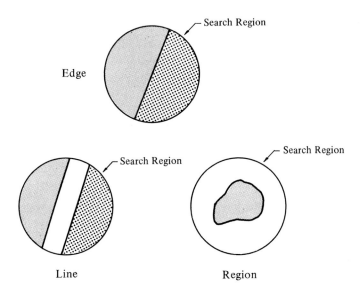

Figure 7.9. Edges, lines, and regions.

As an introduction to template matching, let us consider an image of constant intensity background which contains isolated particles (points) whose intensities are different from the background. Assume that the distance between particles is greater than $[(\Delta x)^2 + (\Delta y)^2]^{1/2}$, where Δx and Δy are the sampling distances in the x and y directions, respectively. Then the particles in the image can be detected by using the template shown in Fig. 7.10. The procedure is as follows. The center of the template (marked 8) is moved around the image from pixel to pixel. At every position, we multiply every point of the image that is inside the template by the number indicated in the corresponding entry of the template, and then add the results. If all image points inside the template area have the same value (constant background), the sum will be zero. If, on the other hand, the center of the mask is located at a particle point, the sum will be different from zero. If the particle is in an off-center location, the sum will also be different from zero, but the magnitude of the response will be weaker than if the particle were in the center of the template. By using a threshold, these weaker responses can be eliminated so that we say a particle has been detected when the sum exceeds the threshold and is, therefore, located in the center of the template.

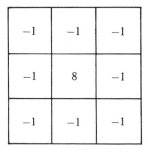

Figure 7.10. Point template.

Let w_1, w_2, ..., w_9 represent the weights (i.e., coefficients) in a 3×3 mask, and let x_1, x_2, ..., x_9 be the gray levels of the pixels inside the mask. From the above discussion it is evident that all we are doing is taking the inner product of the vectors

$$\mathbf{w} = \begin{bmatrix} w_1 \\ w_2 \\ \cdot \\ \cdot \\ \cdot \\ w_9 \end{bmatrix} \tag{7.1-16}$$

and

$$\mathbf{x} = \begin{bmatrix} x_1 \\ x_2 \\ \cdot \\ \cdot \\ \cdot \\ x_9 \end{bmatrix} \tag{7.1-17}$$

where, for example, the first three elements of $\mathbf{w}$ are the elements in the first row of the template, the next three elements are from the second row, and so on. The inner product of $\mathbf{w}$ and $\mathbf{x}$, defined as

$$\mathbf{w}'\mathbf{x} = w_1 x_1 + w_2 x_2 + \cdots + w_9 x_9 \tag{7.1-18}$$

is seen to be identical to the sum of products discussed above. We then say that a particle has been detected if

$$\mathbf{w}'\mathbf{x} > T \tag{7.1-19}$$

where T is a specified threshold. The procedure is easily generalized to an arbitrary window of size $n \times n$, in which case we would work with n^2-dimensional vectors.

Point detection is a fairly straightforward procedure. The next level of complexity involves the detection of lines in an image. Consider the templates shown in Fig. 7.11. If the first template were moved around an image, it would be most strongly responsive to lines (one pixel thick) oriented horizontally. With constant background, the maximum response would result when the line passed through the middle row of the template. The reader can easily verify this by sketching a simple array of 1's with a line of a different gray level (say, 5's) running horizontally through the array. A similar experiment would reveal that the second template in Fig. 7.11 responds best to lines oriented at 45^0; the third template to vertical lines; and the fourth template to lines in the -45^0 direction. These directions can also be established by noting that the preferred direction of each template is weighted with a larger coefficient (i.e., 2) than other possible directions.

Let w_1, w_2, w_3, and w_4 be nine-dimensional vectors formed from the entries of the four templates shown in Fig. 7.11. As discussed above for the point template, the individual responses of the line templates at any point in the image are given by $w_i'x$ for $i = 1, 2, 3, 4$. As before, x is the vector formed from the nine image pixels inside the template area. Given a particular x, suppose that we wish to determine the closest match between the region in question and one of the four line templates. We say that x is closest to the ith template if the response of this template is the largest; in other words, if

$$w_i'x > w_j'x \qquad (7.1\text{-}20)$$

for all values of j, excluding $j = i$. If, for example, $w_1'x$ were greater than $w_j'x$, $j = 2, 3, 4$, we would conclude that the region represented by x is characterized by a horizontal line since this is the feature to which the first template is most responsive.

−1	−1	−1
2	2	2
−1	−1	−1

−1	−1	2
−1	2	−1
2	−1	−1

−1	2	−1
−1	2	−1
−1	2	−1

2	−1	−1
−1	2	−1
−1	−1	2

Figure 7.11. Line templates.

The development of templates for edge detection follows essentially the same reasoning as above, with the exception that we are now interested in detecting transitions between regions. One approach often used for determining such transitions is to implement some form of two-dimensional derivative function. The discrete gradient formulations discussed in Section 4.4.1 in the context of image sharpening are illustrative of this approach. In fact, Fig. 4.27 shows two gradient implementations based on 2 × 2 templates. The following discussion extends the gradient concept to templates of size 3 × 3.

Consider the 3 × 3 image region

a	b	c
d	e	f
g	h	i

Let us define G_x as

$$G_x = (g + 2h + i) - (a + 2b + c) \qquad (7.1\text{-}21)$$

and G_y as

$$G_y = (c + 2f + i) - (a + 2d + c) \qquad (7.1\text{-}22)$$

The gradient at point e is then defined as

$$G = \left[G_x^2 + G_y^2 \right]^{1/2} \qquad (7.1\text{-}23)$$

An alternative definition using absolute values is given by

$$G = |G_x| + |G_y| \qquad (7.1\text{-}24)$$

As indicated in Section 4.4.1, a formulation using absolute values is much easier to implement in a computer.

By comparing the above region and Eq. (7.1-21), we see that G_x is the difference between the first and third rows, where the elements closer to e (b and h) are weighted twice as much as the corner values (this weighting is based on intuitive grounds). Thus, G_x represents an estimate of the derivative in the x direction.

Equation (7.1-23) or (7.1-24) can be implemented by using the two templates shown in Fig. 7.12. Implementation of Eq. (7.1-24), for example, simply requires a sum of the absolute values of the responses of the two templates. It is also important to note that edge detection can be expressed in vector form in exactly the same manner as discussed for line templates. Thus, if x represents the image region in question, we have

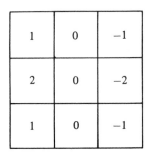

1	2	1
0	0	0
-1	-2	-1

1	0	-1
2	0	-2
1	0	-1

Figure 7.12. Gradient templates

$$G_x = \mathbf{w}_1' \mathbf{x} \tag{7.1-25}$$

and

$$G_y = \mathbf{w}_2' \mathbf{x} \tag{7.1-26}$$

where $\mathbf{w}_1$ and $\mathbf{w}_2$ are the vectors for the two masks shown in Fig. 7.12. The formulations given above for the gradient then become

$$G = \left[(\mathbf{w}_1' \mathbf{x})^2 + (\mathbf{w}_2' \mathbf{x})^2 \right]^{1/2} \tag{7.1-27}$$

and

$$G = |\mathbf{w}_1' \mathbf{x}| + |\mathbf{w}_2' \mathbf{x}| \tag{7.1-28}$$

The vector formulation for the detection of points, lines, and edges has the important advantage that it can be used to detect combinations of these features using a technique developed by Frei and Chen [1977]. In order to see how this can be accomplished, let us consider two hypothetical templates with only three components. In this case, we would have two vectors, $\mathbf{w}_1$ and $\mathbf{w}_2$, which are three-dimensional. Assuming that $\mathbf{w}_1$ and $\mathbf{w}_2$ are orthogonal and normalized so that they have unit magnitude, we have that the terms $\mathbf{w}_1' \mathbf{x}$ and $\mathbf{w}_2' \mathbf{x}$ are equal to the projections of $\mathbf{x}$ onto the vectors $\mathbf{w}_1$ and $\mathbf{w}_2$, respectively. This follows from the fact that, for $\mathbf{w}_1$,

$$\mathbf{w}_1' \mathbf{x} = \|\mathbf{w}_1\| \, \|\mathbf{x}\| \cos \theta \tag{7.1-29}$$

where θ is the angle between the two vectors. Since $\|\mathbf{w}_1\| = 1$,

$$\|\mathbf{x}\| \cos \theta = \mathbf{w}_1' \mathbf{x} \tag{7.1-30}$$

which is the projection of $\mathbf{x}$ onto $\mathbf{w}_1$ (see Fig. 7.13) Similar comments hold for $\mathbf{w}_2$.

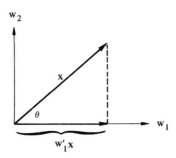

Figure 7.13. Projection of x onto unit vector w_1.

Now suppose that we have three orthogonal vectors of unit magnitude, w_1, w_2, w_3, corresponding to three, 3-point templates. The products $w_1'x$, $w_2'x$, and $w_3'x$ represent the projections of x onto the vectors w_1, w_2, and w_3. According to our earlier discussion, these products also represent the *individual* responses of the three templates. Suppose that templates 1 and 2 are for lines and template 3 is for points. A reasonable question is to ask: Is the region represented by x more like a line or more like a point? Since there are two templates representing lines and we are only interested in the line properties of x, and not on what type of line is present, we could answer the question by projecting x onto the subspace of w_1 and w_2 (which in this case is a plane) and also onto w_3. The angle between x and each of these two projections would tell us whether x is closer to the line or the point subspace. This can be seen from the geometrical arrangement shown in Fig. 7.14. The magnitude of the projection of x onto the plane determined by w_1 and w_2 is given by the quantity $[(w_1'x)^2 + (w_2'x)^2]^{1/2}$, while the magnitude (i.e., norm) of x is

$$\|x\| = \left[(w_1'x)^2 + (w_2'x)^2 + (w_3'x)^2 \right]^{1/2} \tag{7.1-31}$$

The angle between x and its projection is then

$$\theta = \cos^{-1}\left\{ \frac{\left[(w_1'x)^2 + (w_2'x)^2 \right]^{1/2}}{\left[(w_1'x)^2 + (w_2'x)^2 + (w_3'x)^2 \right]^{1/2}} \right\}$$

$$= \cos^{-1}\left\{ \frac{\left[\sum_{i=1}^{2} (w_i'x)^2 \right]^{1/2}}{\left[\sum_{j=1}^{3} (w_j'x)^2 \right]^{1/2}} \right\}$$

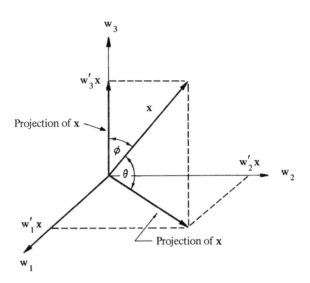

Figure 7.14. Projections of **x** onto subspace (plane) determined by $\mathbf{w}_1$ and $\mathbf{w}_2$, and onto subspace $\mathbf{w}_3$.

$$\theta = \cos^{-1}\left\{ \frac{1}{\|\mathbf{x}\|} \left[\sum_{i=1}^{2} (\mathbf{w}'_i\mathbf{x})^2 \right]^{1/2} \right\} \tag{7.1-32}$$

where the last step follows from Eq. (7.1-31). A similar development would yield the angle of projection onto the $\mathbf{w}_3$ subspace:

$$\phi = \cos^{-1}\left\{ \frac{1}{\|\mathbf{x}\|} \left[\sum_{i=3}^{3} (\mathbf{w}'_i\mathbf{x})^2 \right]^{1/2} \right\}$$

$$= \cos^{-1}\left\{ \frac{1}{\|\mathbf{x}\|} |\mathbf{w}'_3\mathbf{x}| \right\} \tag{7.1-33}$$

Thus, if $\theta < \phi$, we say that the region represented by **x** is closer to the characteristics of a line than of a point.

If we now consider 3 × 3 masks, the problem becomes nine-dimensional, but the above concepts are still valid. We need, however, 9 nine-dimensional orthogonal vectors to form a complete basis. The templates shown in Fig. 7.15 (proposed by Frei and Chen [1977]) satisfy this condition. The first four masks are suitable for detecting edges; the second set of

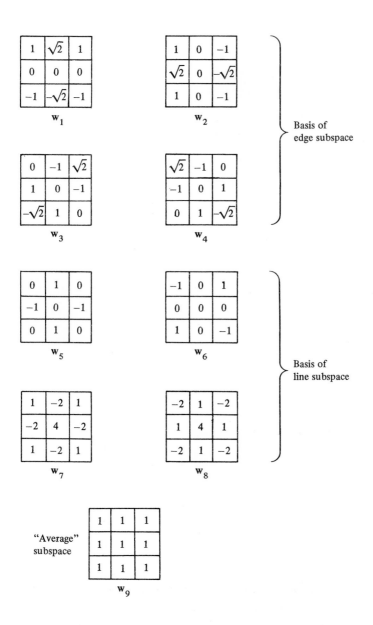

Figure 7.15. Orthogonal templates. (From Frei and Chen [1977].)

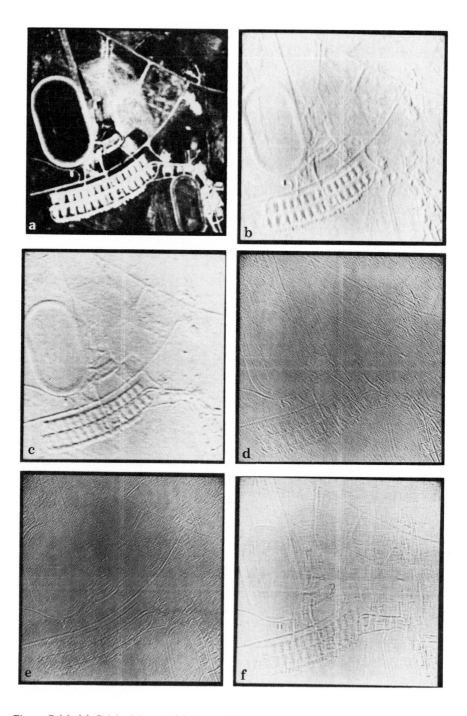

Figure 7.16. (a) Original image. (b) through (f) Projections onto w_1, w_2, w_3, w_4, and w_5 subspaces, respectively. (From Hall and Frei [1976].)

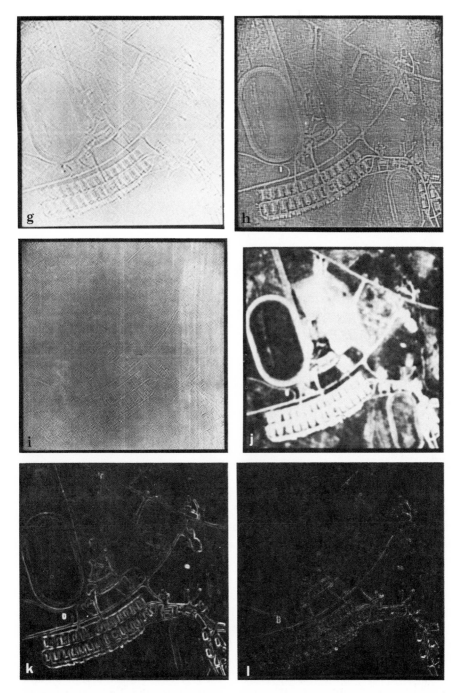

Figure 7.16. (Continued.) (g) through (j) Projections onto w_6, w_7, w_8, and w_9 subspaces. (k) Magnitude of projection onto edge subspace. (l) Magnitude of projection onto line subspace. (From Hall and Frei [1976].)

four masks represents templates suitable for line detection; and the last template (added to complete the bases) is proportional to the average of the pixels in the region at which the mask is located in an image.

Given a 3×3 region represented by $\mathbf{x}$, and assuming that the vectors $\mathbf{w}_i$, $i = 1, 2, ..., 9$, have been normalized, we have from the above discussion that

$$p_e = \left[\sum_{i=1}^{4} (\mathbf{w}_i'\mathbf{x})^2 \right]^{1/2} \tag{7.1-34}$$

$$p_l = \left[\sum_{i=5}^{8} (\mathbf{w}_i'\mathbf{x})^2 \right]^{1/2} \tag{7.1-35}$$

and

$$p_a = |\mathbf{w}_9'\mathbf{x}| \tag{7.1-36}$$

where p_e, p_l, and p_a are the magnitudes of the projections of $\mathbf{x}$ onto the edge, line, and average subspaces, respectively.

Similarly, we have that

$$\theta_e = \cos^{-1}\left\{ \frac{1}{\|\mathbf{x}\|} \left[\sum_{i=1}^{4} (\mathbf{w}_i'\mathbf{x})^2 \right]^{1/2} \right\} \tag{7.1-37}$$

$$\theta_l = \cos^{-1}\left\{ \frac{1}{\|\mathbf{x}\|} \left[\sum_{i=5}^{8} (\mathbf{w}_i'\mathbf{x})^2 \right]^{1/2} \right\} \tag{7.1-38}$$

and

$$\theta_a = \cos^{-1}\left\{ \frac{1}{\|\mathbf{x}\|} |\mathbf{w}_9'\mathbf{x}| \right\} \tag{7.1-39}$$

where θ_e, θ_l and θ_a are the angles between $\mathbf{x}$ and its projections onto the edge, line, and average subspaces, respectively. These concepts are, of course, directly extendable to other bases and dimensions, as long as the basis vectors are orthogonal.

Example: The image shown in Fig. 7.16(a) is a 256×256 aerial photograph of a football stadium site. Figures 7.16(b) through (j) are the magnitudes of the projections along the individual basis vectors obtained by using each of the masks in Fig. 7.15 and, for each position of the ith mask, computing a pixel value equal to $|\mathbf{w}_i'\mathbf{x}|$. Figure 7.16(k) shows the magnitude of the projections onto the edge subspace [Eq. (7.1-34)], and Fig. 7.16(l) was formed from the magnitudes of the projections onto the line subspace [Eq. (7.1-35)]. In this example, the best results were obtained with the edge subspace projections, thus indicating a strong edge content in the original image.   □

7.1.2.2 Region Growing

Since the goal of segmentation is to partition an image or set of images into regions, a direct approach to this problem is to attempt a partitioning of the image frame into regions which satisfy a similarity criterion, i.e., group points into regions. The advantage of this approach is that it results not only in boundary points of regions but also with a similarity criterion satisfied for all points within the regions. The "meaningfulness" of the segmentation may still need to be determined by subjective methods.

In order to perform point groupings, three fundamental issues must be resolved. The number of regions must be determined. The properties or features which distinguish one region from the others must also be determined. Finally, a suitable similarity criterion which will produce a meaningful segmentation must be specified. In the general case, all three problems must be resolved. However, it is often possible to use a priori knowledge in the form of a model to simplify the segmentation task.

One such approach is called *region growing*. Suppose that the number of regions and the location of a single point within each region are known. Then, one may develop an algorithm which starts at the known points and appends all neighboring points which are similar in gray level, color, texture, gradient, or other properties, to the known point in order to form a region. A simple similarity measure is the Euclidean distance of the measured property of the test point from the known point.

A numerical example of region growing is shown in Fig. 7.17. Based on gray level values and the gray level histogram, the original image shown in Fig. 7.17(a) may be assumed to consist of two regions. Starting points of (3, 2) and (3, 4), respectively, are assumed for the two regions. The segmentation results using a Euclidean distance less than 3 are shown in Fig. 7.17 (b). The segmentation resulted in two regions, as expected, and the same segmentation would have resulted from any other starting point in the region. The segmentation results using the same starting points but grouping points with distance less than 4 is shown in Fig. 7.17(c). Note the sensitivity to the similarity threshold.

7.1.2.3 Region Clustering

The technique described in Section 7.1.2.2 is based on the assumption that initial points in each region are available. This requirement is seldom met in practice and it becomes necessary to search for the region directly. One possibility is to use a region clustering algorithm.

The region clustering problem applies directly to image segmentation, and differs from general clustering only by the requirement that points within a cluster must be contiguous in the image plane as well as similar in properties. This requirement may also be used to define contiguous image

	1	2	3	4	5
1	0	0	4	6	7
2	1	1	5	8	7
3	0	<u>1</u>	6	<u>7</u>	7
4	2	0	7	6	6
5	0	1	4	6	4

(a)

a	a	b	b	b
a	a	b	b	b
a	a	b	b	b
a	a	b	b	b
a	a	b	b	b

(b)

a	a	a	a	a
a	a	a	a	a
a	a	a	a	a
a	a	a	a	a
a	a	a	a	a

(c)

Figure 7.17. Example of region growing using known starting point. (a) Original array. (b) Segmentation result using distance less than 3. (c) Segmentation result using distance less than 4.

points with similar properties, as distinguished from a cluster which is simply a set of points with similar properties. The general procedure of region clustering is to first examine the set of image measurements to determine the number and location of clusters in measurement space, then apply these cluster definitions to the image to obtain region clusters. Criteria may be applied to both the clustering procedure and to the region clustering. Different criteria may be required at each step. For example, a color random dot image would provide clusters in the three-dimensional color space, but would not provide regional clusters.

The following procedure is illustrative of a region clustering technique. The reader should be aware, however, that cluster seeking is a major area of automatic information processing and, as such, can only be introduced in our present discussion. The interested reader can consult, for example, the book by Tou and Gonzalez [1974] for additional details and references on this topic.

A partition of a set X is any collection of sets $\{R_1, R_2, ..., R_n\}$ such that the union of the sets R_k is exactly X and the pairwise intersection of the R_k is null unless the two sets are identical. If we define some equivalence relation on the picture – a simple example is: $P(i, j)$ is equivalent to $P(k, l)$ if their values are equal – then this relation induces a natural equivalence relation on the grid of points given by: (i, j) is equivalent to (k, l) if and only if $P(i, j)$ is equivalent to $P(k, l)$.

Any equivalence relation on the grid of points yields a partition of the grid into equivalence classes. For example, if the values of $P(i, j)$ range from 0 to 63, then 64 equivalence class masks could be produced whose values would be equal to 1 if the relation were satisfied and 0 otherwise. The mask images would be pairwise disjoint and the union of the 64 masks would fill the entire grid.

The equivalence classes can be further subdivided into maximally-connected subsets called *connected components*. Connectedness may be defined in terms of the neighbors of a points (i, j). The 4-connected neighbors of a point are the four nondiagonally adjacent neighbors. The 8-connected neighbors are the 8 surrounding neighbor points. Using 4-connectedness, we say that two points, p_1 and p_2, belonging to a subset R of the grid, are connected if there exists a sequence of points in R, the first of which is p_1 and the last of which is p_2, such that consecutive points are 4-connected neighbors. With this definition .of connected points, we may define a region as a subset of R in which any pair of points is connected with respect to R. The equivalence masks may be divided into maximally-connected regions (sometimes called *atomic regions*). These can then be used as building blocks to form regions which respond to natural, meaningful segmentations.

A method for representing a region during segmentation, developed by Brice and Fennema [1970], is as follows. One may consider the picture grid,

G, to be a subgrid of a larger grid, S. In particular, if G is an $n \times m$ grid then let S be a $(2n + 1) \times (2m + 1)$ grid in which the points (i, j) of G, are placed on the points $(2i + 1, 2j + 1)$ of S. The points in the picture correspond to points of S in which each subscript is odd, and the remaining points may be used to represent the boundaries of regions.

Representing regions in this manner yields a simple algorithm for finding atomic regions of a picture. Each point of G, except for its edges, is compared with the one above it and the one to its right. If a difference in gray scale is encountered, the boundary segment is inserted between them. After each picture point is considered, the grid has been partitioned into regions. An example is shown in Fig. 7.18. Note that the equal gray levels along the diagonal produce separate regions because a connectedness of 4 was used for the equivalence relation.

Since the atomic regions produced in the previous step may not correspond to physical or visible boundaries, merging or splitting of regions may be necessary. If the region boundaries are assumed to be oriented such that the region always lies to the left of the boundary, then the operation of merging two regions into one may be accomplished by adding the boundaries. An example is shown in Fig. 7.19. A simple method for splitting regions is to split the region along a straight line as shown in Fig. 7.20. A more general procedure would cut along arbitrary curves, but would be considerably more difficult to implement in a machine.

7.2 REGIONAL DESCRIPTIONS

Once a region of an image has been identified, it is usually of interest to characterize the image by a set of descriptors which are reasonably insensitive to such variations as changes in size, rotation, translation, etc. The motivation for choosing region descriptors is not only to reduce the amount of raw data in the region, but also to bring out features which will aid in differentiating between regions with different attributes. In this section we consider three typical approaches to this problem.

7.2.1 Fourier Descriptors

Once the points on the boundary of a region have been determined, it is often of interest to extract information from these points which will be useful in discriminating between different region shapes. It is shown in this section that the discrete Fourier transform (DFT) can be used for this purpose.

Suppose that M points on the boundary of a region are available. We

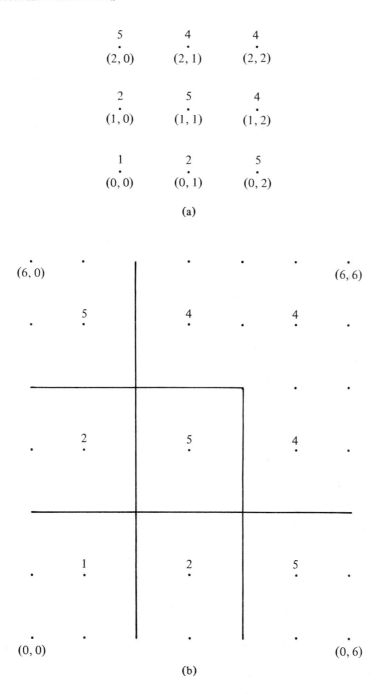

Figure 7.18. Example of Brice and Fennema segmentation method. (a) A 3 X 3 array with gray level and grid values shown. (b) The 7 X 7 supergrid with gray levels placed on odd subscripted points and resulting segmentation.

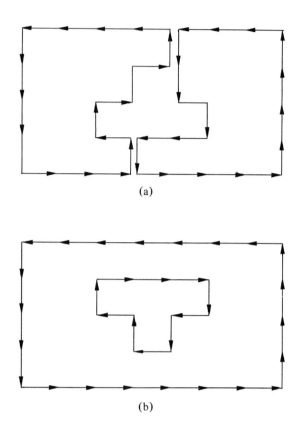

(a)

(b)

Figure 7.19. (a) Two regions with a common boundary. (b) Result of merging by adding
the directed boundary segments.

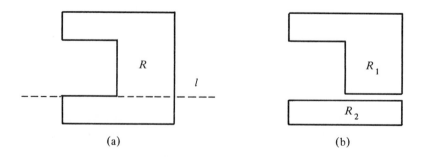

(a)   (b)

Figure 7.20. Splitting a region along a straight line.

may view the region as being in the complex plane, with the ordinate being the imaginary axis and the abscissa being the real axis, as shown in Fig. 7.21. The x-y coordinates of each point in the contour to be analyzed become complex numbers, $x + jy$. Starting at an arbitrary point on the contour, and tracing once around it, yields a sequence of complex numbers. The DFT of this sequence will be referred to in the following discussion as the Fourier Descriptor (FD) of the contour.

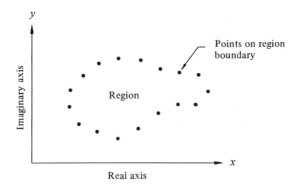

Figure 7.21. Representation of a region boundary in the complex plane.

Since the DFT is a reversible linear transformation, there is no information gained or lost by this process. However, certain simple manipulations of this frequency domain representation of shape can eliminate dependency on position, size, and orientation. Given an arbitrary FD, several successive steps can normalize it so that it can be matched to a test set of FDs regardless of its original size, position, and orientation.

7.2.1.1. Normalization

The frequency domain operations which affect the size, orientation, and starting point of the contour follow directly from properties of the DFT (see Section 3.3). To change the size of the contour, the components of the FD are simply multiplied by a constant. Due to linearity, the inverse transform will have its coordinates multiplied by the same constant.

To rotate the contour in the spatial domain simply requires multiplying each coordinate by $\exp(j\theta)$, where θ is the angle of rotation. Again, by linearity, the constant $\exp(j\theta)$ has the same effect when the frequency domain coefficients are multiplied by it.

To see how the contour starting point can be moved in the frequency domain, recall the periodicity property of the DFT. The finite sequence of numbers in the spatial domain actually represents one cycle of a periodic function. The DFT coefficients are actually coefficients of the Fourier series representation of this periodic function. Remembering these facts, it is easy to see that shifting the starting point of the contour in the spatial domain corresponds to multiplying the kth frequency coefficient in the frequency domain by $\exp(jkT)$, where T is the fraction of a period through which the starting point is shifted. (As T goes from 0 to 2π, the starting point traverses the whole contour once.)

Given the FD of an arbitrary contour, the normalization procedure requires performing the normalization operations such that the contour has a standard size, orientation, and starting point. A standard size is easily defined by requiring the Fourier component $F(1)$ to have unity magnitude. If the contour is a simple closed figure, and it is traced in the counterclockwise direction, this coefficient will be the largest.

The orientation and starting point operations affect only the phases of the FD coefficients. Since there are two allowable operations, the definition of standard position and orientation must involve the phases of at least two coefficients. Let us denote the FD array of length M by $\{F(-M/2 + 1), \ldots F(-1), F(0), F(1), \ldots F(M/2)\}$. One obvious coefficient to use is $F(1)$, already normalized to have unity magnitude. If we start by requiring the phase of $F(1)$ to be some value, say zero, it can be shown that if the kth coefficient is also required to have zero phase, there are $(k-1)$ possible starting point/orientation combinations which satisfy these restrictions.

The obvious procedure is to require $F(1)$ and $F(2)$ to have phases equal to some specified value, thereby achieving a unique standard normalization. This sounds like a solution to the problem, but while $F(1)$ is guaranteed to have unity magnitude after normalizing the FD for size, there is no such guarantee for $F(2)$. A consistent solution to this problem can be obtained by selecting a nonzero coefficient to use for normalization, and then using a third coefficient to resolve the ambiguity which may be caused by the multiple normalization effect.

7.2.1.2. Practical Considerations

The practical implementation of this procedure requires paying attention to a few details not mentioned above. Theoretically, the procedure involves an exact representation of a contour which is sampled at uniform spacing. While nonuniform spacing can result in a frequency domain representation which converges faster, there are some serious difficulties involved in attempting to define a standard sampling strategy using nonuniform spacing.

Remembering that the FFT algorithm requires an input array whose length is an integer power of 2, it is clear that the length of an arbitrary chain representation must be adjusted before the FFT can be used. A procedure for doing this is to compute the perimeter of the contour, divide it by the desired length (desired power of 2), and starting at one point, trace around the contour saving the coordinates of appropriately spaced points. The desired power of 2 might be the smallest power of 2 larger than the length of the chain.

Practically, the input to the shape analysis algorithm will be a contour taken from a sampled picture. The perimeter of this contour will be an approximation to the actual perimeter of the contour. While it can be argued that, for high enough sampling density in the original picture, the chain is an arbitrarily good approximation to the contour, this argument breaks down if one considers the density of points around the approximate contour versus the exact contour.

Consider an equilateral right triangle oriented so that the legs line up with the x and y axes, with the hypotenuse at 45 degrees. The "length" of the contour, if an ordinary four neighbor chain is used, will be four times the length of one leg; the hypotenuse will be as long as both legs combined. Obviously, the density of points on the hypotenuse will depart from the proper value by a factor of $\sqrt{2}$. This error will cause the normalized Fourier descriptors (NFDs) of simple figures such as triangles to differ substantially, and render the algorithm virtually useless.

One solution to this problem is to use an eight-neighbor chain code, in which the four diagonal neighbors of a point can also be the next point in the chain. In the example just considered, this eliminates the point density error. Of course, for different orientations, there will still be a certain amount of error due to the chain code approximation, but this is reduced from a maximum of about 40% to a maximum of about 8%. Experimental results using the eight-neighbor chain code confirm that this error is tolerable. If a picture is contoured using a four-neighbor chain code and it is desired to process the contours using the FD method, the four chain codes can be easily converted to approximate eight chain codes which are suitable for analysis.

Other practical considerations involve the normalization process. While, theoretically, any nonzero coefficient can be used with $F(1)$ to define standard orientation and starting point as outlined above, practical contours show the effects of noise and quantization error. This noise perturbs the phases of the FD coefficients so that the coefficients of lower amplitude can be substantially affected. It can be shown that the mean-square error in the frequency domain corresponds to the point-by-point mean-square error in the spatial domain. It follows from this result that slight shifts in orientation and/or starting point due to noise can have drastic effects on classification of shapes made using this criterion. One way to minimize this effect is to choose the largest magnitude coefficients as normalization coefficients. $F(1)$ is already the largest, so the second largest is chosen to accompany $F(1)$. Generally a third coefficient will be required to decide which of the allowable normalizations is optimum, as explained above. This coefficient can be chosen to be the largest remaining coefficient suitable for resolving the ambiguity.

The normalization procedure tends to reduce the proportion of the information contained in the phase, as compared to that contained in the magnitudes. Also, it can be shown that if the contour under analysis has bilateral symmetry, the resulting NFD will have phases equal to either the normalization phase (phase to which the normalization coefficients are constrained), or that value plus 180 degrees. In view of these results, classification using only the magnitudes of the NFD seems like a reasonable procedure. In this case, the normalization procedure consists of simply dividing each coefficient by the magnitude of $F(1)$.

Example: As an illustration of the above approach for shape description, consider the aircraft silhouttes shown in Fig. 7.22. These silhouttes were obtained by (1) computing the NFDs of the boundary (512 points were used), (2) retaining the 32 lowest frequency components while setting the rest equal to zero, and (3) taking the inverse Fourier transform of the modified 512-array to obtain an approximation to the original data. As shown in Fig. 7.22, the results, although a bit distorted, retained the basic features of the different aircraft. The information present in the lowest 32 components, therefore, was sufficient to differentiate between the shapes of these aircraft.   □

7.2.2 Moments

The Fourier descriptions discussed in the previous section are based on the assumption that a set of boundary points is available. Sometimes a region may be given in the form of interior points and, as indicated earlier, one may be interested in finding descriptors which are invariant to variations in translation, rotation, and size. The moment approach discussed below is often used for this purpose.

Given a two-dimensional continuous function $f(x, y)$ we define the moment of order $(p + q)$ by the relation

$$m_{pq} = \int_{-\infty}^{\infty} \int_{-\infty}^{\infty} x^p y^q f(x, y) \, dx \, dy \qquad (7.2\text{-}1)$$

for $p, q = 0, 1, 2, \ldots$.

A uniqueness theorem (Papoulis [1965]) states that if $f(x, y)$ is piecewise continuous and has nonzero values only in a finite part of the $x\text{-}y$

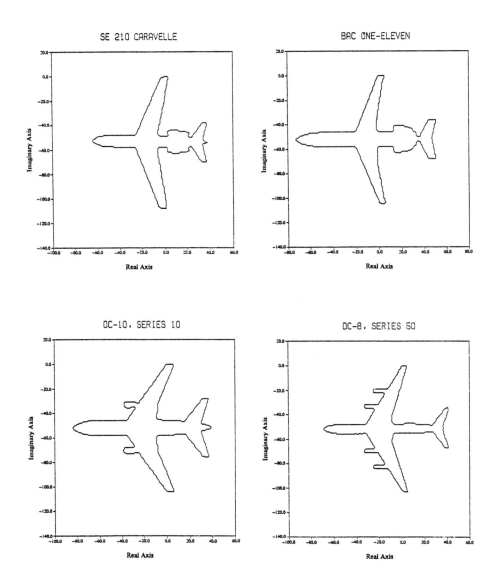

Figure 7.22. Shapes obtained by using Fourier descriptors. (Courtesy of T. Wallace, Electrical Engineering Department, Purdue University.)

plane, then moments of all orders exist and the moment sequence (m_{pq}) is uniquely determined by $f(x, y)$ and, conversely (m_{pq}) uniquely determines $f(x, y)$. The *central moments* can be expressed as

$$\mu_{pq} = \int_{-\infty}^{\infty} \int_{-\infty}^{\infty} (x - \bar{x})^p (y - \bar{y})^q f(x, y) \, dx \, dy \qquad (7.2\text{-}2)$$

where

$$\bar{x} = \frac{m_{10}}{m_{00}}, \qquad \bar{y} = \frac{m_{01}}{m_{00}}$$

For a digital image, Eq. (7.2-2) becomes

$$\mu_{pq} = \sum_x \sum_y (x - \bar{x})^p (y - \bar{y})^q f(x, y) \qquad (7.2\text{-}3)$$

The central moments of order 3 are as follows:

$$\mu_{10} = \sum_x \sum_y (x - \bar{x})^1 (y - \bar{y})^0 f(x, y)$$

$$= m_{10} - \frac{m_{10}}{m_{00}} (m_{00})$$

$$= 0 \qquad (7.2\text{-}4)$$

$$\mu_{11} = \sum_x \sum_y (x - \bar{x})^1 (y - \bar{y})^1 f(x, y)$$

$$= m_{11} - \frac{m_{10} m_{01}}{m_{00}} \qquad (7.2\text{-}5)$$

$$\mu_{20} = \sum_x \sum_y (x - \bar{x})^2 (y - \bar{y})^0 f(x, y)$$

$$= m_{20} - \frac{2 m_{10}^2}{m_{00}} + \frac{m_{10}^2}{m_{00}} = m_{20} - \frac{m_{10}^2}{m_{00}} \qquad (7.2\text{-}6)$$

$$\mu_{02} = \sum_x \sum_y (x - \bar{x})^0 (y - \bar{y})^2 f(x, y)$$

$$= m_{02} - \frac{m_{01}^2}{m_{00}} \qquad (7.2\text{-}7)$$

$$\mu_{30} = \sum_x \sum_y (x - \bar{x})^3 (y - \bar{y})^0 f(x, y)$$

$$= m_{30} - 3\bar{x}m_{20} + 2m_{10}\bar{x}^2 \qquad (7.2\text{-}8)$$

$$\mu_{12} = \sum_x \sum_y (x - \bar{x})(y - \bar{y})^2 f(x, y)$$

$$= m_{12} - 2\bar{y}m_{11} - \bar{x}m_{02} + 2\bar{y}^2 m_{10} \qquad (7.2\text{-}9)$$

$$\mu_{21} = \sum_x \sum_y (x - \bar{x})^2 (y - \bar{y})^1 f(x, y)$$

$$= m_{21} - 2\bar{x}m_{11} - \bar{y}m_{20} + 2\bar{x}^2 m_{01} \qquad (7.2\text{-}10)$$

$$\mu_{03} = \sum_x \sum_y (x - \bar{x})(y - \bar{y})^3 f(x, y)$$

$$= m_{03} - 3\bar{y}m_{02} + 2\bar{y}^2 m_{01} \qquad (7.2\text{-}11)$$

In summary,

$$\mu_{00} = m_{00}, \qquad \mu_{11} = m_{11} - \bar{y}m_{10}$$

$$\mu_{10} = 0, \qquad \mu_{30} = m_{30} - 3\bar{x}m_{20} + 2m_{10}\bar{x}^2$$

$$\mu_{01} = 0, \qquad \mu_{12} = m_{12} - 2\bar{y}m_{11} - \bar{x}m_{02} + 2\bar{y}^2 m_{10} \quad (7.2\text{-}12)$$

$$\mu_{20} = m_{20} - \bar{x}m_{10}, \quad \mu_{21} = m_{21} - 2\bar{x}m_{11} - \bar{y}m_{20} + 2\bar{x}^2 m_{01}$$

$$\mu_{02} = m_{02} - \bar{y}m_{01}, \quad \mu_{03} = m_{03} - 3\bar{y}m_{02} + 2\bar{y}^2 m_{01}$$

The *normalized central moments*, denoted by η_{pq}, are defined as:

$$\eta_{pq} = \frac{\mu_{pq}}{\mu_{00}^{\gamma}} \qquad (7.2\text{-}13a)$$

where

$$\gamma = \frac{p + q}{2} \qquad (7.2\text{-}13b)$$

for $p + q = 2, 3, \dots$.

From the second and third moments, a set of seven *invariant moments* can be derived[†]. They are given by

$$\varphi_1 = \eta_{20} + \eta_{02} \tag{7.2-14}$$

$$\varphi_2 = (\eta_{20} - \eta_{02})^2 + 4\eta_{11}^2 \tag{7.2-15}$$

$$\varphi_3 = (\eta_{30} - 3\eta_{12})^2 + (3\eta_{21} + \eta_{03})^2 \tag{7.2-16}$$

$$\varphi_4 = (\eta_{30} + \eta_{12})^2 + (\eta_{21} + \eta_{03})^2 \tag{7.2-17}$$

$$\varphi_5 = (\eta_{30} - 3\eta_{12})(\eta_{30} + \eta_{12})\left[(\eta_{30} + \eta_{12})^2 - 3(\eta_{21} + \eta_{03})^2\right]$$
$$+ (3\eta_{21} - \eta_{03})(\eta_{21} + \eta_{03})\left[3(\eta_{30} + \eta_{12})^2 - (\eta_{21} + \eta_{03})^2\right] \tag{7.2-18}$$

$$\varphi_6 = (\eta_{20} - \eta_{02})\left[(\eta_{30} + \eta_{12})^2 - (\eta_{21} + \eta_{03})^2\right]$$
$$+ 4\eta_{11}(\eta_{30} + \eta_{12})(\eta_{21} + \eta_{03}) \tag{7.2-19}$$

$$\varphi_7 = (3\eta_{12} - \eta_{30})(\eta_{30} + \eta_{12})\left[(\eta_{30} + \eta_{12})^2 - 3(\eta_{21} + \eta_{03})^2\right]$$
$$+ (3\eta_{21} - \eta_{03})(\eta_{21} + \eta_{03})\left[3(\eta_{30} + \eta_{12})^2 - (\eta_{21} + \eta_{03})^2\right] \tag{7.2-20}$$

This set of moments has been shown to be invariant to translation, rotation, and scale change Hu [1962].

Example: The image shown in Fig. 7.23(a) was reduced to half size [Fig. 7.23(b)], mirror-imaged [Fig. 7.23(c)], and rotated by 2 degrees and 45 degrees, as shown in Figs. 7.23(d) and (e). The seven moment invariants given in Eqs. (7.2-14) through (7.2-20) were then computed for each of these images, and the logarithm of the results taken to reduce the dynamic range. As shown in Table 7.1, the results for Figs. 7.23(b) through (e) are in reasonable agreement with the invariants computed for the original image. The major cause of error can be attributed to the digital nature of the data.□

7.2.3 Topological Descriptors

Topological properties are useful for global descriptions of regions in the image plane. Simply defined, topology is the study of properties of a figure which are unaffected by any deformation without tearing or joining. Consider, for example, the region shown in Fig. 7.24. If we define as a

---

[†]Derivation of these results involves concepts which are beyond the scope of the present discussion. The interested reader should consult the book by Bell [1965] and the paper by Hu [1962] for a detailed discussion.

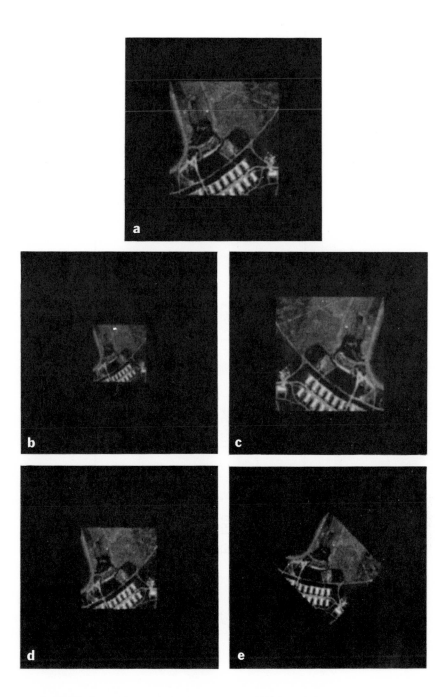

Figure 7.23. Images used to demonstrate properties of moment invariants.

Table 7.1. Moment invariants for the images in Figs. 7.23(a) through (e).

Invariant (log)	Original	Half size	Mirrored	Rotated(2°)	Rotated(45°)
φ_1	6.249	6.226	6.919	6.253	6.318
φ_2	17.180	16.954	19.955	17.270	16.803
φ_3	22.655	23.531	26.689	22.836	19.724
φ_4	22.919	24.236	26.901	23.130	20.437
φ_5	45.749	48.349	53.724	46.136	40.525
φ_6	31.830	32.916	37.134	32.068	29.315
φ_7	45.589	48.343	53.590	46.017	40.470

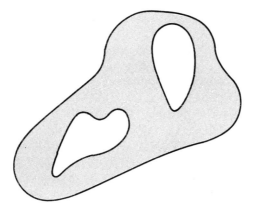

Figure 7.24. A region with two holes.

topological descriptor the number of holes in the region, it is evident that this property will not be affected by a stretching or rotation transformation. The number of holes, however, will in general change if we tear or fold the region. It is noted that, since stretching affects distance, topological properties do not depend on any notion of distance or any properties which are implicitly based on the concept of a distance measure.

Another topological property useful for region description is the number of connected components. A connected component of a set is a subset of maximal size such that any two of its points can be joined by a connected curve lying entirely within the subset. Figure 7.25 shows a region with three connected components.

The number of holes (H) and connected components (C) in a figure can be used to define the *Euler* number, E, as follows:

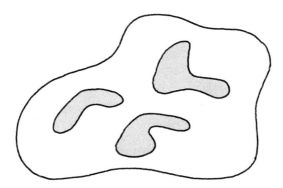

Figure 7.25. A region with three connected components.

$$E = C - H \tag{7.2-21}$$

The Euler number is also a topological property. The regions shown in Fig. 7.26, for example, have Euler numbers equal to 0 and -1, respectively, since the "A" has one connected component and one hole and the "B" one connected component but two holes.

Regions represented by straight line segments (referred to as *polygonal networks*) have a particularly simple interpretation in terms of the Euler number. A polygonal network is shown in Fig. 7.27. It is often important to classify interior regions of such a network into faces and holes. If we denote the number of vertices by W, the number of edges by Q, and the number of faces by F, we have the following relationship, called the *Euler formula*:

$$W - Q + F = C - H \tag{7.2-22}$$

which, in view of Eq. (7.2-21), is related to the Euler number:

$$W - Q + F = C - H = E \tag{7.2-23}$$

The network shown in Fig. 7.27 has 7 vertices, 11 edges, 2 faces, 1 connected region, and 3 holes; thus,

$$7 - 11 + 2 = 1 - 3 = -2$$

Although topological concepts are rather general, they provide an additional feature which is often useful in characterizing regions in a scene.

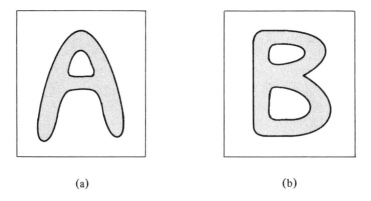

(a)   (b)

Figure 7.26. Regions with Euler number equal to 0 and -1, respectively.

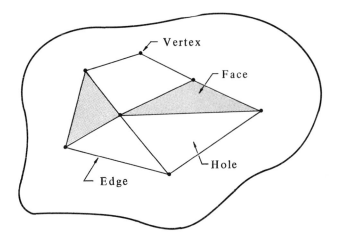

Figure 7.27. A region containing a polygonal network.

7.3 RELATIONAL DESCRIPTIONS

Once an image has been segmented into regions or components, the next level of complexity in an image description task is to organize these elements into a meaningful relational structure. Although a unified body of theory dealing with techniques for relating components of an image has yet to be developed, the idea of using techniques based on grammatical concepts is emerging as a promising approach to the structural description problem.

As an introduction to this concept, consider the simple staircase structure shown in Fig. 7.28(a). Assuming that this structure has been segmented out of an image, let us suppose that we wish to describe it in some formal way. By defining the two *primitive elements a* and *b* shown, we may code Fig. 7.28(a) in the form shown in Fig. 7.28(b). The most obvious property of the coded structure is the repetitiveness of the elements *a* and *b*. A simple description approach, therefore, would be to formulate a recursive relationship involving these primitive elements. One possibility is to use the following *rewriting rules*:

    1) $S \rightarrow aA$
    2) $A \rightarrow bS$
    3) $A \rightarrow b$

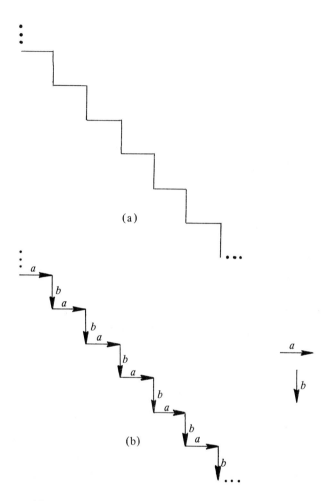

Figure 7.28. (a) A simple staircase structure. (b) Coded structure.

where S and A are variables and the elements a and b are constants corresponding to the primitives defined above. The first rule indicates that S can be replaced by primitive a and variable A. This variable, in turn, can be replaced by b and S or by b alone. If we replace A by bS, this takes us back to the first rule and the procedure can be repeated. If A is replaced by b, the procedure terminates because no variables are left in the expression. Figure 7.29 illustrates some sample derivations of these rules, where the numbers below the structures represent the order in which rules 1, 2, and 3 were applied. It is noted that the relationship between a and b is preserved by the fact that these rules force an a always to be followed by a b.

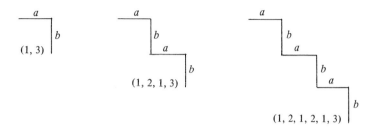

Figure 7.29. Sample derivations for the rules $S \rightarrow aA$, $A \rightarrow bS$, $A \rightarrow b$.

In the following discussion it is assumed that all derivations start with a special symbol, S, (called the *starting symbol*). Based on this convention, the first element in the structures generated by the above rules is always an a and the last element a b. Other variations of starting and ending elements are easily incorporated by adding more rules. The key point of this illustration, however, is that three simple rewriting rules can be used to generate (or describe) an infinite number of "similar" structures. As shown in the following sections, this approach also enjoys the advantage of having a solid theoretical foundation.

7.3.1 String Grammars and Languages

The coded structure illustrated in Fig. 7.28 is composed of connected strings of symbols. In this section we introduce some concepts from formal language theory for handling such strings. This field deals with the study of mathematical models used for the generation, translation, or other processes involving strings of an artificial language.

The origin of formal language theory may be traced to the middle 1950's with the development by Noam Chomsky of mathematical models of grammars related to his work in natural languages. One of the original goals of the work of linguists working in this area was to develop computational grammars capable of describing natural languages such as English. The hope was that, if this could be done, it would be a relatively simple matter to "teach" computers to interpret natural languages for the purposes of translation and problem solving. Although it is generally agreed that these expectations have been unrealized thus far, spin-offs of research in this area have had significant impact on other fields such as compiler design, computer languages, automata theory, and, more recently, pattern recognition and image processing.

We begin the development with some basic definitions.

An *alphabet*, *V*, is any finite set of symbols.

A *sentence*, *string*, or *word* over an alphabet *V* is any string of finite length composed of symbols from the alphabet. For example, given the alphabet $V = \{0, 1\}$, the following are valid sentences: $\{0, 1, 00, 01, 10, 11, 000, 001, ...\}$.

The sentence with no symbols is called the *empty sentence*, which we shall denote by λ. For an alphabet *V*, we will use V^* to denote the set of all sentences composed by symbols from *V*, including the empty sentence. The symbol V^+ will denote the set of sentences $V^* - \lambda$. For example, given the alphabet $V = \{a, b\}$, we have $V^* = \{\lambda, a, b, aa, ab, ba, ...\}$ and $V^+ = \{a, b, aa, ab, ba, ...\}$.

A *language* is any set (not necessarily finite) of sentences over an alphabet.

As is true in natural languages, a serious study of formal language theory must be focused on grammars and their properties. We define a *formal string grammar* (or simply *grammar*) as the four-tuple $G = (N, \Sigma, P, S)$, where *N* is a set of *nonterminals* (variables); Σ is a set of terminals (constants); *P* is a set of productions or rewriting rules; and *S* is the *start* or *root* symbol. It is assumed that *S* belongs to the set *N*, and that *N* and Σ are disjoint sets. The alphabet *V* is the union of *N* and Σ.

The *language* generated by *G*, denoted by *L(G)*, is the set of strings which satisfy two conditions: (1) each string is composed only of terminals (i.e., each string is a *terminal* sentence), and (2) each string can be derived from *S* by suitable applications of productions from the set *P*.

The following notation will be used throughout this section. Nonterminals will be denoted by capital letters: *S*, *A*, *B*, *C*, ... Lower-case letters at the beginning of the alphabet will be used for terminals: *a*, *b*, *c*, ... Strings of terminals will be denoted by lower-case letters toward the end of the alphabet: *v*, *w*, *x*, ... Strings of mixed terminals and nonterminals will be represented by lower-case Greek letters: $\alpha, \beta, \gamma, \delta, ...$.

The set *P* of productions consists of expressions of the form $\alpha \to \beta$, where α is a string in V^+ and β is a string in V^*. In other words, the symbol $\to$ indicates replacement of the string α by the string β. The symbol $\underset{G}{\Rightarrow}$ will be used to indicate operations of the form $\gamma\alpha\delta \underset{G}{\Rightarrow} \gamma\beta\delta$ in the grammar *G*; that is, $\underset{G}{\Rightarrow}$ indicates the replacement of α by β by means of the production $\alpha \to \beta$, γ and δ being left unchanged. It is customary to drop the *G* and simply use the symbol $\Rightarrow$ when it is clear which grammar is being considered.

Example: Consider the grammar $G = (N, \Sigma, P, S)$, where $N = \{S\}$, $\Sigma = \{a, b\}$ and $P = \{S \to aSb, S \to ab\}$. If the first production is applied $m - 1$ times, we obtain

$$S \Rightarrow aSb \Rightarrow aaSbb \Rightarrow a^3Sb^3 \Rightarrow \cdots a^{m-1}Sb^{m-1}$$

Applying now the second production results in the string

$$a^{m-1}Sb^{m-1} \Rightarrow a^m b^m$$

The language generated by this grammar is seen to consist solely of strings of this type, where the length of a particular string depends on m. We may express $L(G)$ in the form $L(G) = \{a^m b^m \mid m \geqslant 1\}$. It is noted that the simple grammar of this example is capable of producing a language with an infinite number of strings.   □

7.3.1.1 Types of Phrase Structure Grammars

Grammars of the form described above, in which the productions have the general form $\alpha \rightarrow \beta$, are called *phrase structure grammars*. It is common practice to categorize these grammars based on the type of restrictions placed on the productions.

A grammar in which the general form of productions $\alpha \rightarrow \beta$ (α in V^+ and β in V^*) is allowed is called an *unrestricted grammar*.

A *context-sensitive grammar* has productions of the form $\alpha_1 A \alpha_2 \rightarrow \alpha_1 \beta \alpha_2$, where α_1 and α_2 are in V^*, β is in V^+, and A is in N. This grammar allows replacement of the nonterminal A by the string β only when A appears in the context $\alpha_1 A \alpha_2$ of strings α_1 and α_2.

A *context-free grammar* has productions of the form $A \rightarrow \beta$, where A is in N and β is in V^+. The name "context free" arises from the fact that the variable A may be replaced by a string β regardless of the context in which A appears.

Finally, a *regular* (or *finite-state*) *grammar* is one with productions of the form $A \rightarrow aB$ or $A \rightarrow a$, where A and B are variables in N and a is a terminal in Σ. Alternative valid productions are $A \rightarrow Ba$ and $A \rightarrow a$. However, once one of the two types has been chosen, the other set must be excluded.

These grammars are sometimes called *type* 0, 1, 2, and 3 *grammars*, respectively. It is interesting to note that all regular grammars are context free, all context-free grammars are context sensitive, and all context-sensitive grammars are unrestricted.

Although unrestricted grammars are considerably more powerful than the other three types, their generality presents some serious difficulties from both a theoretical and a practical point of view. To a large extent, this is also true of context-sensitive grammars. For these reasons, most of the work dealing with the use of grammatical concepts for image description and pattern recognition has been limited to context-free and regular grammars.

Example: The grammar given in the previous example is context free because its productions are of the form $S \rightarrow \beta_1$, and $S \rightarrow \beta_2$, with S being a single nonterminal and $\beta_1 = aSb$, $\beta_2 = ab$ being strings in V^+. It is interesting to note that the language generated by this context-free grammar,

$L(G) = \{a^m b^m \mid m \geq 1\}$, cannot be generated by a regular grammar. In other words, the types productions allowed under the definition of a regular grammar are not capable of generating *only* strings of the form $a^m b^m$. For instance, the regular grammar $G = (N, \Sigma, P, S)$, with $N = \{S\}$, $\Sigma = \{a, b\}$, $P = \{S \rightarrow aS, S \rightarrow bS, S \rightarrow a, S \rightarrow b\}$, can generate the strings $a^m b^m$, but it is also capable of generating other types of strings such as a^m and b^m. $\quad\square$

7.3.1.2 Use of Positional Operators

Since strings are one-dimensional structures, it is necessary when applying them to image description to establish an appropriate method for reducing two-dimensional positional relations to one-dimensional form.

Most applications of string grammars to image description are based on the idea of extracting connected line segments from the objects of interest. One approach is to follow the contour of an object and code the result with segments of specified direction and/or length. This procedure is illustrated in Fig. 7.30.

Another, somewhat more general, approach is to describe sections of an image (such as small homogeneous regions) by directed line segments which can be joined in other ways besides head-to-tail connections. This approach is illustrated in Fig. 7.31(a); Fig. 7.31(b) shows some typical operations that can be defined on the extracted line segments.

The two approaches just described, although not exhaustive, are typical of procedures used for reducing two-dimensional information to string form. The following examples should further clarify these concepts.

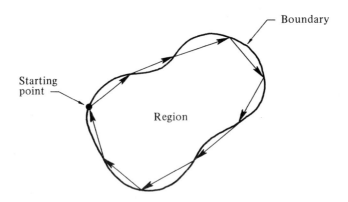

Figure 7.30. Coding a region boundary with directed line segments.

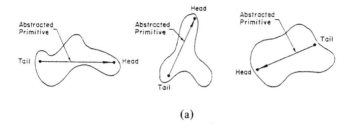

(a)

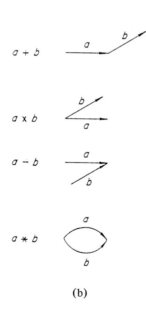

(b)

Figure 7.31. (a) Abstraction of regions by directed line segments. (b) Some operations involving the abstracted primitives.

Example: An interesting illustration of image description by boundary tracking is the grammar proposed by Ledley [1964, 1965] to characterize submedian and telocentric chromosomes. This grammar utilizes the primitive elements shown in Fig. 7.32(a), which are detected as a chromosome boundary is tracked in a clockwise direction. Typical submedian and telocentric chromosome shapes are shown in Fig. 7.32(b), along with the string representation obtained by tracking the boundary of each chromosome.

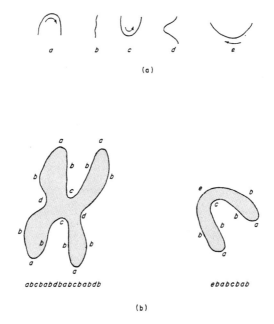

Figure 7.32. (a) Primitives of a chromosome grammar. (b) Submedian and telocentric chromosomes. (From Ledley [1964].)

The complete grammar is given by $G = (N, \Sigma, P, S)$, where $\Sigma = \{a, b, c, d, e\}$, $N = \{S, T, A, B, C, D, E, F\}$, and

$$
\begin{array}{lll}
P: & 1) & S \rightarrow C \cdot C \\
   & 2) & T \rightarrow A \cdot C \\
   & 3) & C \rightarrow B \cdot C \\
   & 4) & C \rightarrow C \cdot B \\
   & 5) & C \rightarrow F \cdot D \\
   & 6) & C \rightarrow E \cdot F
\end{array}
$$

$$\begin{array}{ll}
7)\ \ E \to F \cdot c & 13)\ \ B \to B \cdot b \\
8)\ \ D \to c \cdot F & 14)\ \ B \to b \\
9)\ \ A \to b \cdot A & 15)\ \ B \to d \\
10)\ \ A \to A \cdot b & 16)\ \ F \to b \cdot F \\
11)\ \ A \to e & 17)\ \ F \to F \cdot b \\
12)\ \ B \to b \cdot B & 18)\ \ F \to a
\end{array}$$

The operator "$\cdot$" is used to describe simple connectivity of the terms in a production as the boundary is tracked in a clockwise direction. □

The above grammar is in reality a combination of two grammars with starting symbols S, and T, respectively. Thus, starting with S allows generation of structures which correspond to submedian chromosomes. Similarly, starting with T produces structures which correspond to telocentric chromosomes.

Example: As a second illustration, let us discuss in some detail the Picture Description Language (PDL) proposed by Shaw [1970] for describing objects using operators of the form shown in Fig. 7.31(b).

Consider the following simple PDL grammar:

$$G = (N, \Sigma, P, S)$$

with

$$N = \{S, A_1, A_2, A_3, A_4, A_5\}$$

$$\Sigma = \{a \nearrow, b \searrow, c \to, d\downarrow\}$$

$$P: \quad \begin{aligned}
& S \to d + A_1 \\
& A_1 \to c + A_2 \\
& A_2 \to \sim d * A_3 \\
& A_3 \to a + A_4 \\
& A_4 \to b * A_5 \\
& A_5 \to c
\end{aligned}$$

where ($\sim d$) indicates the primitive d with its direction reversed, and a, b, c, and d are elements whose directions are shown in the set Σ.

Application of the first production yields a primitive d followed by a variable A_1 not yet defined. All we know at this point is that the tail of the structure represented by A_1 will be connected to the head of d because this primitive is followed by the "+" operator. The variable A_1 resolves into

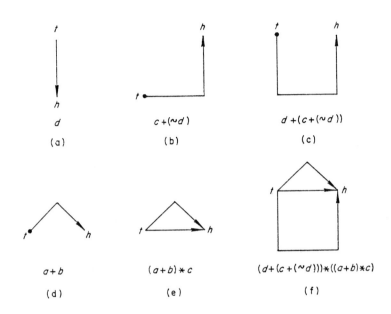

Figure 7.33. Steps in the construction of a PDL structure. Note the heads and tails of composite structures.

$d + A_2$, where A_2 is not yet defined. Similarly, A_2 resolves into $\sim d * A_3$. The results of applying the first three productions are shown in Figs. 7.33(a), (b), and (c). From the definition of the operator "$*$" we know that when A_3 is resolved it will be connected to the composite structure shown in Fig. 7.33 (c) in a tail-to-tail and head-to-head manner. The final result obtained by applying all the productions is shown in Fig. 7.33(f).

The PDL grammar described above can generate only one structure. However, the scope of structures generated by this grammar can be extended by introducing recursiveness — the capability of a variable to replace itself — into the productions. For example, suppose that we define the following productions:

$$S \rightarrow d + A_1$$
$$A_1 \rightarrow c + A_1$$
$$A_1 \rightarrow \sim d * A_2$$
$$A_2 \rightarrow a + A_2$$

$$A_2 \rightarrow b*A_2$$
$$A_2 \rightarrow c$$

If these productions were applied in the order shown, they would produce Fig. 7.33(f). However, this new set of productions allows, for instance, the application of the first production followed by the third, completely omitting the second production. If the remaining productions were applied in order, we would obtain a triangular structure. Furthermore, these productions allow the generation of infinite structures by repeated substitutions of a variable by itself. The variety of structures generated by the above grammar can be increased further by letting A_1 and A_2 equal S. This substitution would yield the maximum capability of this grammar. □

7.3.2 Higher-Dimensional Grammars

The types of string grammars discussed in the previous section are best suited for applications where the connectivity of primitives can be expressed in a head-to-tail or other continuous manner. In this section we consider a more general approach to the grammatical description problem by allowing a higher level of primitive description capability. The grammars required to handle this added capability are, as should be expected, more complex and difficult to analyze on a formal basis. As an illustration of these concepts, we consider below two typical generalizations of string grammars.

7.3.2.1 Tree Grammars

A *tree T* is a finite set of one or more nodes such that
(1) there is a unique node designated the root; and
(2) the remaining nodes are partitioned into m disjoint sets
   $T_1, ..., T_m$, each of which in turn is a tree called a *subtree* of T.

The *tree frontier* is the set of nodes at the bottom of the tree (the *leaves*), taken in order from left to right. For example, the tree shown below has root $ and frontier xy.

Generally, two types of information in a tree are important, namely: (1) information about a node stored as a set of words describing the node, and (2) information relating a node to its neighbors stored as a set of pointers to those neighbors. As used in image descriptions, the first type of information identifies a pattern primitive, while the second type defines the physical relationship of the primitive to other substructures.

Example: The structure shown in Fig. 7.34(a) can be represented by a tree by using the relationship "inside of". Thus, denoting the root of the tree by the symbol $, we see from Fig. 7.34(a) that the first level of complexity involves *a* and *c* inside $. This produces two branches emanating from the root, as shown in Fig. 7.34(b). The next level involves *b* inside *a* and *d* and *e* inside *c*. Finally, we complete the tree by noting that *f* is inside *e*.   □

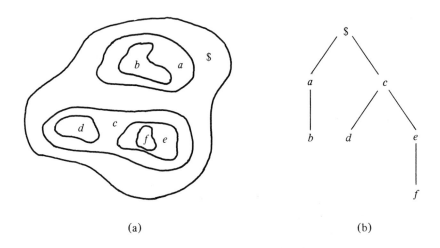

(a)   (b)

Figure 7.34. (a) A simple composite region. (b) Tree representation obtained by using the relationship "inside of."

A *tree grammar* is defined as the five-tuple $G = (N, \Sigma, P, r, S)$, where N and Σ are, as before, sets of nonterminals and terminals, respectively; S is the start symbols which can, in general, be a tree; P is a set of productions of the form $\Omega \rightarrow \Psi$, where Ω and Ψ are trees; and r is a *ranking function* which denotes the number of direct descendants of a node whose label is a terminal in the grammar.

The form of production $\Omega \rightarrow \Psi$ is analogous to that given for unrestricted string grammars and, as such, is usually too general to be of much practical use. A type of production which has found wide acceptance in the study of tree systems is an *expansive* production, which is of the form,

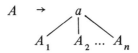

where $A, A_1, A_2, ..., A_n$ are nonterminals, and a is a terminal. A tree grammar that has only productions of this form is called an *expansive tree grammar*.

Example: As an illustration of a tree grammar, consider the circuit structure shown in Fig. 7.35(a). The tree representation shown in Fig. 7.35(b) was

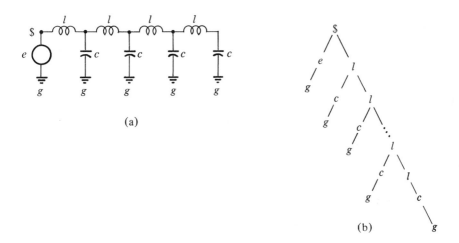

Figure 7.35. Tree representation of a connected figure.

obtained by defining the root at the left-most node and using the relationship "connected to".

A tree grammar which generates only trees of this form is given by $G = (N, \Sigma, P, r, S)$, where $N = \{S, A\}$, $\Sigma = \{e, g, l, c, \$\}$, and P is the following set of productions:

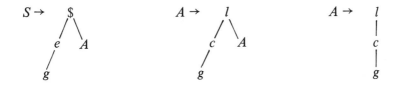

The ranking functions are in this case $r(e) = 1$, $r(g) = 0$, $r(l) = \{2, 1\}$, $r(c) = 1$,

$r(\$) = 2$. It is noted that only three simple productions can generate an infinite number of structures by the recursiveness defined on A.

An expansive tree grammar which generates the same structures is given by $G = (N, \Sigma, P, r, S)$, where $N = \{S, A_1, A_2, A_3, A_4, A_5\}$, $\Sigma = \{e, g, l, c, \$\}$, and the set P consists of the following productions:

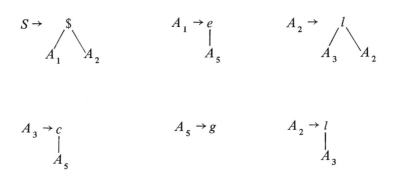

The ranking functions are $r(e) = 1, r(g) = 0, r(l) = \{1, 2\}, r(c) = 1, r(\$) = 2$. □

7.3.2.2 Web Grammars

As illustrated in Fig. 7.36, webs are undirected graph structures whose nodes are labeled. When used for image description, webs allow representations at a level considerable more abstract than that afforded by string or tree formalisms.

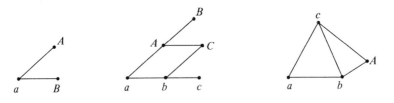

Figure 7.36. Some simple webs.

In a conventional phrase-structure string grammar rewriting rules of the form $\alpha \rightarrow \beta$ are used to replace one string by another. Such a rule is completely specified by specifying strings α and β; any string $\gamma \alpha \delta$ which contains α as a substring can be immediately rewritten as $\gamma \beta \delta$. Similarly, the productions of expansive tree grammars are interpreted without difficulty. The definition of rewriting rules involving webs, however, is much more complicated. Thus, if we want to replace a subweb α of the web ω by another subweb β, it is necessary to specify how to *embed* β in ω in place of α. As will be seen below this can be done by using embedding rules. An important point, however, is that the definition of an embedding rule must not depend on the "host web" ω because we want to be able to replace α by β in any web containing α as a subweb.

Let V be a set of labels and N_α and N_β the set of nodes of webs α and β, respectively. Based on the above concepts we define a *web rewriting rule* as a triple (α, β, ϕ), where ϕ is a function from $N_\beta \times N_\alpha$[†] into 2^V (the set of subsets of labels). This function specifies the embedding of β in place of α; that is, it specifies how to join to nodes of β to the neighbors of each node of the *removed* subweb α. Since ϕ is a function from the set of ordered pairs $N_\beta \times N_\alpha$, its argument is of the form (n, m), for n in N_β, and m in N_α. The values of $\phi(n, m)$ specify the allowed connections of n to the neighbors of m. For example, $\phi(B, A) = \{C, D\}$ means "join node B (in β) to the neighbors of node A (in α) whose labels are either C or D". We will omit the embedding specification and instead use the term "normal" to denote situations in which there is no ambiguity in a rewriting rule.

A *web grammar* is defined as a four-tuple $G = (N, \Sigma, P, S)$ where N is the nonterminal vocabulary, Σ is the terminal vocabulary, P is a set of web productions, and S is the starting symbol. As usual, S is in N, and the vocabulary V is the union of N and Σ.

Example: Consider the web grammar $G = (N, \Sigma, P, S)$, where

$$N = \{S\},$$

$$\Sigma = \{a, b, c\}, \text{ and}$$

$$P \text{ is the following set of triples}$$

[†]The symbol $\times$ is used in this context to denote the cartesian product of the sets N_β and N_α (i.e., the set of ordered pairs (n, m) such that n is an element of N_β and m is an element of N_α).

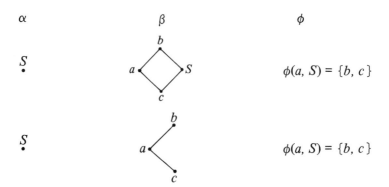

The embedding specified by ϕ indicates that α ($\alpha = S$ in this case) can be rewritten as β by connecting node a of β to the neighbors of S labeled b and c. It is noted that the rule does not apply in this case to the first execution of a production because the generation starts with a single point web without neighbors. In situations like this, it will be implicitly understood that ϕ is null for the first application of the production, and that the embedding rule describes replacements during the course of a derivation in which the subweb to be rewritten is embedded in a host web.

It is easily verified that this web grammar produces structures of the form

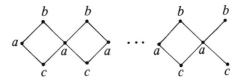

$\square$

The above definition of a web grammar is analogous to that of an unrestricted string grammar and, as such, is too broad to be of much practical use. As in the case of string grammars, however, it is possible to define restricted types of web grammars by limiting the generality of the productions.

We shall call a web rewriting rule (α, β, ϕ) *context sensitive* if there exists a nonterminal point A of α such that ($\alpha - A$) is a subweb of β. In this case the rule rewrites only a single point of α, regardless of how complex α is. If α contains a single point, we will call (α, β, ϕ) a *context-free* rule. It is noted that this is a special case of a context-sensitive rule since ($\alpha - A$) is empty when α contains a single point. By representing strings as webs (e.g., the string $aAbc...$ may be expressed as the web $\overset{a}{\cdot} \rightarrow \overset{A}{\cdot} \rightarrow \overset{b}{\cdot} \rightarrow \overset{c}{\cdot}$ ...) it is easily shown that the above rewriting rules are analogous to context-sensitive and context-free string productions, as defined in Section 7.3.1.1.

If the terminal vocabulary of a web grammar consists of a single symbol, every point of every web generated by the grammar will have the same label. In this case we can ignore the labels and identify the webs by their underlying graphs. This special type of web grammar is sometimes referred to as a *graph grammar*.

Example: Consider the context-sensitive graph grammar $G = (N, \Sigma, P, S)$, where,

$$N = \{A, B, C, S\}$$
$$\Sigma = \{a\}, \text{ and}$$
P is the following set of triples:

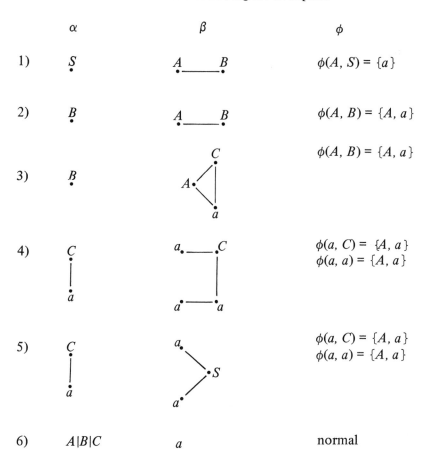

	α	β	ϕ
1)	S	$A \quad B$	$\phi(A, S) = \{a\}$
2)	B	$A \quad B$	$\phi(A, B) = \{A, a\}$
3)	B	(see figure)	$\phi(A, B) = \{A, a\}$
4)	C over a	(see figure)	$\phi(a, C) = \{A, a\}$, $\phi(a, a) = \{A, a\}$
5)	C over a	(see figure)	$\phi(a, C) = \{A, a\}$, $\phi(a, a) = \{A, a\}$
6)	$A|B|C$	a	normal

This grammar generates graph structures which consist of an arbitrary number of series and parallel sections. The parallel segments are separated by at least one series element, and all structures begin and end with at least one such element. Two simple derivations are as follows:

As in the previous example, the single terminal *a* has been omitted from the final patterns.   □

Example: The following context-sensitive web grammar generates some simple geometrical figures: $G = (N, \Sigma, P, S)$, where

$$N = \{S, A, B\}$$
$$\Sigma = \{a, b, c\}, \text{ and}$$

P is the following set of triples:

	α	β	φ
1)	S	$\overset{\textstyle A}{\underset{\textstyle \dot{a}}{\displaystyle \mid}}$	normal
2)	$\overset{\textstyle A}{\underset{\textstyle \dot{a}}{\displaystyle \mid}}$	$\begin{matrix} b. & \underline{\quad} & .A \\ \mid & & \mid \\ a\cdot & \underline{\quad} & \cdot a \end{matrix}$	$\phi(b, A) = \{b, a\}$ $\phi(a, a) = \text{normal}$
3)	$b\underline{\qquad}A$	$\begin{matrix} b. & \underline{\quad} & .A \\ \mid & & \mid \\ b\cdot & \underline{\quad} & \cdot a \end{matrix}$	$\phi(a, A) = \{b, a\}$ $\phi(b, b) = \text{normal}$
4)	$\overset{\textstyle A}{\underset{\textstyle a}{\displaystyle \mid}}$	$\begin{matrix} b. & \underline{\quad} & .b \\ \mid & & \mid \\ a\cdot & \underline{\quad} & \cdot B \end{matrix}$	$\phi(b, A) = \{b, a\}$ $\phi(a, a) = \text{normal}$

5)   a _____ B $\phi(b, B) = \{b, a\}$
                                                        $\phi(a, a) = \text{normal}$

6)   b $\phi(a, B) = \{b, a\}$
       | $\phi(b, b) = \text{normal}$
       |
       B

7)   A   b   normal

8)   B   a   normal

As an example, consider the following derivation:

 S $\overset{1}{\Rightarrow}$ 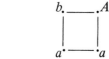 $\overset{2}{\Rightarrow}$

$\overset{2}{\Rightarrow}$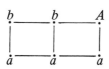

$\overset{3}{\Rightarrow}$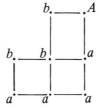

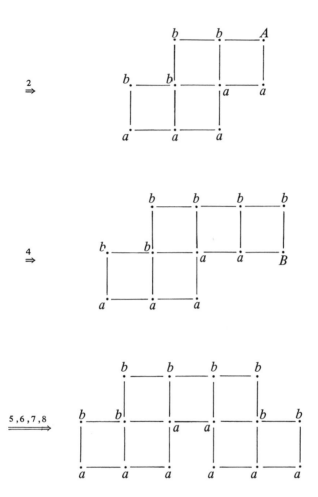

It is noted that the upper edge of the structure is labeled with b's and the lower edge with a's. The structures generated by this grammar, all of which share this property, consist of the set of geometrical structures which can be drawn by stacking squares to the right, above, or below the previously drawn square. It is also noted that a transition upward or downward can be made only from a square that was drawn to the right.   □

7.4 DESCRIPTIONS OF SIMILARITY

Measures of similarity may be established at any level of complexity of an image, ranging from the trivial case of comparing two pixels to the highly complex problem of determining in some meaningful way how similar two or more scenes are.

7.4.1 Distance Measures

Some of the techniques discussed previously can be used as the basis of comparison between two image regions. Consider, for example, the moment descriptors introduced in Section 7.2.2. Suppose that the moments for two regions are arranged in the form of two vectors, x_1 and x_2. The distance between x_1 and x_2, given by

$$D(x_1, x_2) = \|x_1 - x_2\|$$

$$= \sqrt{(x_1 - x_2)'(x_1 - x_2)} \tag{7.4-1}$$

may be used as a measure of similarity between these two descriptors. This becomes a particularly attractive approach if we are interested in comparing a given descriptor of unknown origin against two or more descriptors whose characteristics have been previously established. If the known descriptors are denoted by x_1, x_2, ..., x_L and the unknown descriptor by x, then we say that x is more similar to the ith descriptor if x is closer to x_i than to any other vector; that is, if

$$D(x, x_i) < D(x, x_j) \tag{7.4-2}$$

for $j = 1, 2, ..., L, j \neq i$. This approach can be used with a variety of descriptors, as long as they can be meaningfully expressed in vector form.

7.4.2 Correlation

Given a digital image $f(x, y)$ of size $M \times N$, suppose that we wish to determine if it contains a region which is similar to some region $w(x, y)$ of size $J \times K$, where $J < M$ and $K < N$. One of the methods most often used for the solution of this problem is to perform a correlation between $w(x, y)$ and $f(x, y)$.

In its simplest form, the correlation between these two functions is given by

$$R(m, n) = \sum_x \sum_y f(x, y) w(x - m, y - n) \tag{7.4-3}$$

where $m = 0, 1, 2, ..., M - 1, n = 0, 1, 2, ..., N - 1$, and the summation is taken over the image region where $w(x, y)$ is defined. The procedure is illustrated in Fig. 7.37; for any value of (m, n) inside $f(x, y)$ we apply Eq. (7.4-3) to obtain one value of R. As m and n are varied, $w(x, y)$ moves around the image area and we obtain the function $R(m, n)$. The maximum value of $R(m, n)$ then indicates the position where $w(x, y)$ best matched $f(x, y)$. It is noted that accuracy will be lost for values of m and n near the edges of $f(x, y)$, with the amount of error being proportional to the size of $w(x, y)$.

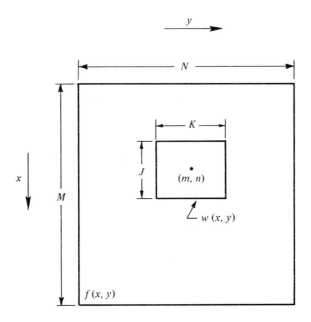

Figure 7.37. Arrangement for obtaining the correlation of $f(x, y)$ and $w(x, y)$ at a given point (m, n).

Figure 7.38 illustrates the concepts just discussed. Figure 7.38(a) is $f(x, y)$, Fig. 7.38(b) is $w(x, y)$, and Fig. 7.38(c) is $R(m, n)$ displayed as an intensity function. Note the higher intensity of $R(m, n)$ in the position where a match with $w(x, y)$ was found.

The correlation function defined in Eq. (7.4-3) is really nothing more than template matching, as defined in Section 7.1.2.1, with the exception that, in the present context, the template [i.e., $w(x, y)$] is in general a sub-image and not a function specifically designed to detect edges, lines, etc.

A slightly more complicated definition of correlation which adapts to the characteristics of the image is given by the expression

$$R(m, n) = \frac{\sum_x \sum_y f(x, y)w(x - m, y - n)}{\left[\sum_x \sum_y f^2(x, y)\right]^{1/2}} \tag{7.4-4}$$

Introduction of the normalizing factor in the denominator of this expression has a tendency to sharpen the peaks of $R(m, n)$. It is important to note that the normalizing factor is computed over the area over which $w(x, y)$ is defined and, therefore, varies as a function of displacement.

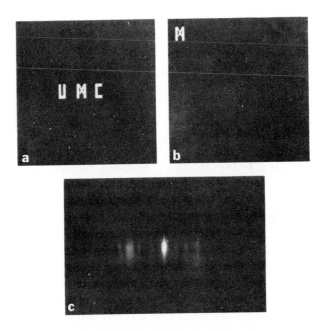

Figure 7.38. Example of correlation. Note the brightness of $R(m, n)$ at the position where the two letters match. (From Hall *et al* [1971].)

As mentioned in Section 3.3.8.2, correlation can also be carried out in the frequency domain via an FFT algorithm. If the functions are of the same size, this approach can be more efficient that a direct implementation of correlation in the spatial domain. In implementing Eq. (7.4-3), however, it is important to note that $w(x, y)$ is usually of much smaller dimension than $f(x, y)$. A trade-off estimate performed by Campbell [1969] indicates that if the number of nonzero terms in $w(x, y)$ is less than 132, then a direct implementation of Eq. (7.4-3) is more efficient than using the FFT approach. This figure, of course, depends on the machine and algorithms used, but it gives an indication of approximate image size when one should begin to consider the frequency domain as an alternative.

Similar comments apply to the numerator of Eq. (7.4-4). The denominator, however, requires a computation which is proportional to the average of $f^2(x, y)$ over the region of definition of $w(x, y)$. Since this term is in the form of a correlation of $f(x, y)$ with itself for zero displacement ($m = n = 0$) we could use the FFT approach. It is noted, however, that this would yield the entire correlation function for all allowed values of m and n in the region of interest. Since only the result for $m = n = 0$ is of interest, this approach is not warranted for most applications of Eq. (7.4-4).

7.4.3 Structural Similarity

Descriptions of structural similarity are, in general, more difficult to formulate and apply than the approaches discussed in the previous two subsections because the concept of a meaningful structure is highly problem dependent. If, for example, we characterize a region by the number of holes it contains, then we would say that a picture of a number "8" is more similar to a "B" than to an "A". This descriptor, however, would be meaningless for establishing similarity between pictures of human faces.

Typical structural descriptors which may be used for similarity measures are lengths of line segments, angles between segments, brightness characteristics, areas of regions, positions of regions in an image with respect to one another, etc. It should be stressed, however, that extracting (automatically) some of these components from an image is one of the most challenging parts of the problem.

Measures of similarity based on relationships between structural components can often be formulated by using the grammatical concepts introduced in Section 7.3. Suppose that we have two types of objects which can be generated by the productions of two grammars, G_1 and G_2, respectively. Given a particular object, we say that it is more similar to the objects of the first category if the given object can be generated by G_1, but not by G_2. If the object can be generated by both grammars, then we obviously have an ambiguity which cannot be resolved with the given information. If neither grammar can generate the object, we may use as a measure of similarity the closeness with which the grammars can approximate the given structure.

7.5 CONCLUDING REMARKS

Image segmentation and description are essential preliminary steps in most automatic pictorial pattern recognition and scene analysis problems. As indicated by the range of examples presented in the previous sections, the choice of one segmentation or description technique over another is dictated mostly by the peculiar characteristics of the problem being considered. The methods discussed in this chapter, although far from exhaustive, are representative of techniques commonly used in practice. The references cited below can be used as the basis for further study of this topic.

REFERENCES

Edge extraction by gray-level thresholding is one of the oldest techniques developed for digital image processing. Some typical early references on this topic are the papers by Doyle [1962], Narasimhan and Fornango [1963], and Rosenfeld *et al* [1965]. The optimum thresholding technique discussed in Section 7.1.1.2 is based on the papers by Chow and Kaneko [1972], and Hall [1972]. Additional reading on the use of multivariable histograms may be found in Ohlander [1975], Price [1976], and Riseman and Arbib [1977].

Template matching techniques (Section 7.1.2.1) have also received considerable attention in the literature. Additional mask structures which complement the ones discussed in this chapter may be found in the papers by Roberts [1965], Prewitt [1970], Kirsch [1971], and Robinson [1976]. The review article by Fram and Deutsch [1975] contains several templates and a comparison of their performance. The paper by VanderBrug and Rosenfeld [1977] considers template matching as a two-stage process. The vector formulation given in Section 7.1.2.1 is based on the work of Frei and Chen [1977].

Additional details on region growing may be found in the paper by Muerle and Allen [1968]. The paper by Pavlidis [1972] is also of interest. The region clustering technique discussed in Section 7.1.2.3 is based on the article by Brice and Fennema [1970]. For additional details and references on the general clustering problem see the book by Tou and Gonzalez [1974].

The material in Section 7.2.1 was contributed by T. Wallace (Elec. Eng. Dept., Purdue University). See also the paper by Brill [1968]. Additional information on the moment approach for region description may be found in Hu [1962], Bell [1965], and Wong and Hall [1977]. The material in Section 7.2.3 is based on a similar discussion by Duda and Hart [1973].

The use of formal language theory for structural description and recognition has received considerable attention in the literature, especially in the last few years. Additional material and references for the topics discussed in Section 7.3 may be found in the books by Fu [1974] and Gonzalez and Thomason [in preparation].

The material in Section 7.4.1 is based on Chapter 3 of the book by Tou and Gonzalez [1974]. For additional details and references on correlation see Horowitz [1957], Harris [1964], Anuta [1969], Pratt [1974], and Rosenfeld and Kak [1976]. References for Section 7.4.3 are the same as those given above for Section 7.3.

APPENDIX A

IMAGE DISPLAY SUBROUTINE

The following FORTRAN subroutine outputs to a line printer a 32-level image. The maximum array size is a 64 × 64, which produces a full, standard computer output sheet. Larger images can be printed by subdividing them into 64 × 64 mosaics.

Usage

IA - A 64 × 64 image. This integer array is formed in the calling program where it is set up as follows:
COMMON IA (64, 64)
INTEGER*2 IA
The first subscript of IA refers to a *row* of the image and the second to a *column*. Care must be taken to keep this convention in mind when building the array.

The following subroutine displays IA on a line printer.

CALL DSP(NX,NY,LAW,IL,IH,NEG,LG)

The arguments are as follows:

NX - Number of rows of IA to be printed; maximum NX is 64.
NY - Number of columns of IA to be printed; maximum NY is 64. If NX = NY = 64 a full page is output.
LAW - Gray-level scale translation variable.
LAW = 1: linear scale,
LAW = 2: square-root scale,
LAW = 3: logarithmic scale,
LAW = 4: "absorption" scale.

IL - minimum gray level in IA, calculated in the calling program.

IH - maximum gray level in IA, calculated in the calling program.

NEG - a value equal to 1 gives the normal image; a value equal to 0 gives the negative of the image.

LG - logical unit number for the line printer.

The characters used in the program to obtain the thirty-two gray levels are shown in Fig. A.1. The characters in a column, when over-printed, produce the gray level indicated.

```
MMMMMMHHHHHXHXOZWMNOS=I*++=:-.-
WWWWWW###*++----        =   -    -
####OO+-
OOO
+
```

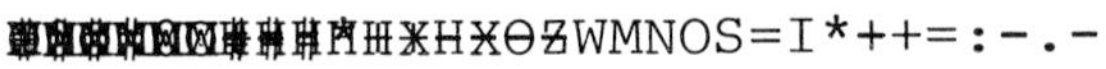

▉▉▉▉▉▉▉#▉#▉#XHXOZWMNOS=I*++=:-.-   G ray Levels

Figure A.1. Over-print characters used to obtain thirty–two gray levels. The 32nd character is a blank.

```
      SUBROUTINE DSP(NX,NY,LAW,IL,IH,NEG,LG)
C
C     **LINE PRINTER IMAGE OUTPUT SUBROUTINE**
C
C     ADAPTED BY B. A. FITTES, ELECTRICAL ENG.
C     DEPT., UNIVERSITY OF TENNESSEE, FROM
C     A PROGRAM WRITTEN BY J. L. BLANKENSHIP,
C     INSTRUMENTATION AND CONTROLS DIV., OAK
C     RIDGE NATIONAL LABORATORY, OAK RIDGE,
C     TN.  THE OVERPRINT METHOD USED IS FROM
C     "CONSIDERATIONS FOR EFFICIENT PICTURE
C     OUTPUT VIA LINE PRINTER," BY P. HENDERSON
C     AND S. TANIMOTO, REPORT NO. 153, 1974,
C     COMPUTER SCIENCE LAB., ELEC. ENG. DEPT.,
C     PRINCETON UNIVERSITY.
      COMMON IA(64,64)
      INTEGER*2 IA
      INTEGER*2 IB(64,64),LEV(32),BLANK(5)
```

```
      LOGICAL*1 LINE(128,5),GRAY(32,5)
C     SPECIFY GRAY-LEVEL CHARACTERS
      DATA GRAY /6*'M',5*'H','X','H','X','O',
     1 'Z','W','M','N','O','S','=','I','*','+',
     2 '+','=',':','-','.','-',' ',6*'W',3*'#',
     3 '*','+','+',4*'-',5*' ','=',' ',' ',' ','-',
     4 3*' ','-',3*' ',4*'#','O','O','+','-',
     5 24*' ','O','O','O',29*' ','+',31*' '/
      GN=32.0
      FL=IL
      FH=IH
      IF ((FH-FL).GT.0.0) GO TO 100
      T=FL
      FL=FH
      FH=T
  100 RANGE=(FH-FL+1)/GN
      AA=(SQRT(FH)-SQRT(FL))/GN
      EE=(FH-FL)/ALOG(GN+1.0)
      T=AMAX1(FL,1.0)
      SS=-(1.0/GN)*ALOG(FH/T)
C
C
C     A VECTOR LEV IS COMPUTED NEXT.  THIS
C     VECTOR IS A SET OF BREAK POINTS USED TO
C     DETERMINE THE SCALED VALUE OF IA(I,J).
C     THE MATRIX IB IS THE RESULT OF SCALING
C     IA.  IB(I,J)=K IF IA(I,J) IS LESS THAN
C     LEV(K+1) BUT GREATER THAN OR EQUAL TO
C     LEV(K).
C
      DO 160 I=1,32
      GO TO (110,120,130,140),LAW
  110 FLEV=FL+(I-1)*RANGE+0.5
      GO TO 150
  120 FLEV=(SQRT(FL)+(I-1)*AA)**2+0.5
      GO TO 150
  130 FLEV=FL+EE*ALOG(FLOAT(I))+0.5
      GO TO 150
  140 FLEV=FH*EXP(SS*(GN-I))+0.5
  150 LEV(I)=FLEV
  160 CONTINUE
      IF (NX.GT.64) NX=64
      IF (NY.GT.64) NY=64
      DO 180 I=1,NX
      DO 180 J=1,NY
      KLT=1
      DO 170 K=1,32
```

```
        IF (1A(I,J).GE.LEV(K)) KLT=K
170 CONTINUE
        IB(1,J)=KLT
180 CONTINUE
C
C
C       ONCE IB HAS BEEN COMPUTED, THE PICTURE CAN
C       BE PRINTED.  EACH POINT IN THE PICTURE CAN
C       CONSIST OF UP TO FIVE CHARACTERS OVERPRINTED
C       ON ONE ANOTHER.  SINCE THERE ARE 32 POSSIBLE
C       GRAY LEVELS, THERE IS A 32X5 MATRIX, GRAY,
C       THAT CONTAINS ALL OF THE COMBINATIONS. SINCE
C       EACH ELEMENT OF IB IS AN INTEGER BETWEEN
C       1 AND 32, IT CAN BE USED AS AN INDEX ON
C       GRAY TO OBTAIN THE CORRECT COMBINATION.
C       THE OUTPUT BUFFER, LINE, IS A 128X5 MATRIX.
C       EACH POINT IS OUTPUT TWICE HORIZONTALLY AND
C       ONCE VERTICALLY TO ATTEMPT TO COMPENSATE
C       FOR THE SPACING OF THE PRINTER.  HENCE, THE
C       FULL OUTPUT BUFFER REPRESENTS ONE ROW OF IB.
C       AS THE ROW IS GENERATED THERE IS A VECTOR,
C       BLANK, THAT INDICATES WHETHER OR NOT ANY
C       NON-BLANK CHARACTERS ARE PRESENT IN THE
C       BUFFER.  IF THERE ARE NOT ANY, THAT ROW IS
C       NOT PRINTED.  THIS SPEEDS UP THE PRINTING
C       PROCESS.
C
        WRITE(LG,1)
        IX=NX
        IY=2*NY
        DO 210 I=1,IX
        DO 190 J=1,5
        BLANK(J)=0
190 CONTINUE
        DO 200 K=2,IY,2
        J=K/2
        NG=IB(I,J)
        IF (NEG.EQ.0) NG=33-NG
        DO 200 L=1,5
        LINE(K-1,L)=GRAY(NG,L)
        LINE(K,L)=GRAY(NG,L)
        IF (NG.NE.32) BLANK(L)=1
200 CONTINUE
        WRITE (LG,2)
        DO 210 L=1,5
        IF (BLANK(L).EQ.0) GO TO 210
        WRITE (LG,3) (LINE(M,L), M=1,IY)
```

```
  210 CONTINUE
    1 FORMAT (1H1)
    2 FORMAT (1H )
    3 FORMAT (1H+,3X,128A1)
      RETURN
      END
```

APPENDIX B
CODED IMAGES

The following 64 × 64, 32-level images can be used as test data for many of the image processing concepts developed in the text. Along with each image is shown a coded array which contains an alphanumeric character for each pixel in the image. The range of these characters is from 0 through 9 and A through V, which corresponds to thirty-two gray levels. These characters can, for example, be punched on sixty-four computer cards, one row per card, and input into any image processing program. The first step after reading the coded image into the computer is to convert the alphanumeric characters into numerical levels in a range such as 0 to 31. The resulting numerical array can be used in its original form or it can be corrupted by, for example, adding noise to each pixel. This flexibility allows generation of a variety of input data which can be used to illustrate the effects of image processing algorithms. The results before and after processing can be displayed on a standard line printer by using the subroutine given in Appendix A. When using this routine, the gray-tone images may look slightly different from the ones shown in the following pages, depending on the type of lineprinter used.

Figure B.1. Mona Lisa.

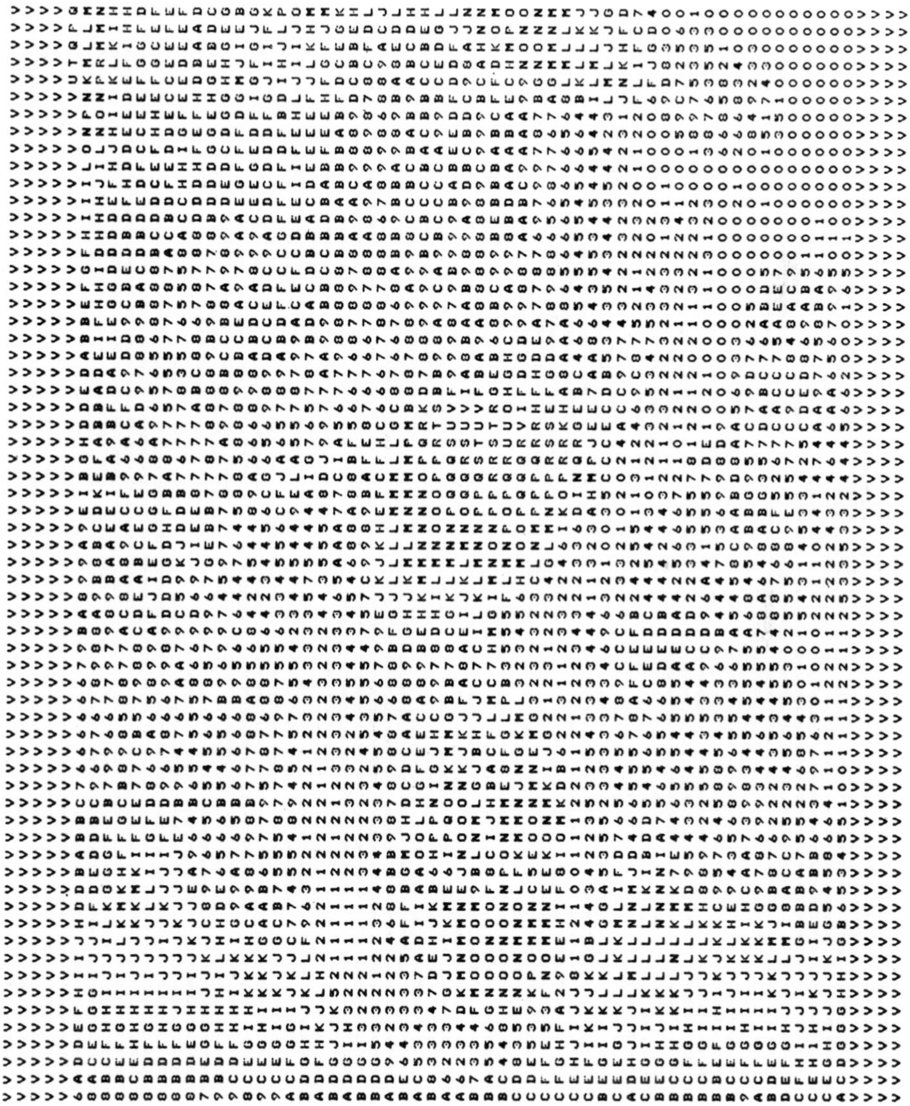

Figure B.2. Coded Mona Lisa.

Figure B.3. Characters.

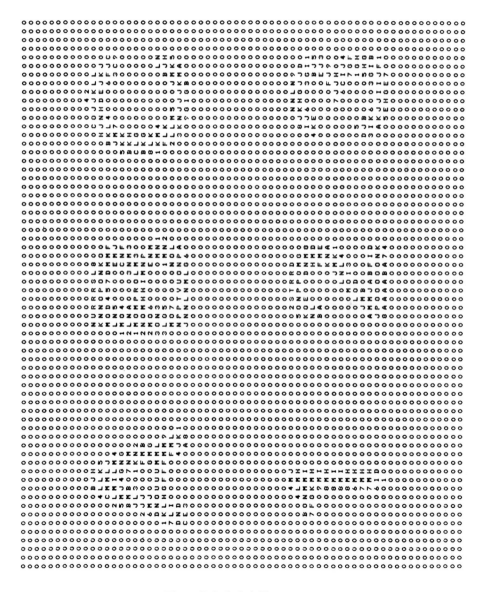

Figure B.4. Coded Characters.

Figure B.5. Jet.

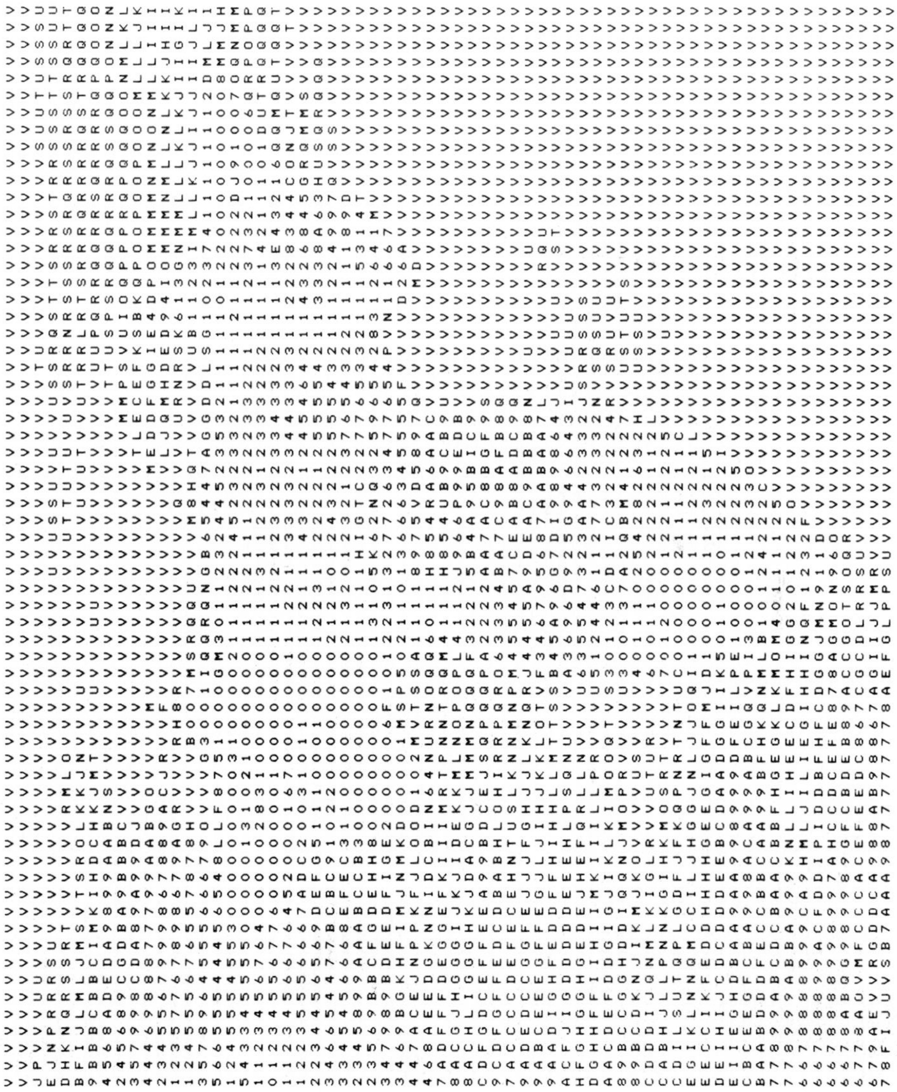

Figure B.6. Coded Jet.

Figure B.7. Lincoln.

Figure B.8. Coded Lincoln.

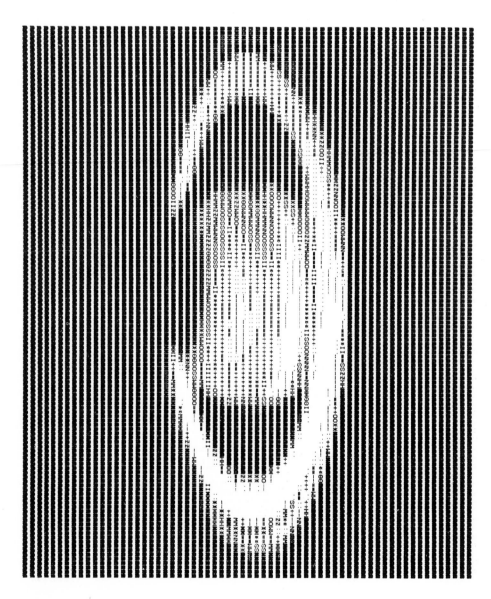

Figure B.9. Saturn.

Figure B.10. Coded Saturn.

Figure B.11. Chromosomes.

Figure B.12. Coded Chromosomes.

Figure B.13. Fingerprint.

Figure B.14. Coded Fingerprint.

Figure B.15. Statue of Liberty.

Figure B.16. Coded Statue of Liberty.

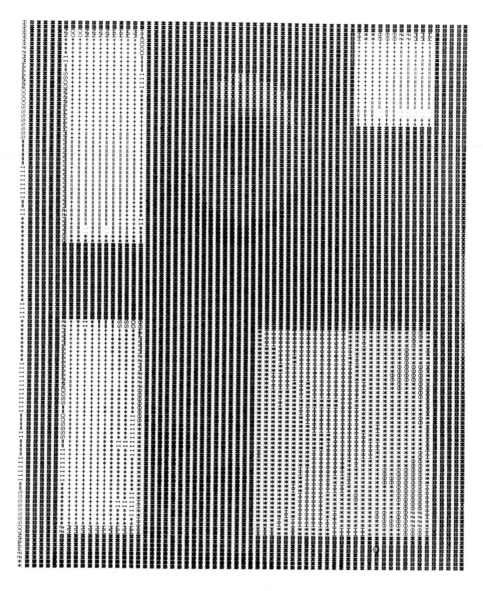

Figure B.17. Geometrical Figures.

Figure B.18. Coded Geometrical Figures.

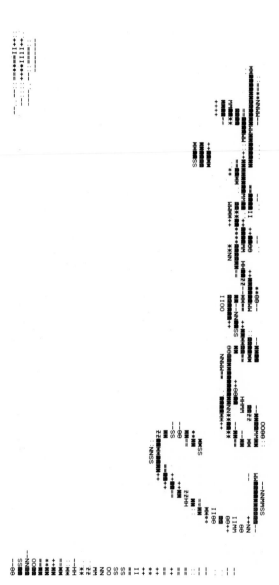

Figure B.19. Word.

Figure B.20. Coded Word.

Figure B.21. Boy.

Figure B.22. Coded Boy.

Figure B.23. Byplane.

Figure B.24. Coded Byplane.

BIBLIOGRAPHY

Abramson, A. [1963]. *Information Theory and Coding*, McGraw-Hill, New York.

Ahmed, N., Natarajan, T., and Rao, K. R. [1974]. "Discrete Cosine Transforms," *IEEE Trans. Comp.*, vol. C-23, pp. 90-93.

Ahmed, N. and Rao, K. R. [1975]. *Orthogonal Transforms for Digital Signal Processing*, Springer-Verlag, New York.

Anderson, G. L. and Netravaly, A. N. [1976]. "Image Restoration Based on a Subjective Criterion," *IEEE Trans. Syst. Man. Cyb.*, vol. SMC-6, no. 12, pp. 845-853.

Andrews, H. C. [1970]. *Computer Techniques in Image Processing*, Academic Press, New York.

Andrews, H. C., Tescher, A. G., and Kruger, R. P. [1972]. "Image Processing by Digital Computer," *IEEE Spectrum*, vol. 9, no. 7, pp. 20-32.

Andrews, H. C. [1974]. "Digital Image Restoration: A Survey," *Computer J.*, vol. 7, no. 5, pp. 36-45.

Andrews, H. C. and Hunt B. R. [1977]. *Digital Image Restoration*, Prentice-Hall, Englewood Cliffs, New Jersey.

Anuta, P. F. [1969]. "Digital Registration of Multispectral Video Imagery," *Soc. Photo-Optical Instrum. Engs.*, vol. 7, pp. 168-175.

Baumert, L. D., Golomb, S. W., and Hall, M., Jr. [1962]. "Discovery of a Hadamard Matrix of Order 92," *Bull. Am. Math. Soc.*, vol. 68, pp. 237-238.

Bell, E. T. [1965]. *Men of Mathematics*, Simon and Schuster, New York.

Bellman, R. [1970]. *Introduction to Matrix Analysis* (2nd ed.), McGraw-Hill, New York.

Biberman, L. M. [1973]. "Image Quality," in *Perception of Displayed Information*, (L. M. Biberman, ed.), Plenum Press, New York.

Billingsley, F. C., Goetz, A. F. H., and Lindsley, J. N. [1970]. "Color Differentiation by Computer Image Processing," *Photo. Sci. Eng.*, vol. 14, no. 1, pp. 28-35.

Blackman, E. S. [1968]. "Effects of Noise on the Determination of Photographic System Modulation Transfer Function," *Photogr. Sci. Eng.*, vol. 12, pp. 244-250.

Blackman, R. B. and Tukey, J. W. [1958]. *The Measurement of Power Spectra*, Dover Publications, New York.

Brice, C. R. and Fennema, C. L. [1970]. "Scene Analysis Using Regions," *Artificial Intelligence*, vol. 1, pp. 205-226.

Brigham, E. O. [1974]. *The Fast Fourier Transform*, Prentice-Hall, Englewood Cliffs, New Jersey.

Brill, E. L. [1968]. "Character Recognition Via Fourier Descriptors," WESCON, Paper 25/3, Los Angeles, California.

Brown, J. L., Jr. [1960]. "Mean-Square Truncation Error in Series Expansions of Random Functions," *J. SIAM*, vol. 8, pp. 18-32.

Budak, A. [1974]. *Passive and Active Network Analysis and Synthesis,* Houghton-Mifflin, Boston.

Campbell, J. D. [1969]. "Edge Structure and the Representation of Pictures," Ph.D. dissertation, Dept. of Elec. Eng., University of Missouri, Columbia.

Carlson, A. B. [1968]. *Communication Systems,* McGraw-Hill, New York.

Chen, P. H. and Wintz, P. A. [1976]. "Data Compression for Satellite Images," TR-EE-76-9, School of Electrical Engineering, Purdue University, West Lafayette, Indiana.

Chow, C. K. and Kaneko, T. [1972]. "Automatic Boundary Detection of the Left Ventricle from Cineangiograms," *Comp. and Biomed. Res.,* vol. 5, pp. 388–410.

Cochran, W. T., Cooley, J. W., *et al* [1967]. "What is the Fast Fourier Transform?," *IEEE Trans. Audio and Electroacoustics,* vol. AU-15, no. 2, pp. 45–55.

Cooley, J. W. and Tukey, J. W. [1965]. "An Algorithm for the Machine Calculation of Complex Fourier Series," *Math. of Comput.,* vol. 19, pp. 297–301.

Cooley, J. W., Lewis, P. A. W., and Welch, P. D. [1967a]. "Historical Notes on the Fast Fourier Transform," *IEEE Trans. Audio and Electroacoustics,* vol. AU-15, no. 2, pp. 76–79.

Cooley, J. W., Lewis, P. A. W., and Welch, P. D. [1967b]. "Application of the Fast Fourier Transform to Computation of Fourier Integrals," *IEEE Trans. Audio and Electroacoustics,* vol. AU-15, no. 2, pp. 79–84.

Cooley, J. W., Lewis, P. A. W., and Welch, P. D. [1969]. "The Fast Fourier Transform and its Applications," *IEEE Trans. Educ.,* vol. E-12, no. 1, pp. 27–34.

Cornsweet, T. N. [1970]. *Visual Perception,* Academic Press, New York.

Cutrona, L. J., Leith, E. N., and Palermo, C. J. [1960]. "Optical Data Processing and Filtering Systems," *IRE Trans. Info. Theory,* vol. IT-6, no. 3, pp. 386–400.

Cutrona, L. J. and Hall, W. D. [1968]. "Some Considerations in Post-Facto Blur Removal," in *Evaluation of Motion-Degraded Images,* NASA Publ. SP-193, pp. 139–148.

Danielson, G. C. and Lanczos, C. [1942]. "Some Improvements in Practical Fourier Analysis and Their Application to X-Ray Scattering from Liquids," *J. Franklin Institute,* vol. 233, pp. 365–380 and 435–452.

Davenport, W. B. and Root, W. L. [1958]. *An Introduction to the Theory of Random Signals and Noise,* McGraw-Hill, New York.

Deutsch, R. [1965]. *Estimation Theory,* Prentice-Hall, Englewood Cliffs, New Jersey.

Digital Image Processing [1974]. Special issue of *Computer,* vol. 7, no. 5.

Digital Picture Processing [1972]. Special issue of the *Proceedings of the IEEE,* vol. 60, no. 7.

Doyle, W. [1962]. "Operations Useful for Similarity-Invariant Pattern Recognition," *J. ACM,* vol. 9, pp. 259–267.

Duan, J. R. and Wintz, P. A. [1974]. "Information Preserving Coding for Multispectral Scanner Data," TR-EE-74-15, School of Electrical Engineering, Purdue University, West Lafayette, Indiana.

Duda, R. O. and Hart, P. E. [1973]. *Pattern Classification and Scene Analysis,* Wiley, New York.

Elias, P. [1952]. "Fourier Treatment of Optical Processes," *J. Opt. Soc. Am.,* vol. 42, no. 2, pp. 127–134.

Elsgolc, L. E. [1962]. *Calculus of Variations,* Addison-Wesley, Reading, Mass.

Essman, J. and Wintz, P. A. [1973]. "The Effects of Channel Errors in DPCM Systems and Comparison with PCM Systems," *IEEE Trans. on Comm.,* vol. COM-21, no. 8, pp. 867–877.

Evans, R. M. [1959]. *An Introduction to Color,* John Wiley & Sons, Inc., New York.

Falconer, D. G. [1970]. "Image Enhancement and Film Grain Noise," *Opt. Acta,* vol. 17, pp. 693–705.

Fine, N. J. [1949]. "On the Walsh Functions," *Trans. Am. Math. Soc.,* vol. 65, pp. 373–414.

Fine, N. J. [1950]. "The Generalized Walsh Functions," *Trans. Am. Math. Soc.,* vol. 69, pp. 66–77.

Fram, J. R. and Deutsch, E. S. [1975]. "On the Quantitative Evaluation of Edge Detection Schemes and Their Comparison with Human Performance," *IEEE Trans. Computers*, vol. C-24, no. 6, pp. 616–628.

Frendendall, G. L. and Behrend, W. L. [1960]. "Picture Quality – Procedures for Evaluating Subjective Effects of Interference," *Proc. IRE*, vol. 48, pp. 1030–1034.

Freeman, H. [1961]. "On the Encoding of Arbitrary Geometric Configurations," *IEEE Trans. Elec. Computers*, vol. EC-10, pp. 260–268.

Frei, W. and Chen, C. C. [1977]. "Fast Boundary Detection: A Generalization and a New Algorithm," *IEEE Trans. Computers*, (in press).

Frieden, B. R. [1972]. "Restoring with Maximum Likelihood and Maximum Entropy," *J. Opt. Soc. Amer.*, vol. 62, pp. 511–518.

Frieden, B. R. [1974]. "Image Restoration by Discrete Deconvolution of Minimal Length," *J. Opt. Soc. Amer.*, vol. 64, pp. 682–686.

Fu, K. S. [1974]. *Syntactic Methods in Pattern Recognition*, Academic Press, New York.

Fu, K. S. and Rosenfeld A. [1976]. "Pattern Recognition and Image Processing," *IEEE Trans. Computers*, vol. C-25, no. 12, pp. 1336–1346.

Gattis, J. and Wintz, P. A. [1971]. "Automated Techniques for Data Analysis and Transmission," TR-EE-71-37, School of Electrical Engineering, Purdue University, West Lafayette, Indiana.

Gaven, J. V., Jr., Tavitian, J., and Harabedian, A. [1970]. "The Informative Value of Sampled Images as a Function of the Number of Gray Levels Used in Encoding the Images," *Phot. Sci. Eng.*, vol 14, no. 1, pp. 16–20.

Gentleman, W. M. and Sande, G. [1966]. "Fast Fourier Transform for Fun and Profit," *Fall Joint Computer Conf.*, vol. 29, pp. 563–578, Spartan, Washington, D. C.

Gentleman, W. M. [1968]. "Matrix Multiplication and Fast Fourier Transformations," *Bell System Tech. J.*, vol. 47, pp. 1099–1103.

Gish, H. and Pierce, J. N. [1968]. "Asymptotically Efficient Quantizer," *IEEE Trans. Info. Theory*, vol. IT-14, pp. 676–683.

Golomb, S. W. and Baumert, L. D. [1963]. "The Search for Hadamard Matrices," *Am. Math. Monthly*, vol. 70, pp. 27–31.

Gonzalez, R. C. [1972]. "Syntactic Pattern Recognition – Introduction and Survey," *Proc. Nat. Elec. Conf.*, vol. 27, pp. 27–31.

Gonzalez, R. C. and Fittes, B. A. [1975]. "Gray-Level Transformations for Interactive Image Enhancement," *Proc. Second Conf. on Remotely Manned Systems*, pp. 17-19.

Gonzalez, R. C., Edwards, J. J., and Thomason, M. G. [1976]. "An Algorithm for the Inference of Tree Grammars," *Internat. J. Computers and Info. Sci.*, vol. 5, no. 2, pp. 145–163.

Gonzalez, R. C. and Fittes, B. A. [1977]. "Gray-Level Transformations for Interactive Image Enhancement," *Mechanism and Machine Theory*, vol. 12, pp. 111–112.

Gonzalez, R. C. and Thomason, M. G. [in preparation]. *Syntactic Pattern Recognition: An Introduction*, Addison-Wesley, Reading, Mass.

Good, I. J. [1958]. "The Interaction Algorithm and Practical Fourier Analysis," *J. Roy. Statist. Soc.* (London), vol. B20, pp. 361–367; *Addendum*, vol. 22, 1960, pp. 372–375.

Goodman, J. W. [1968]. *Introduction to Fourier Optics*, McGraw-Hill, New York.

Graham, C. H., ed. [1965]. *Vision and Visual Perception*, John Wiley & Sons, New York.

Graham, D. N. [1967]. "Image Transmission by Two-Dimensional Contour Coding," *Proc. IEEE*, vol. 55, pp. 336–346.

Habibi, A. [1971]."Comparison of N*th* Order DPCM Encoder with Linear Transformations and Block Quantization Techniques," *IEEE Trans. Commun. Tech.*, vol. COM-19, no. 6.

Habibi, A. [1972]. "Two-Dimensional Bayesian Estimate of Images," *Proc. IEEE*, vol. 60, pp. 878–883.

Habibi, A. and Wintz, P. A. [1971]. "Image Coding by Linear Transformations and Block Quantization," *IEEE Trans. Comm. Tech.*, vol. COM-19, pp. 50–62.

Habibi, A. and Wintz, P. A. [1974]. "Hybrid Coding of Pictorial Data," *IEEE Trans. Comm. Tech.*, vol. COM-22, no. 5, pp. 614–624.

Hadamard, J. [1893]. "Resolution d'une Question Relative aux Determinants," *Bull. Sci. Math.*, Ser. 2, vol. 17, Part I, pp. 240–246.

Hall, E. L. [1972]. "Automated Computer Diagnosis Applied to Lung Cancer," *Proc. of the 1972 Int. Conf. on Cybernetics and Society*, New Orleans, La.

Hall, E. L. [1974]. "Almost Uniform Distributions for Computer Image Enhancement," *IEEE Trans. Computers*, vol. C-23, no. 2, pp. 207–208.

Hall, E. L. *et al.* [1971]. "A Survey of Preprocessing and Feature Extraction Techniques for Radiographic Images," *IEEE Trans. Computers*, vol. C-20, no. 9, pp. 1032–1044.

Hall, E. L. and Frei, W. [1976]. "Invariant Features for Quantitative Scene Analysis," Final Report, Contract F 08606-72-C-0008, Image Processing Institute, University of Southern California.

Hammond, J. L. and Johnson, R. S. [1962]. "Orthogonal Square-Wave Functions," *J. Franklin Inst.*, vol. 273, pp. 211–225.

Harmuth, H. F. [1968]. "A Generalized Concept of Frequency and Some Applications," *IEEE Trans. Info. Theory*, vol. IT-14, no. 3, pp. 375–382.

Harris, J. L. [1964]. "Resolving Power and Decision Theory," *J. Opt. Soc. Amer.*, vol. 54, pp. 606–611.

Harris, J. L. [1966]. "Image Evaluation and Restoration," *J. Opt. Soc. Amer.*, vol. 56, pp. 569–574.

Harris, J. L. [1968]. "Potential and Limitations of Techniques for Processing Linear Motion-Degraded Images," in *Evaluation of Motion Degraded Images*, NASA Publ. SP-193, pp. 131–138.

Hecht, E. and Zajac, A. [1975]. *Optics,* Addison-Wesley Publishing Co., Reading, Mass.

Helstrom, C. W. [1967]. "Image Restoration by the Method of Least Squares," *J. Opt. Soc. Amer.*, vol. 57, no. 3, pp. 297–303.

Henderson, K. W. [1964]. "Some Notes on the Walsh Functions," *IEEE Trans. Electronic Computers*, vol. EC-13, no. 1, pp. 50–52.

Horner, J. L. [1969]. "Optical Spatial Filtering with the Least-Mean-Square-Error Filter," *J. Opt. Soc. Amer.*, vol. 59, pp. 553–558.

Horowitz, M. [1957]. "Efficient Use of a Picture Correlator," *J. Opt. Soc. Amer.*, vol. 47, pg. 327.

Hotelling, H. [1933]. "Analysis of a Complex of Statistical Variables into Principal Components," *J. Educ. Psychol.*, vol. 24, pp. 417–441 and 498–520.

Hu, M. K. [1962]. "Visual Pattern Recognition by Moment Invariants," *IRE Trans. Info. Theory*, vol. IT-8, pp. 179–187.

Huang, T. S. [1965]. "PCM Picture Transmission," *IEEE Spectrum*, vol. 2, no. 12, pp. 57–63.

Huang, T. S. [1966]. "Digital Picture Coding," *Proc. Nat. Electron. Conf.*, pp. 793–797.

Huang, T. S. [1968]. "Digital Computer Analysis of Linear Shift-Variant Systems," in *Evaluation of Motion-Degraded Images*, NASA Publ. SP-193, pp. 83–87.

Huang, T. S., ed. [1975]. *Picture Processing and Digital Filtering*, Springer, New York.

Huang, Y. and Schultheiss, P. M. [1963]. "Block Quantization of Correlated Gaussian Random Variables," *IEEE Trans. Commun. Syst.*, vol. CS-11, pp. 289–296.

Huffman, D. A. [1952]. "A Method for the Construction of Minimum Redundancy Codes," *Proc. IRE*, vol. 40, no. 10, pp. 1098–1101.

Hummel, R. A. [1974]. "Histogram Modification Techniques," Technical Report TR-329, F-44620-72C-0062, Computer Science Center, University of Maryland, College Park, Maryland.

Hunt, B. R. [1971]. "A Matrix Theory Proof of the Discrete Convolution Theorem," *IEEE Trans. Audio and Electroacoust.*, vol. AU-19, no. 4, pp. 285–288.

Hunt, B. R. [1973]. "The Application of Constrained Least Squares Estimation to Image Restoration by Digital Computer," *IEEE Trans. Computers*, vol. C-22, no. 9, pp. 805–812.

IEEE Trans. Circuits and Syst. [1975]. Special issue on digital filtering and image processing, vol. CAS-2, pp. 161–304.

IEEE Trans. Computers [1972]. Special issue on two-dimensional signal processing, vol. C-21, no. 7.

IES Lighting Handbook [1972]. Illuminating Engineering Society Press, New York.

Jain, A. K. and Angel, E. [1974]. "Image Restoration, Modeling, and Reduction of Dimensionality," *IEEE Trans. Computers*, vol. C-23, pp. 470–476.

Kahaner, D. K. [1970]. "Matrix Description of the Fast Fourier Transform," *IEEE Trans. on Audio and Electroacoustics*, vol. AU-18, no. 4, pp. 442–450.

Karhunen, K. [1947]. "Über Lineare Methoden in der Wahrscheinlichkeitsrechnung," *Ann. Acad. Sci. Fennicae*, Ser. A137 (translated by I. Selin in "On Linear Methods in Probability Theory," T-131, 1960, The RAND Corp., Santa Monica, Calif.)

Kirsch, R. [1971]. "Computer Determination of the Constituent Structure of Biological Images," *Comp. and Biomed. Res.*, vol. 4, pp. 315–328.

Kiver, M. S. [1955]. *Color Television Fundamentals*, McGraw-Hill, New York.

Kodak Plates and Films for Scientific Photography [1973]. Publication no. P-315, Eastman Kodak Co., Rochester, New York.

Kohler, R. J. and Howell, H. K. [1963]. "Photographic Image Enhancement by Superposition of Multiple Images," *Phot. Sci. Eng.*, vol. 7, no. 4, pp. 241–245.

Koschman, A. [1954]. "On the Filtering of Nonstationary Time Series," *Proc. 1954 Nat. Electron. Conf.*, pg. 126.

Kramer, H. P. and Mathews, M. V. [1956]. "A Linear Coding for Transmitting a Set of Correlated Variables," *IRE Trans. Info. Theory*, vol. IT-2, pp. 41–46.

Lawley, D. N. and Maxwell, A. E. [1963]. *Factor Analysis as a Statistical Method*, Butterworth & Co., Ltd., London.

Ledley, R. S. [1964]. "High-Speed Automatic Analysis of Biomedical Pictures," *Science*, vol. 146, no. 3461, pp. 216–223.

Ledley, R. S., *et al.* [1965]. "FIDAC: Film Input to Digital Automatic Computer and Associated Syntax-Directed Pattern Recognition Programming System," in *Optical and Electro-Optical Information Processing Systems* (J. Tippet, D. Beckowitz, L. Clapp, C. Koester, and A. Vanderburgh, Jr., eds.), MIT Press, Cambridge, Mass., Chapter 33.

Legault, R. R. [1973]. "The Aliasing Problems in Two-Dimensional Sampled Imagery," in *Perception of Displayed Information*, (L. M. Biberman, ed.), Plenum Press, New York.

Lipkin, B. S. and Rosenfeld, A. eds. [1970]. *Picture Processing and Psychopictorics*, Academic Press, New York.

Loève, M. [1948]. "Fonctions Aléatoires de Seconde Ordre," in P. Lévy, *Processus Stochastiques et Mouvement Brownien*, Hermann, Paris, France.

Lohman, A. W. and Paris, D. P. [1965]. "Space-Variant Image Formation," *J. Opt. Soc. Amer.*, vol. 55, pp. 1007–1013.

MacAdam, D. P. [1970]. "Digital Image Restoration by Constrained Deconvolution," *J. Opt. Soc. Amer.*, vol. 60, pp. 1617–1627.

Max, J. [1960]. "Quantizing for Minimum Distortion," *IRE Trans. Info. Theory*, vol. IT-6, pp. 7–12.

McFarlane, M. D. [1972]. "Digital Pictures Fifty Years Ago," *Proc. IEEE*, vol. 60, no. 7.

McGlamery, B. L. [1967]. "Restoration of Turbulence-Degraded Images," *J. Opt. Soc. Amer.*, vol. 57, no. 3, pp. 293–297.

Mees, C. E. K. and James, T. H. [1966]. *The Theory of the Photographic Process*, Macmillan, New York.

Meyer, H., Rosdolsky, H. G., and Huang, T. S. [1973]. "Optimum Run Length Codes," *IEEE Trans. on Comm.*, vol. COM-22, no. 6, pp. 826–835.

Moon, P. [1961]. *The Scientific Basis of Illuminating Engineering*, Dover Publications, New York.

Mueller, P. F. and Reynolds, G. O. [1967]. "Image Restoration by Removal of Random Media Degradations," *J. Opt. Soc. Amer.*, vol. 57, pp. 1338–1344.

Muerle, J. L. and Allen, D. C. [1968]. "Experimental Evaluation of Techniques for Automatic Segmentation of Objects in a Complex Scene," in *Pictorial Pattern Recognition*, (G. C. Cheng *et al.*, eds.), Thompson Book Co., Washington, D. C.

Narasimhan, R. and Fornango, J. P. [1963]. "Some Further Experiments in the Parallel Processing of Pictures," *IEEE Trans. Elec. Computers*, vol. EC-12, pp. 748–750.

Nelson, C. N. [1971]. "Prediction of Densities in Fine Detail in Photographic Images," *Photo. Sci. Eng.*, vol. 15, pp. 82–97.

Noble, B. [1969]. *Applied Linear Algebra*, Prentice-Hall, Englewood Cliffs, New Jersey.

Ohlander, R. B. [1975]. "Analysis of Natural Scenes," Ph.D. dissertation, Dept. of Computer Science, Carnegie-Mellon Univ., Pittsburgh, Penn.

O'Neill, E. L. [1956]. "Spatial Filtering in Optics," *IRE Trans. Info. Theory*, vol. IT-2, no. 2, pp. 56–65.

Oppenheim, A. V., Schafer, R. W., and Stockham, T. G., Jr. [1968]. "Nonlinear Filtering of Multiplied and Convolved Signals," *Proc. IEEE*, vol. 56, no. 8, pp. 1264–1291.

Oppenheim, A. V. and Schafer, R. W. [1975]. *Digital Signal Processing*, Prentice-Hall, Englewood Cliffs, New Jersey.

Panter, P. F. and Dite, W. [1951]. "Quantization Distortion in Pulse Code Modulation with Nonuniform Spacing of Levels," *Proc. IRE*, vol. 39, pp. 44–48.

Papoulis, A. [1962]. *The Fourier Integral and its Applications*, McGraw-Hill, New York.

Papoulis, A. [1965]. *Probability, Random Variables, and Stochastic Processes*, McGraw-Hill, New York.

Papoulis, A. [1968]. *Systems and Transforms with Applications in Optics*, McGraw-Hill, New York.

Pattern Recognition [1970]. Special issue on pattern recognition in photogrammetry, vol. 2, no. 4.

Pavlidis, T. [1972]. "Segmentation of Pictures and Maps Through Functional Approximation," *Comp. Graph. Image Proc.*, vol. 1, pp. 360–372.

Pearson, D. E. [1975]. *Transmission and Display of Pictorial Information*, John Wiley & Sons (Halsted Press), New York.

Perrin, F. H. [1960]. "Methods of Appraising Photographic Systems," *Journal of the SMPTE*, vol. 49, pp. 151–156 and 239–249.

Phillips, D. L. [1962]. "A Technique for the Numerical Solution of Certain Integral Equations of the First Kind," *J. Assoc. Comp. Mach.*, vol. 9, pp. 84–97.

Pratt, W. K. [1971]. "Spatial Transform Coding of Color Images," *IEEE Trans. Comm. Tech.*, vol. COM-19, no. 6, pp. 980–991.

Pratt, W. K. [1974]. "Correlation Techniques of Image Registration," *IEEE Trans. Aerospace and Elec. Syst.*, vol. AES-10, no. 3, pp. 353–358.

Prewitt, J. M. S. [1970]. "Object Enhancement and Extraction," in *Picture Processing and Psychopictorics*, (B. S. Lipkin and A. Rosenfeld, eds.), Academic Press, New York.

Price, K. E. [1976]. "Change Detection and Analysis in Multispectral Images," Dept. of Computer Science, Carnegie-Mellon Univ., Pittsburgh, Penn.

Proc. IEEE [1967]. Special issue on redundancy reduction, vol. 55, no. 3.

Proc. IEEE [1972]. Special issue on digital picture processing, vol. 60, no. 7.

Proctor, C. W. and Wintz, P. A. [1971]. "Picture Bandwidth Reduction for Noisy Channels," TR-EE 71-30, School of Electrical Engineering, Purdue University, West Lafayette, Indiana.

Ready, P. J. and Wintz, P. A. [1973]. "Information Extraction, SNR Improvement, and Data Compression in Multispectral Imagery," *IEEE Trans. Comm.*, vol. COM-21, no. 10, pp. 1123–1131.

Rino, C. L. [1969]. "Bandlimited Image Restoration by Linear Mean-Square Estimation," *J. Opt. Soc. Amer.*, vol. 59, pp. 547–553.

Riseman, E. A. and Arbib, M. A. [1977]. "Computational Techniques in Visual Systems. Part II: Segmenting Static Scenes," IEEE Computer Society Repository, R77-87.

Robbins, G. M. and Huang, T. S. [1972]. "Inverse Filtering for Linear Shift-Variant Imaging Systems," *Proc. IEEE*, vol. 60, pp. 862–872.

Roberts, L. G. [1965]. "Machine Perception of Three-Dimensional Solids," in *Optical and Electro-Optical Information Processing*, (J. T. Tippet, ed.), MIT Press, Cambridge, Mass.

Robinson, G. S. [1976]. "Detection and Coding of Edges Using Directional Masks," University of Southern Cal., Image Processing Institute, Report no. 660.

Rosenfeld, A. [1969]. *Picture Processing by Computer*, Academic Press, New York.

Rosenfeld, A. [1972]. "Picture Processing," *Comp. Graph. Image Proc.*, vol. 1, pp. 394–416.

Rosenfeld, A. [1973]. "Progress in Picture Processing: 1969-71," *Computer Surv.*, vol. 5, pp. 81–108.

Rosenfeld, A. [1974]. "Picture Processing: 1973," *Comp. Graph. Image Proc.*, vol. 3, pp. 178–194.

Rosenfeld, A. *et al.* [1965]. "Automatic Cloud Interpretation," *Photogrammetr. Eng.*, vol. 31, pp. 991–1002.

Rosenfeld, A. and Kak, A. C. [1976]. *Digital Picture Processing*, Academic Press, New York.

Roth, W. [1968]. "Full Color and Three-Dimensional Effects in Radiographic Displays," *Investigative Radiology*, vol. 3, pp. 56–60.

Rudnick, P. [1966]. "Note on the Calculation of Fourier Series," *Math. of Comput.*, vol. 20, pp. 429–430.

Runge, C. [1903]. *Zeit. für Math. and Physik*,vol. 48, pg.433.

Runge, C. [1905]. *Zeit. für Math. and Physik*, vol.53, pg. 117.

Runge, C. and König, H. [1924]. "Die Grundlehren der Mathematischen Wissenschaften," *Vorlesungen über Numerisches Rechnen*, vol. 11, Julius Springer, Berlin, Germany.

Rushforth, C. K. and Harris, R. W. [1968]. "Restoration, Resolution, and Noise," *J. Opt. Soc. Amer.*, vol. 58, pp. 539–545.

Sakrison, D. J. and Algazi, V. R. [1971]. "Comparison of Line-by-Line and Two-Dimensional Encoding of Random Images," *IEEE Trans. Info. Theory*, vol. IT-17, no. 4, pp. 386–398.

Sawchuk, A. A. [1972]. "Space-Variant Image Motion Degradation and Restoration," *Proc. IEEE*, vol. 60, pp. 854–861.

Schreiber, W. F. [1956]. "The Measurement of Third Order Probability Distributions of Television Signals," *IRE Trans. Info. Theory*, vol. IT-2, pp. 94–105.

Schreiber, W. F. [1967]. "Picture Coding," *Proc. IEEE*, (Special issue on Redundancy Reduction), vo.. 55, pp. 320–330.

Schwartz, J. W. and Barker, R. C. [1966]. "Bit-Plane Encoding: A Technique for Source Encoding," *IEEE Trans. Aerosp. Elec. Systems*, vol. AES-2, no. 4, pp. 385–392.

Schwarz, R. E. and Friedland, B. [1965]. *Linear Systems*, McGraw-Hill, New York.

Scoville, F. W. [1965]. "The Subjective Effect of Brightness and Spatial Quantization," *Quarterly Rept., no. 78*, MIT Research Laboratory of Electronics.

Seidman, Joel [1972]. "Some Practical Applications of Digital Filtering in Image Processing," *Proc. of the Conf. on Computer Image Processing and Recognition*, University of Missouri, Columbia, vol. 2, pp. 9–1–1 through 9–1–16.

Selin, I. [1965]. *Detection Theory*, Princeton University Press, Princeton, New Jersey.

Shack, R. V. [1964]. "The Influence of Image Motion and Shutter Operation on the Photographic Transfer Function," *Appl. Opt.*, vol. 3, pp. 1171–1181.

Shanks, J. L. [1969]. "Computation of the Fast Walsh-Fourier Transform," *IEEE Trans. Computers*, vol. C-18, no. 5, pp. 457–459.

Shaw, A. C. [1970]. "Parsing of Graph-Representable Pictures," *J. ACM*, vol. 17, no. 3, pp. 453–481.

Sheppard, J. J., Jr. [1968]. *Human Color Perception*, Americal Elsevier Publishing Co., New York.

Sheppard, J. J., Jr., Stratton, R. H., and Gazley, C., Jr. [1969]. "Pseudocolor as a Means of Image Enhancement," *Am. J. Optom. Arch. Am. Acad. Optom.*, vol. 46, pp. 735–754.

Slepian, D. [1967a]. "Linear Least-Squares Filtering of Distorted Images," *J. Opt. Soc. Amer.*, vol. 57, pp. 918–922.

Slepian, D. [1967b]. "Restoration of Photographs Blurred by Image Motion," *BSTJ*, vol. 46, pp. 2353–2362.

Slepian, D. and Pollak, H. O. [1961]. "Prolate Spheroidal Wave Functions, Fourier Analysis, and Uncertainty – I," *Bell Sys. Tech. J.*, vol. 40, pp. 43–64.

Smith, S. L. [1963]. "Color Coding and Visual Separability in Information Displays," *J. Applied Psychology*, vol. 47, pp. 358–364.

Snider, H. L. [1973]. "Image Quality and Observer Performance," in *Perception of Displayed Information*, (L. M. Biberman, ed.), Plenum Press, New York.

Som, S. C. [1971]. "Analysis of the Effect of Linear Smear," *J. Opt. Soc. Amer.*, vol. 61, pp. 859–864.

Sondhi, M. M. [1972]. "Image Restoration: The Removal of Spatially Invariant Degradations," *Proc. IEEE*, vol. 60, no. 7, pp. 842–853.

Stevens, S. S. [1951]. *Handbook of Experimental Psychology*, Wiley, New York.

Stockham, T. G., Jr. [1972]. "Image Processing in the Context of a Visual Model," *Proc. IEEE*, vol. 60, no. 7, pp. 828–842.

Stumpff, K. [1939]. *Tafeln und Aufgaben zur Harmonischen Analyse und Periodogrammrechnung*, Julius Springer, Berlin, Germany.

Tasto, M. and Wintz, P. A. [1971]. "Image Coding by Adaptive Block Quantization," *IEEE Trans. Comm. Tech.*, vol. COM-19, pp. 957–972.

Tasto, M. and Wintz, P. A. [1972]. "A Bound on the Rate-Distortion Function and Application to Images," *IEEE Trans. Info. Theory*, vol. IT-18, pp. 150–159.

Thomas, J. B. [1969]. *Statistical Communication Theory*, Wiley, New York.

Thomas, L. H. [1963]. "Using a Computer to Solve Problems in Physics," *Application of Digital Computers*, Ginn, Boston, Mass.

Thomason, M. G. and Gonzalez, R. C. [1975]. "Syntactic Recognition of Imperfectly Specified Patterns," *IEEE Trans. Computers*, vol. C-24, no. 1, pp. 93–96.

Titchmarsh, E. C. [1948]. *Introduction to the Theory of Fourier Integrals*, Oxford Univ. Press, New York.

Tou, J. T. and Gonzalez, R. C. [1974]. *Pattern Recognition Principles*, Addison-Wesley, Reading, Mass.

Twomey, S. [1963]. "On the Numerical Solution of Fredholm Integral Equations of the First Kind by the Inversion of the Linear System Produced by Quadrature," *J. Assoc. Computer Mach.*, vol. 10, pp. 97–101.

VanderBrug, G. J. and Rosenfeld, A. [1977]. "Two-Stage Template Matchings," *IEEE Trans. Computers*, vol. C-26, no. 4, pp. 384–394.

Van Valkenburg, [1955]. *Network Analysis*, Prentice-Hall, Englewood Cliffs, New Jersey.

Walsh, J. L. [1923]. "A Closed Set of Normal Orthogonal Functions," *Am. J. Math.*, vol. 45, no. 1, pp. 5–24.

Weinberg, L. [1962]. *Network Analysis and Synthesis*, McGraw-Hill, New York.

Whelchel, J. E., Jr. and Guinn, D. F. [1968]. "The Fast Fourier-Hadamard Transform and its Use in Signal Representation and Classification," *Eascon 1968 Convention Record*, pp. 561–573.

Wilkins, L. C. and Wintz, P. A. [1970]. "Studies on Data Compression, Part I: Picture Coding by Contours, Part II: Error Analysis of Run-Length Codes," TR-EE 70-17, School of Electrical Engineering, Purdue University, Lafayette, Indiana.

Williamson, J. [1944]. "Hadamard's Determinant Theorem and the Sum of Four Squares," *Duke Math. J.*, vol. 11, pp. 65–81.

Wintz, P. A. [1972]. "Transform Picture Coding," *Proc. IEEE*, vol. 60, no. 7, pp. 809–820.

Wong, R. Y. and Hall, E. L. [1977]. "Scene Matching with Invariant Moments," *Computer Graph. Image Proc.* (in press).

Wood, R. C. [1969]. "On Optimum Quantization," *IEEE Trans. Info. Theory*, vol. IT-15, pp. 248–252.

Yates, F. [1937]. "The Design and Analysis of Factorial Experiments," Commonwealth Agricultural Bureaux, Farnam Royal, Burks, England.

INDEX